GASOLINE ENGINE A

Gasoline engine analysis

for Computer Aided Design

Edited by John Fenton

MECHANICAL ENGINEERING PUBLICATIONS LTD
LONDON

ISBN 0 85298 634 3

Printed in Great Britain at the University Press, Cambridge

Contents

The Editor thanks Cecil French of Ricardo and Richard Jones of CAE International for their considerable help, without which compilation of source material for this volume would have been impossible.

Preface

Aimed at filling the gap between theoretical works on thermodynamics and descriptive works on engine construction, this book is the latest of those texts setting out to provide design calculations for internal combustion (IC) engines – a pioneer of which was V.L. Maleev's *Internal-combustion engines: theory and design* published by McGraw- Hill in 1933. That author was a native of pre-1917 Russia and shared the Eastern tenet of having a very sound base in theoretical engineering. His subsequent move to the USA involved 'circumstances [which] made necessary a change to industrial engineering'. Here he recognised the need for a book giving '. . . proper understanding of the principles involved in the design . . .'. Existing books were unsatisfactory, he argued; 'they dealt with design alone or were too theoretical or of a purely descriptive nature'. His book recognised the need for 'what, how and why'. Design calculations were provided on engine performance, fuels, cooling, inertia loads, gas loads, flywheels, balancing, and logical procedures in detail design.

In the post war years Liverpool University's T.D. Walshaw wrote his study of the best in British and American Practice; *Diesel engine design,* published by George Newnes in 1949. While drawing on experience of mostly larger, slower-speed engines, automobile engines were firmly part of the brief. He calculated inlet air consumption and exhaust emission weight, provided a logical approach to sizing, quantified a variety of conditions in detail design, introduced connecting-rod, crank-shaft and flywheel stressing, and also the calculation of balancing inertias. Cam design and valve gear fundamentals were tackled with a discussion, albeit non-quantitative, of engine structures.

In the mid 1960s A.W. Judge began to divide texts on diesel and petrol engines. His work *Modern petrol engines* (dealing with automobile, aircraft, and stationary types), was published by Chapman and Hall in 1965 (from a first edition in 1956) alongside companion volumes in diesels. While the work was largely descriptive, valuable analytical treatments were provided for performance considerations, supercharging, and cooling.

In 1965, M.H. Howarth in his book *The design of high speed diesel engines,* published by Constable, broke new ground on tackling detail design in a particularly thorough way. Valve and port layouts were treated, combustion chambers examined, along with coolant water passages and crankshaft dimensioning, and there was a look at the fundamentals of crankshaft bearing design; this was alongside an overall view of noise and vibration problems. While the book was descriptive rather than analytical, in the main, focus was on components rather than overall engines.

Meanwhile the American, Charles Fayette Taylor's work *The internal-combustion engine in theory and practice,* published by the MIT Press in 1968, had a useful section on design within Volume 2. Here a systematic approach to design was introduced for a wide range of engine types up to those for heavy locomotives. Comparative design concepts (design ratios) were introduced; expressions were provided for such factors as inlet port swirl coefficient, inlet valve flow capacity, and crankshaft overlap, and an approach to crankcase structural design was suggested. Geometrical factors governing combustion chamber surface to volume ratio with respect to valve seat diameter, camshaft dynamics, and an analysis of valve gear motion, were provided (the latter introducing a computer method of evaluation). Design of geared timing drives was also considered.

The same year, 1968, saw the UK publication by Iliffe Books of *Engine design,* edited by J.G. Giles, using the practice of having separate authors for specific sections. Of the five included, three were concerned with piston engines; one on valve gear and breathing; another on the piston and connecting rod assembly, and the third on overall design considerations.

Finally, mention should be made of the *Diesel engine reference book* published in 1984 by Butterworth, a 32-section work by individual authors and edited by a noted Ricardo engineer, L.R.C. Lilley. This comprehensive volume inevitably overlaps into mechanical design of gasoline piston engines. Much of the text is descriptive, but important analytical sections provide useful data for design and research projects (including the use of computer analysis). Sections on torsional vibration balance and structural-dynamics/noise is particularly useful from the standpoint of design calculation. This, and other specialist subjects dealt with by the above texts, have not been duplicated in this volume, the contents of which are summarised below.

Chapter one is an edited version of a ten-part series of articles by the late S.S. Tresilian of Rolls Royce, on comparative design. Of the companies involved in motor manufacture Rolls Royce was probably alone in the extent to which it drew upon the more disciplined approach to aeronautical engine design, in conceiving 'clean-sheet' designs of car petrol engines. The author considers the fundamentals of performance and efficiency and relates engine breathing parameters and mechanical stresses to the dimensions and proportions of the major engine components, concluding with an introduction to the several different types of piston engine. Chapters two, three, and four are comprised of articles contributed by a number of specialists covering the broad headings of Performance and combustion, Mechanical Performance and Noise Reduction, and Cooling system design. The final chapter, five, CAD for gasoline engines, provides examples of components designed with CAD techniques culminating with case studies of performance enhanced by computer analysis and a total engine system design using computer techniques.

Editor's note

Where the individual contents of this work cannot be used directly in design, it is offered as a start-point for studying particular aspects of engine technology. In certain cases, for example, more recent mathematical techniques have changed the way in which fundamental principles are analysed. However, the exponents of computer aided engineering (CAE) maintain that modern test and re-evaluation techniques can quickly verify the analytical method concerned. Alternative techniques can often be found through literature searches, for which purpose the editor has provided a list of all original source material in the two journals concerned, as well as secondary references in an appendix. Volume one of *Automotive Design Engineering* commenced in September 1962, the last volume becoming volume one of *Automotive Engineer* from end of 1975, and including 1976.

Numerous analytical papers have been written since the renewed interest in exhaust-pollution control and fuel-economy which followed the Californian smog crisis of the 1960s and the fuel crisis of the 1970s. These have extended the variety of student textbooks on thermodynamics available and have given an insight into applied research now being carried out. While many of these papers have thrown new light on the analytical techniques available, comparatively few are of direct assistance in design calculations. By contrast articles commissioned by *Automotive Design Engineering* (and, selectively, *Automotive Engineer,* which succeeded it) were specifically aimed at exposing design engineers to design analysis techniques. These form the core material of the book whose context among existing texts on engine design is described in the Preface.

Foreword

This collection of some fifty articles, previously published in *Automotive Design Engineering* and *Automotive Engineer* is intended to provide the basis for gasoline engine design using computer-based techniques. Where possible, design examples are provided; otherwise analytical and descriptive texts have been chosen to meet the needs of engineering analysis of engine design. The most commonly used terms in CAE (Computer Aided Engineering) are shown in the figure below.

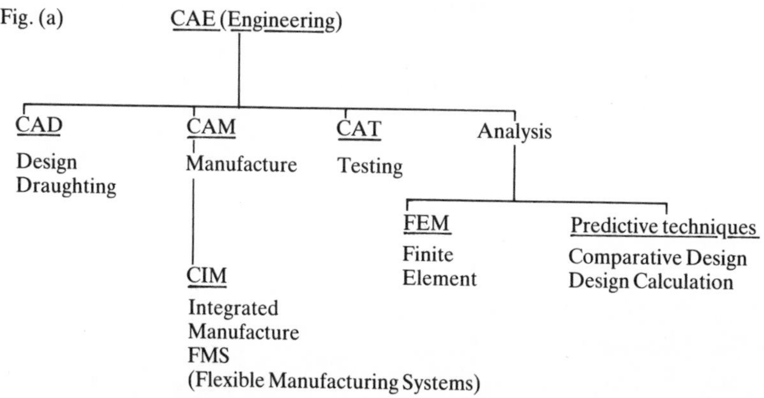

An important feature is that draughting and analysis are considered separately – and that analysis is made up of two distinct areas: Finite Element Analysis, for structural and analogous problems (heat transfer, fluid flow), and Predictive techniques.

At the concept stage of engine design, the major components (crankshaft, connecting rod, piston, camshaft, and valves) may be stressed more efficiently and economically using computer aided Predictive techniques, based on traditional theories, than using FEM. This is because the computer can carry out the repetitious calculations based on analytical statements (such as design calculations) or graphical interpolation (comparative design), whereas FEM often requires, by necessity, well defined geometry and a detailed, time consuming, and sometimes costly approach.

In addition, Predictive techniques are valuable in the form of performance prediction programmes. These can be tailored for conventional designs, to provide a rapid picture of dimensional requirements to suit, say, a broad scatter of performance characteristics, using a process of digital extrapolation, much quicker than 'reading' characteristic curves. If ram tuning is to be employed a 'volumetric-efficiency' program-package can be added, or for multi-fuel operation, a fuels package; there are now programs, for example, which will simulate different combustion cycles.

Richard Jones,
SDRC Engineering Services Ltd

CHAPTER ONE
Comparative design

Based on a single-author article series by the former Chief Designer (Engines) at Rolls Royce, this introductory chapter deals with design ratios applying to most of the principal parameters of the piston engine.

The author's respect for the piston engine as an automotive power unit is based on its widely variable cylinder-assembly configuration, which can satisfy many vehicle packaging problems. After discussing the related criteria for engine performance he deals with the three fundamental efficiencies. Losses resulting in reduced mechanical efficiency form his start point for design-ratio comparisons, considering gas velocity, piston area and speed, and engine speed. The concept of similar and dissimilar engines is introduced and their design ratios are used to investigate stresses in the reciprocating parts; a section on choosing engine proportions then follows.

Valve gear design is next examined in detail with a study of effects in changes of valve speed and valve opening area followed by a discussion of valve gear similarity and stressing. A design approach to the valving of engines is presented with consideration of duration of opening, number of valves, and attainable valve speed. Next, design and development approaches are compared and particular configurations of valve train examined.

Stroke/bore ratio significance is discussed and its relationship with piston speed, number of valves, increased valve speed, crankshaft vibrations, number of cylinders, bearing inertia load, engine speed, and combustion. The chapter concludes with consideration of engine mechanisms as affecting big-end bearings, valve gear comparisons, and compression ratio.

S. S. Tresilian[1]† maintained that the important characteristics of a road vehicle engine are performance, smoothness and quietness, weight, cost, and economy; although the order of importance varies, cost is frequently at the head. Here we are chiefly concerned with power/weight ratio and smoothness of operation. Savings in engine weight can show nearly a

† References are given in the Appendix at the end of the book.

Table 1.1

Car A	*Units of weight*
Power unit, including items whose weight varies with engine displacement – flywheel, clutch, gearbox, battery, starter and radiator	20
Body shell, including seats and closure panels	40
Sub-frames, springs, brakes, axles, steering, wheels, and tyres	40
Total weight	100
Car B	
Power unit	13
Body shell	40
Running gear (40% of total weight)	35
Total weight	88
Weight saving	12
Car C	
Power unit (85% of that for Car B)	11
Body	40
Running gear (40% of total weight)	34
Total weight	85
Weight saving	15

double saving in total car weight. To illustrate this, three imaginary cars will be considered. The first of them (Car A) has the weight breakdown shown in Table 1.1.

Now let us assume that, in Car B, the weight of the power unit has been reduced by a third – i.e., its power/weight ratio improved 50 per cent – and that the 'chassis' weight is a fixed proportion (40 per cent) of the total weight, including the running gear. The weight of this is reduced by its having less engine weight to carry, and further by having less 'chassis' weight to carry – the running gear being part of its own load.

Compared with Car A, this vehicle has an improved power/weight ratio of over 100/88, or 14 per cent. Suppose now that the engine size and power are adjusted to keep the power/weight ratio constant in Car C.

A reduction of nine units of power-unit weight could therefore permit a further saving of six units of 'chassis' weight. It follows that Car C should show a possible fuel-economy gain of 18 per cent over Car A – proportional to the total weights. Since the materials in a 'cheap' engine cost several times as much as the machining, it is worth taking a fresh look at the economics of engine design. The cheapest engine is not necessarily the simplest, nor is that made from the cheapest materials.

The three efficiencies

Of any engine, it is substantially true that, using the notation in Table 1.2

$$\text{Specific output (bhp/litre)} \propto N \times \eta_{th} \times \eta_{vol} \times \eta_{mech}$$

Observed figures for the three efficiencies vary between engines, but much less between those in the same usage categories – for example, trucks, touring cars, sports cars, motor-cycles, or racing cars.

Thermal efficiency

The figure attainable is fixed between narrow limits, for each vehicle category, by the fuels and compression ratio customarily used in that

Table 1.2 Notation for comparative design

Symbol	Meaning	Symbol	Meaning
A	Total piston area	Q	Airflow capacity
A_p	Piston area per piston	R	Ratio θ_a/θ_d
A_o	Equivalent fixed-orifice area per valve	r	Crank radius
		S	Stroke
A_t	Time-area per valve opening cycle	S/B	Stroke/bore ratio
		V_g	Gas velocity
A_v	Valve area per cylinder	V_p	Piston speed
a	Acceleration	V_v	Valve speed
B	Cylinder bore	W_{vo}	Valve spring load, valve open
D	Valve spring diameter	θ_a	Acceleration period of cam (crankshaft angle)
d	Valve head diameter		
h	Valve lift	θ_d	Deceleration period of cam (crankshaft angle)
K, k	Constants		
L	Engine total displacement	ϕ	Total cam opening period (crankshaft angle)
L_c	Displacement per cylinder		
M_v	Equivalent weight of valve gear at valve	η_{th}	Engine thermal efficiency on air basis (per cent)
mep	Mean effective pressure	η_{vol}	Engine volumetric efficiency (per cent)
N	Revolutions per minute		
n	Number of cylinders, or number of valves per cylinder	η_{mech}	Engine mechanical efficiency (per cent)
P	Power		

category. Its variation, therefore, does not concern the mechanical designer, and changes can be made during the development stages. The effects of cylinder size on thermal efficiency are normally too small to concern the designer, since the adverse effects of smaller cylinders can be partially offset by the higher compression ratios that can be used.

Volumetric efficiency

It is probable that if enough engines were investigated, and the correct measurements could be made, we should find that volumetric efficiency is a function of the physical conditions of the ingoing charge. Some research done in America has already correlated the volumetric efficiency with the inlet pipe Mach number in one engine. It is more likely to be controlled in fact by valve-throat Mach number, if that quantity could be measured. If it is not, then it ought to be, because this should always be the principal limitation on the gas flow rate. Upstream restrictions, in the carburettor or inlet pipe, have an effect on the mass flow rate at critical Mach number in the value throat.

It is assumed in this chapter that the volumetric efficiency is a function of gas velocity, and will be the same in engines of the same category at the same gas velocity; also, that other factors affecting the Mach number (such as charge pressure and temperature) will be the same since the engines are similar. The designer's problem is simplified if he can assume – for purposes of comparison – that when the carburettors, inlet and exhaust pipes, and valves are scaled up, the gas velocities will remain the same, and therefore, that the same volumetric efficiency will be attained.

Mechanical efficiency

Engines with 'similar' cylinders – i.e., to scale, but of different size, and running at the same mean piston speeds – will have

(1) the same mep by assumption, and the same output per unit of projected piston area, since

$$\text{Power} \propto \text{mep} \times A_p \times V_p$$

(2) the same bearing rubbing velocities;
(3) the same unit inertia loadings on their bearings;
(4) bearing areas in the same proportion as their piston areas, and so as their engine power.

There is, consequently, no reason why their mechanical efficiencies should differ.

The two most important mechanical losses in piston engines – piston friction and pumping losses – appear as lost mep. This is equivalent to a pressure loss in the power stroke. Therefore, as a percentage of the power developed, they are independent of scale. Pumping losses are a function of gas flow and flow conditions. An engine may be running at a high piston speed with mep pumping losses no greater than for one at low piston speed, if appropriate porting and valves are provided and gas speeds are the same. For this reason the volumetric efficiency can have a bigger effect on the mechanical efficiency than the actual mechanical losses. Improving the volumetric efficiency can improve the mechanical efficiency. Engines with high peak-power volumetric efficiencies show high mechanical efficiencies, even at high piston speeds. Mechanical losses that increase as an over-unity power of the piston speed form a small percentage of the total losses, and can be neglected by the mechanical designer.

The pumping work is done by the pistons, and must therefore carry two lots of piston friction losses, so an engine showing 70 per cent overall mechanical efficiency may be doing its pumping work at only 50 per cent mechanical efficiency. Useful improvements might be made in the part-load consumption of engines if power could be reduced in some other way than by throttling the charge – for example, by controlling the injection of fuel in the neighbourhood of the sparking-plug, with a stratified charge.

The mechanical designer may thus be faced with designing a small engine to achieve the same performance as a larger one in the same usage category, where he has the same limitations on fuel, noise level, materials, and cost. In this context, the variations in these efficiencies are unimportant. The car engine is primarily an air pump, refined to high standards of quietness and smoothness. Its performance is decided mainly by this capacity; it has been responsible for most of the past improvements in engine performance, and it offers the greatest scope for the future.

Gas velocity

For continuity, $V_g \times A_v = V_p \times A_p$, therefore

$$\text{gas velocity } V_g \propto \frac{A_p}{A_v} \cdot v_p$$

This classical definition of gas velocity, a mechanical one, is used throughout. It enables valve area and gas velocity to be expressed in mechanical units, and permits performances and stresses of valve gears to be simply connected with those of pistons. The definition assumes that valves, valve gear, and ports are similar in the cases being compared. It also assumes that, where the number of cylinders varies, the induction-pipe and carburettor-choke areas are in proportion to the product $A_p \times V_p$, or $L \times N$, to preserve the same gas velocities.

Valve areas vary considerably. In the touring, sports, and racing car categories, known variations in the ratio of valve-head diametral area to piston area range from 20 per cent to 40 per cent. This does not mean that gas velocities in these engines vary by this amount; rather the piston speeds at which they will develop their peak power. At this performance, their gas velocities will be substantially the same, because their airflows are governed by physical laws which do not depend on engine design.

Piston area and piston speed

A consequence of the previous definition is that the piston speed attainable at peak power is governed by the valve/piston area ratio. In other words, $V_p \propto A_v/A_p \cdot V_g$.

We assume that V_g is substantially constant, at peak output, particularly in similar engines in the same usage category. This means, also, that if we compare engines of the same displacement, and with the same valves and valve gear (i.e., A_v) but varying S/B ratio, then

$$V_p \propto (A_v/A_p)V_g \propto 1/A_p \propto S$$

(V_g is assumed constant, and N is constant since $V_p = 2S \cdot N$). Therefore, piston area and piston speed are inversely proportional and interchangeable. Because of this, A_p and V_p are no longer acceptable as indications of the potential performance of an engine. With S/B a variable, then for one cylinder

$$Q = A_v \cdot V_g = N \cdot L_c$$

and with V_g constant

$$N \propto A_v L_c$$

Significance of engine speed

In engines at peak power

$$\text{power } P \propto \text{mep} \times L \times N$$

It is a basic assumption here that engines in the same usage category develop substantially the same peak-power mep. Therefore, for comparative purposes

$$P/L \propto N$$

The power/litre – and, therefore, to a first approximation, the power/weight ratio for the complete power unit of engine plus flywheel, clutch, and gearbox – is proportional to the designed peak-power engine speed. This simple rule embraces all the principal variables in an engine's specification – swept volume, number of cylinders, bore, stroke, S/B ratio, piston area, or designed piston speed.

As a measure of the specific performance of engines, the designed engine speed forms a more important indication than piston area or piston speed. Mechanical and thermal stresses should not be the limitations on power output; if they are, the designer has not finished his job. The ultimate limitations should be aerodynamic – the physical laws of gas flow.

It will be shown later that engine speed provides a connexion between power output per litre and the design performance of the valve gear. Although there are, admittedly, other ways of improving engine power/weight ratio, such as light alloys or supercharging, the unit designed for high engine speed will always be lighter than a lower-speed one, irrespective of the construction.

ENGINE PROPORTIONS FOR COMPARATIVE DESIGN

Similar engines

It can be shown that engines having similar cylinders and valving will develop the same mep and have the same piston speed at peak power. The following assumptions are made

constant: mep, V_p, B/S, A_v/A_p
varying: L, L_c, N, P, B, S, n

(In the following, subscripts refer to the bracketed variables, considered constant, and P signifies engine power.) Then

$$P \propto n \cdot B^2 \propto A \propto 1/N^2$$
$$(P/L) \propto N; \qquad (P/L)_n \propto 1/L_c^{1/3}$$
$$(P/L) \propto n^{1/3}; \qquad (P/L)_A \propto n^{1/2}$$

Such engines will have:

(1) the same gas speeds at the same piston speeds;
(2) the same inertia loading per unit area of bearing area or piston area, and the same stresses;
(3) the same bearing rubbing velocities.

Dissimilar engines

Effects of varying some of the principal proportions of some hypothetical engines will now be considered. The design difficulties likely to arise, and the steps to be taken to overcome them, are the subject of later sections. In general, we can assume the following (the actual situation differs, though, in each of the succeeding sub-sections)

constant: mep
varying: $B, S, L, L_e, n, B/S, A_v/A_p, N$

Then

$$P \propto N; \qquad V_p \propto A_v/A_p$$

(a) Engines of the same power but varying piston speed

These engines have geometrically similar cylinders. Assumptions are

constant: mep, $P, n, B/S, A_v$
varying: $V_p, B, S, N, L, A_v, A_p$

Then

$$A_p \propto 1/V_p; \qquad L \propto (A_p)^{3/2} \propto (1/V_p)^{3/2}; \qquad (P/L)_p \propto (V_p)^{3/2}$$

The advantages sometimes claimed for the large low-speed engine are greater smoothness and quietness, but the reverse is probably truer. Improved fuel consumption is also claimed in favour of engines with low piston speeds. However, careful tests with aircraft power units, where facts and not hearsay are essential, show that this is not so. Identical consumptions are shown by tests on large and small engines both running full-throttle at the same altitude cruise power but at different engine speeds. The reason for this is that the two chief losses, pumping work and piston friction (both equivalent to lost mep), are the same in the two cases at the same power. If the engines are sufficiently similar, and the meps are the same at the same altitudes and powers, then product $A_p \cdot V_p$ will also be the same, as will be the lost work due to the two causes mentioned. There is no saving at all at cruising power conditions to offset the great weight of the larger engine.

(d) Engines of the same displacement and piston speed, but varying B/S *ratio*

It is assumed that there are no mechanical limitations – such as valve gear – to prevent the engines developing the same mep. Other assumptions are

constant: L, n, V_p, mep, A_v/A_p
varying: B/S, B, S, N, A_v, P

Then

$$L \propto B^2 \cdot S \propto B^3 \cdot (S/B) = \text{constant}$$

Therefore

$$B^3 \propto B/S \quad \text{and} \quad B \propto (B/S)^{1/3}$$

Also

$$P \propto N \propto B^2 \propto (B/S)^{2/3}; \qquad (P/L)_L \propto (B/S)^{2/3}$$

Table 1.3 compares three four-cylinder 2 litre engines, each developing 690 kN/m^2 mep at 900 m/min piston speed.

Table 1.3

	Engine 1	*Engine 2*	*Engine 3*
S/B	1.25	1.00	0.75
N (r/min)	4580	5310	6430
Relative power per litre	1.00	1.160	1.405

(c) *Engines of the same output and firing at the same top-gear frequency, but with different cylinder numbers and displacement*

Here, let us compare three engines of the same total piston area and piston speed. (See Table 1.4.)

These engines, if geared for the same power and piston speed at the same road speed, would have the same displacement per kilometre and the same potential road performance; they would also fire with the same frequency. It might be hard for the driver to tell them apart, except that Engine 1 would be much lighter than the other two.

Table 1.4

	Engine 1	*Engine 2*	*Engine 3*
L	2	3	4
n	4	6	8
S/B	0.75	1.38	2.13
N at 900 m/min	6430	4300	3220

(d) Exchange of piston area for piston speed
The following assumptions are made

constant: L, n, A_v, N, P
varying: $S/B, V_p, B, D, A_v/A_p$

For engines of the same displacement and with similar valve gear, the power, r/min, and P/L will be the same

$$V_p \propto S \propto (S/B)^{2/3} \propto (1/B^2) \propto (1/A_p)$$

This illustrates the current trend towards 'oversquare' engines, with the aim of reduced piston speed; smoothness and lower fuel consumptions are claimed as advantages for this layout.

Since A_v is unchanged, $V_p \propto (A_v/A_p)$.

(e) Constant displacement but varying number of cylinders
In this case our assumptions are

constant: L, V_p
varying: $n, S/B, B, S$

Comparing three $2\frac{1}{2}$ litre engines, all of the same stroke, total piston area, and total valve area, see Table 1.5. The potential power of the four-cylinder unit is as great as that of the eight. It has the same piston area and stroke as the latter; if of the four-valve type, it can have exactly the same valves and gear. Consequently, it can be regarded as an eight-cylinder engine shortened to a four of the same potential output.

Table 1.5

	Engine 1	*Engine 2*	*Engine 3*
n	4	6	8
S/B	0.75	0.92	1.06
L_c (/cyl)	0.61	0.40	0.30

Stresses in the reciprocating parts

It is appropriate to follow our general consideration of engine proportions with an investigation of the stresses in the pistons and connecting rods of the hypothetical engines examined in the previous section. Piston motion, though not simple harmonic, is of harmonic type. The stresses in different designs, or in any given design at different running conditions, can be compared by means of formulae based on harmonic motion.

$$\text{Piston inertia force } F \propto M \cdot r \cdot \omega^2 \propto M \cdot S \cdot N^2$$

In engines having substantially similar piston and rod assemblies, the following is true of any stress in those assemblies. This stress

$$\propto (F/A_p) \propto (M \cdot S \cdot N^2/A_p) \propto (B^3 \cdot S \cdot N^2/B^2) \propto B \cdot S \cdot N^2 \propto B/S \cdot V_p^2$$

since $M \propto B^3$ and $V_p \propto S \cdot N$. The comparison is exact if the geometrical similarity between the engines is exact. It is still substantially true for variations in B/S if the piston and rod dimensions are scaled transversely with the bore, and similarity is maintained except for rod length.

A designer who evolves a range of substantially similar engines can therefore write, for the whole range

$$\text{Stress} = k \cdot B \cdot S \cdot N^2 \quad \text{or} = K(B/S)V_p^2$$

This enables him to make quick estimates of the effects of changes in proportions or running conditions. Different constants must be used, though, for different types of rod (for example, light alloy as against steel).

The formula just stated can be modified as follows to allow more accurately for differences in the ratio of connecting rod length to crank radius

$$\text{Stress} \propto B \cdot S \cdot N^2(1 + r/L_r), \quad \text{or} \quad (B/S) \cdot V_p^2(1 + r/L_r)$$

where r is the crank radius and L_r the connecting rod length.

If changes have to be made to the proportions or to the piston speed of an engine, and the stresses are not to be altered, then an allowance can be made for changed rod or bearing proportions – for example, a change in big-end bearing area in proportion to piston area

$$\text{Stress} \propto (B/S \cdot V_p^2)(1 + r/L_r) \cdot A_p/A_b$$

The change required in the bearing area to keep the bearing loading constant can be assessed approximately from the following

$$\frac{A_b}{A_p} \propto \frac{B}{S} \cdot V_p^2\left(1 + \frac{r}{L}\right)$$

This formula can be used to estimate the necessary alteration in any other rod dimensions, such as neck area in relation to piston area, or big-end bolt area. All these estimates are, of course, approximate, for the guidance of the designer in the early stages. It is still necessary to do a complete stress investigation of the design finally adopted, allowing for the probable extra weight involved, or for improved materials.

It has been proposed by other writers that piston-ring stresses or running conditions should be compared on the basis of acceleration. In the authors view, this is wrong, because acceleration is a meaningless number when used by itself, without reference to scale. Similar pistons and rods running at the same piston speed have the same stresses, but not the same accelerations.

$$\text{Maximum acceleration} \propto r\omega^2 \propto S \cdot N^2$$

If piston speed $2S \cdot N$ is constant, then

$$\text{Maximum acceleration} \propto N \propto \frac{1}{B}$$

Small pistons have greater accelerations than big ones, though they may have the same velocity at each crank angle.

From the above-mentioned formulae, similar piston rings running in similar engines at the same piston speeds will have the same inertia pressures on their top and bottom faces expressed in pressure over the projected ring area. We can compare the inertia pressures beneath rings of varying proportions, running in engines of maybe differently varying proportions; on such rings

$$\text{Inertia force} \propto M \cdot r \cdot N^2 \propto w \cdot h \cdot B \cdot S \cdot N^2$$

where w = radial width of ring and h = height of ring

$$\text{Inertia pressure} \propto \frac{w \cdot h \cdot b \cdot S \cdot N^2}{w \cdot B} \propto h \cdot S \cdot N^2 \propto \frac{h}{S} \cdot V_p^2$$

CHOOSING ENGINE PROPORTIONS

Two lines of approach are open to the designer.

(a) He can start with a definite displacement in view. It is then possible for him to investigate simultaneously the effect on the power output of various combinations of mep, n, B/S, and V_p, by combining the previous individual comparison formulae.

In any engines, in actual figures

$$\text{Power} = \text{mep} \cdot N \cdot L/K_1$$

where K depends on units

$$L = n \cdot \pi/4 \cdot B^2 \cdot S = n \cdot \pi/4 \cdot (B/S)^2 \cdot S^3$$

therefore

$$\text{Power} = (\text{mep}/K_2) V_p (\pi \cdot n/4)^{1/3} \cdot (B/S \cdot L)^{2/3}$$

(b) He can start with a certain required power output, when from the first equation above it can be shown that

$$L = K_3 \cdot S/B \cdot (P/\text{mep} \cdot V_p)^{3/2} \cdot (4/\pi \cdot n)^{1/2}$$

For simple comparison purposes

$$\text{Power} \propto V_p \cdot n^{1/3} \cdot (B/S \cdot L)^{2/3}$$
$$(L/P) \propto S/B \cdot (1/V_p)^{2/3} \cdot (P/n)^{1/2}$$

K_1, K_2, K_3 vary according to the chosen system of units. These formulae involve assessment of stress after trying the various combinations of proportions. B/S should be put into these as a ratio, not as actual dimensions.

VALVE GEAR DESIGN

Only poppet valve gear will be considered. Sleeve valves have long fallen out of favour for automobile engines, largely because of the high first and replacement costs of the sleeves and their driving gear. They show no advantage on port area, and the stroke cannot be reduced below about 85 per cent of the bore because then the piston begins to mask the ports. Also, when run by themselves on test rigs, sleeve valves are far from silent. There would be a case for them, though, if air-cooled engines were to be seriously considered for general use, since they constitute a valve gear whose enclosure and lubrication do not compromise the provisions for the cooling air flow and its access to the heat-dissipating surfaces.

Before any change is contemplated to valve size or valve-gear design, as a means of increasing engine performance, the whole breathing system should be investigated to avoid restrictions in inlet and exhaust tracts, and the best possible port and valve shapes should be evolved by flow tests. The coefficient of discharge of the average valve is about 0.65 – i.e., its effective area is about two-thirds of its nominal peripheral area. This means that the valve area has in fact to be about 50 per cent oversize to pass the desired flow. For this a high price is paid in the design and performance of the operating gear. The actual valve weighs about 85 per cent more than an ideal one of 100 per cent aerodynamic efficiency (discharge coefficient of 1.0), if this were possible. It needs 25 per cent more lift, and operating forces nearly $2\frac{1}{2}$ times as great. This puts port design in perspective by showing that improvements here are more worth seeking than increases in valve size and area, and should come first. Good aerodynamic design is even more important than good mechanical design.

Alternatively, the valve should be used more efficiently. This can be done by avoidance of carburettor or tract restrictions, or by the use of fuel injection whereby the fuel is atomized by pump pressure.

Valve speed

This quality is analogous to piston speed; every valve gear has a maximum figure which can be used as a measure of its dynamic performance. Valve speed is defined as

$$V_v = \text{lift} \times \text{r/min/opening period}$$

or

$$V_v = h \cdot N/\phi$$

the opening period being measured in crankshaft degrees.

Valve speed can be used to compare the performance of different designs of valve gear, or of valve gears in different engines. However, comparisons must be made at some constant condition, either a peak-power engine speed or valve-bounce engine speed, or at the same piston speed or valve-gear stress.

As explained earlier, the choice of valve sizes against required piston speed is made early in the design of an engine. Also the designer presumably knows his usual or preferred h/d ratio; therefore, since

$$N = 6V_p/S; \quad \text{then} \quad V_v = (6/\phi)(h/S)V_p$$

where V_p = piston speed and S = stroke.

In the first stages of the design, an estimate can be made of the sort of valve-gear performance needed to reach the required engine performance. The type of valve gear can be selected with the certainty that it will meet the engine performance with the valve size necessary. Valve gear stress will be known, as shown later.

Valve opening area

If we take as an example the simple constant-acceleration cam, in which the lift curve is calculated by assuming constant values for the acceleration and deceleration of the valve, then for one lift cycle the product 'time-area' is given by the following

$$A_t = (\pi/3)[(R + 2/R + 1)]h \cdot d \cdot \phi \cdot \omega$$

Where R = ratio acceleration period/deceleration period = θ_a/θ_d, h = lift of valve, d = diameter of valve, ϕ = total duration of opening, ω = crankshaft speed, $\phi/\omega = t$ = time of opening, or

$$A_t = (\pi/18)[(R + 2)(R + 1)]h \cdot d \cdot \phi^\circ/N$$

where ϕ° = opening period and N = crankshaft r/min.

Since the camshaft is rotating continuously and its cycle frequency equals half crankshaft speed, a valve has in fact a constant average area of opening which does not vary with speed. Multiplying the above figures for one cycle by the frequency gives this equivalent orifice area for a valve

$$A_0 = \pi[(R + 2)/(R + 1)]h \cdot d \cdot \phi^\circ/K_4$$

where K_4 depends on units chosen.

This mean area is approximately 20 per cent of the full-life peripheral area, $\pi \cdot d \cdot h$.

A mechanically operated poppet valve is equivalent to a constant-area fixed orifice. Consequently, piston and cylinder with valves can be visualized as a continuous flow pump, of the type of a Roots blower; these pumps operate between two fixed orifices, one representing the inlet valve and system and the other the exhaust valve and system. Gas conditions of pressure, temperature, and volume are different on each side. The pumping capacity of the piston is large, and the flow is always controlled by the orifices and gas conditions.

Let us now suppose that the cam is not of the constant-acceleration type, with a calculated lift curve, but is built up from arcs of circles. Then if the time-area of one lift is plotted and integrated, the mean orifice area with other sizes of valve and similar cams follow the relationship

$$A_v \propto h \cdot d \cdot \phi$$

So far, valve and piston sizes have been compared merely by the ratio of their diameters squared; this ratio is more often available than any other, and is a quick estimate of proportions. To express valve area in units of valve gear mechanical performance, it is necessary to consider the product $h \cdot d \cdot \phi$, since these are interdependent in the design of the valve gear when the influence of engine speed is taken into account.

Valve gear similarity

For valve gears to be 'similar' they must:

(1) be geometrically scaled linearly – i.e., valves, springs, operating gear, and cams all have the same linear size relationship with, say, the valve head;
(2) run at 'scale speed' – revolutions per minute varying inversely with valve head diameter;
(3) have the same design of cam, one in which the lift is the same function of the angle – or, alternatively, the cams are scaled;
(4) have the same ratios h/d and θ_a/θ_d and the same opening period.

Such gears then have:

(a) at their 'scale speeds', the same maximum valve velocity, the same velocity at the same cam angles, and the same valve speed;
(b) bouncing speeds inversely proportional to scale, and bounce occurring at the same valve speed;
(c) accelerations inversely proportional to scale;
(d) inertia forces proportional to d^2; however, valve springs to scale and at the same stress will give the requisite spring loads, in proportion to the square of their dimensions;
(e) constant spring loads in units of valve head area and inertia forces/ per cam base projected area (diameter × width); surface stresses of cams and tappets will be the same;
(f) the same tensile, compressive, bending, or torsion stresses;
(g) the same bearing loads and rubbing speeds.

Valve gear stresses

As in the case of piston motion, valve gear motion is of harmonic type and can be compared by means of comparison formulae based on s.h.m.

In similar valve gears

$$\text{Stress} \propto \frac{F}{A} \propto \frac{M \cdot r \cdot \omega^2}{A} \propto \frac{M \cdot h \cdot N^2}{A} \propto \frac{d^3 \cdot h \cdot N^2}{d^2}$$

and

$$d \cdot h \cdot N^2 = \frac{d}{h}(h^2 \cdot N^2) = \frac{d}{h} V_v^2$$

The accuracy of this method of stress comparison depends upon the degree of similarity in the valve gears, and the designer has to use his judgment in applying it. Valve gears must be of the same type – e.g., pushrod or ohc – and of substantially similar mass relative to the valve; also they must have similar values for the ratio θ_a/θ_d.

For a family of similar valve gears to any one design or set or proportions, we can write for all of them

$$\text{Stress} = k \cdot d \cdot h \cdot N^2; \quad \text{or} \quad = k \cdot \frac{d}{d} \cdot V_v^2$$

This formula can be used to compare the required stress in springs, or in tappets, rockers, rocker shafts, etc., over the intended range of speeds. The constants will differ for different designs or types of valve gear.

Should all members of such a family of similar valve gears be required to work at the same stress figures everywhere – as is normal practice – then the above formula can be written

$$k = d \cdot h \cdot N^2; \quad \text{or} \quad d \cdot h = k/N^2$$

In this form it demonstrates two important principles:

(1) for a given design, if stress is a limitation, d is interchangeable with h (there are well-known aerodynamic limitations on this, but mechanically it is true);
(2) if d is interchangeable with h, only the choice of engine speed (not that of valve diameter) determines the peripheral opening area that can be provided.

For the same stresses and bouncing r/min, the required valve area can be provided by a small valve with a large lift or by a large valve with a small lift. The size of valve shown on a drawing does not in fact determine the area; only the choice of bouncing r/min does that. It can be argued that the large valve can always be made to lift high enough by changing the size of spring, and so on, but this applies also to the small valve. Consequently, it is also possible to write for a family of similar valve gears

$$A_v = d \cdot h = k/N^2$$

This explains why engines of widely differing *B/S* ratios, and with different sizes of valves, sometimes give the same powers, the output not being connected in any way with that ratio. Valve area is controlled not by the chosen valve size but by the valve-gear dynamic performance – in fact by the attainable valve speed, as will be shown in the next section. An increase in valve area commensurate with a change in *B/S* ratio can only be provided by improving the attainable dynamic performance of the valve gear.

Valving of engines

A prior knowledge of what valve-gear performance will be required and what design changes must be made to the gear to give this with certainty, avoids the usual and generally familiar engine design sequence.

(1) Selecting hoped-for power.
(2) Choosing small displacement and high speed.
(3) Drawing largest possible valves in the head; choosing 'tight' timing for good low-speed performance.
(4) Designing valve springs and cams, and then finding that, due to excessive stresses, the lift is hopelessly inadequate.

It is worth repeating here that good aerodynamic efficiency of the existing valves should be verified by tests before the complete valve gear is subjected to any redesign or uprating likely to involve trial-and-error mechanical development. If more area and better valve-gear performance really must be designed in, it is usually preferable to achieve them without changing existing satisfactory practice as regards stresses, materials, hardness, etc. Larger or faster-running valves, and therefore a higher potential specific power output, can be provided by altering the duration of opening, the number of valves, or the attainable valve speed.

Duration of opening

Since $V_v = h \cdot N/\phi$, then $N = \theta.\ V_v/h$. It follows that r/min can be increased in direct proportion to duration, at constant lift, with no increase in valve speed and stress. Alternatively, duration and lift can both be increased in the same proportion, with no change in the valve-bounce speed or stress. Modified springs will be necessary to provide the same loads at higher lifts. The change of ϕ only is equivalent to a reduction in camshaft speed relative to crankshaft speed: the ω in the equivalent valve gear s.h.m. is reduced.

Number of valves

If one inlet valve with its gear is replaced by two similar ones with their gear, scaled down to give the same total area, then the diameter and lift of each of the smaller valves will be 0.707 of those of the original single valve. Again, the equivalent weight of each of the small valves and its gear will be only 35 per cent of that of the original valve (weight varies as d^3). The inertia forces involved in operating the smaller valves at the same revolution per minute will be

$$F \propto M \cdot r \cdot \omega^2 \propto d^3 \cdot d \propto d^4 = \tfrac{1}{4}$$

while stresses

$$\propto \frac{M \cdot r \cdot \omega^2}{A} \propto \frac{d^3 \cdot d}{d^2} \propto d^2 = \tfrac{1}{2}$$

Together the two valves only require half the previous total force to operate them, and will give rise to only half the stresses of the original valve.

Alternatively, to operate at the same stresses as the original single valve, the two smaller valves can be speeded up $\sqrt{2}$ times – or about 40 per cent. They will still pass the same amount of air, since this is proportional to the product $n \cdot d \cdot h \cdot \phi$ and is independent of the speed of operation. The engine will be turning 40 per cent faster, so the piston displacement can be cut down to $1/\sqrt{2}$ (or 0.707) of the original value (for example, by reducing the stroke) without reducing the pumping capacity and potential power. Though running 40 per cent faster, the two valves will still require only the same total force to operate them as the original single valve.

If the original displacement is retained, however, the single inlet valve could be replaced by two similar ones of greater combined area, and running faster, without altering the valve speed and stress. Assuming the ratio d/h to be invariable, the airflow capacity is given by

$$Q \propto n \cdot d \cdot h \propto n \cdot h^2$$

If $V_v \propto h \cdot N = \text{constant}$, then $N \propto 1/h$

$$N \propto n \cdot h^2 \propto n/N^2; \quad \text{or} \quad \propto \sqrt[3]{n}$$

Therefore, a 26 per cent increase in area and speed is possible with double valves – or 44 per cent with triple valves – for the same stress. Since single inlet valves are sometimes 20 per cent larger in diameter than exhaust valves, it may only be necessary to lengthen the exhaust period slightly to balance the valves' speeds and stresses for the case of two inlets with one exhaust.

Attainable valve speed

If in a given engine no change is proposed in the number of valves, overall valve timing durations, or the ratio *d*/*h*, how would the dynamic performance of the valve gear have to be altered to meet a required increase in engine performance? Valves might have to be bigger, and at the same time work faster, if the rise in power involves higher crankshaft speeds. For a given cylinder, the airflow capacity

$$Q \propto N \propto A_v \propto d \cdot h \propto \frac{d}{h} \cdot h^2 \propto h^2$$

or

$$d \propto h \propto \sqrt{N}$$
$$V_v \propto h \cdot N \propto N^{3/2} \propto Q^{3/2}$$

If no change is made to the design, and the valves and gear are merely scaled up,

$$\text{Stresses} \propto V_2^2 \propto Q^3 \propto (P)^3$$

On the other hand, if a change is made to the valve-gear design, to enable it to work at higher valve speed with no increase in stress, then

$$Q \propto N \propto V_v^{2/3}$$

That is, a design alteration to double the attainable valve speed, say a change from heavy pushrods to a light ohc layout, permits 26 per cent larger diameter valves and 60 per cent greater potential air flow.

One method of achieving the change in valve speed is to reduce the effective weight. If the calculations for the valve motion are done in respect of the valve itself and not the cam, then the effective weight at the valve is the weight of the valve parts plus the weight of the tappet-side parts (if there is a rocker), reduced as the rocker arm ratio squared should the tappet-side arm be the shorter.

Spring diameter can be changed in proportion to that of the valve. A spring scaled up in both coil and wire diameters will provide a force, at constant stress, in proportion to the square of its diameter. Then to raise the bouncing speed of a valve gear

$$V_v^2 \propto a \cdot h \propto \frac{W_{vo}}{M_v} \cdot h$$

where W_{vo} = spring force, valve open, and M_v = valve gear effective weight. The lift h is retained in these formulae to make them dimensionally correct regardless of scale.

With spring diameter a variable, the preceding statement should be written

$$V_v^2 \propto (D/d)^2 \cdot h/M_v$$

or

$$V_v \propto (D/d)(h/M_v)^{1/2}$$

where D = spring coil diameter and d = valve head diameter.

Therefore, to produce a required change in valve speed, the design must be such that

$$M_v \propto (D/d \cdot V_v)^2 \cdot h$$

If the cam is derived from a calculated lift curve, then it may be possible when changing the valve gear design to change also the ratio $R = \theta_a/\theta_d$. This has only a small influence on the effective area of opening, but a large effect on the bouncing valve speed attainable. Pushrod valve gears sometimes have a rocker-arm leverage ratio unfavourable to the cam – to ease the duty of the valve spring – and an R value of about 1/4 to avoid overloading the cam by heavy initial acceleration. With light ohc valve gear this ratio may be as low as 1/8. This is part of the gain in changing to ohc: the bouncing speed attainable from a given valve spring is proportional to the time period θ_d, and would be increased in the proportion of 4/5 to 8/9, or by 11 per cent in this case, without change in ϕ.

It follows that $V_v \propto \theta_d \propto 1/(R+1)$. The previously stated formula then becomes

$$M_v \propto [(D/dV_v)\{1/(R+1)\}]^2 \cdot h$$

or

$$V_v \propto (D/d) \cdot \{1/(R+1)\} \cdot (h/M_v)^{1/2}$$

Development or change?

It is not always necessary to change the design to get better valve-gear performance. This mechanism responds to 'development', as does the rest of the engine. Most designers are familiar with the routine which starts with stronger valve springs, followed by improved design or materials for the tappets, pushrods, and other components, until reliability is attained at a new and higher valve speed.

The acceleration diagram of the cam should be checked against the lift curve, to make sure that the acceleration is sufficiently harmonic in form to use the spring force fully. An ideal cam form has part of a sine curve for the lift during the acceleration period, so that the shape of the deceleration period is part of another sine curve, and the spring force can be correct at all lifts at maximum speed.

Finally, the means available to improve the breathing of an engine can be combined into one formula. It is assumed as a first approximation that the power is proportional to the air consumption – a simplification that the mechanical designer must make if he is to reduce the problem to usable terms and that

$$(P/L_c) \propto (Q/L_c) \propto N$$

Then, if L_c, V_v, ϕ, and n are all variable, but the ratio d/h is fixed

$$Q \propto P \propto L_c \cdot N \propto n \cdot h^2 \cdot \phi$$

therefore

$$h \propto (L_c \cdot N/n \cdot \phi)^{1/2}$$

also

$$V_v \propto (h \cdot N/\phi) \propto (N/\phi)(L_c N/n \cdot \phi)^{1/2} \propto (L_c/n)^{1/2} \cdot (N/\phi)^{3/2}$$

therefore

$$Q/L_c \propto P/L_c \propto N \propto V_v^{2/3} \cdot \phi \cdot (n/L_c)^{1/3}$$

This is a general equation connecting power per litre with poppet valve-gear performance required, in any size of cylinder. If engines compared under these formulae have valve gears of similar design, then in those valve gears

$$\text{stress} \propto V_v^2 \propto (L_c/n)(N/\phi)^3$$

The valving equation

$$\frac{P}{L_c} \propto \frac{Q}{L_c} \propto N \propto V_v^{2/3} \cdot \phi \cdot \left(\frac{n}{L_c}\right)^{1/3}$$

can be written, multiplying through by $L_c^{1/3}$, as follows

$$\frac{p}{L_c^{2/3}} \propto N \cdot L_c^{1/3} \propto V_v^{2/3} \cdot \phi \cdot n^{1/3}$$

where $p/L_c^{2/3}$ is a 'non-dimensional' form of power per unit of piston area and $N \cdot L_c^{1/3}$ a 'non-dimensional' piston speed. Both are independent of S/B ratio. They are units of comparison of engine performance, but not so significant to the engine designer as p/L_c or N, which are measures of power/weight ratio.

The equation can also be written

$$Q \propto P \propto L_c \cdot N \propto (V_v \cdot L_c)^{2/3} \phi \cdot n^{1/3}$$

increase in ϕ allows useful increases in valving and potential airflow. It is equivalent to reducing the speed of the camshaft relative to that of the crankshaft. At constant displacement

$$Q \propto P \propto N \propto \phi \propto V_p$$

Although it is well known that there is a useful limit to ϕ, its increase is worth investigation, because of the consequently-permitted relaxation in valve-gear duty or increase in effective valve area. This point is now being exploited by companies fitting automatic gearboxes to their cars, the engines of which are no longer required to pull hard at low speeds in top gear. Again, it must be emphasized that a mere change in ϕ in an existing engine is not enough; the full power increase cannot be hoped for unless the inlet and exhaust tract areas are increased in proportion to maintain conditions of gas velocity and pressure. If no increase in power is required, then an increase in ϕ could be a means of reducing the necessary engine displacement

$$d \cdot h \cdot \phi \propto Q \propto \text{hp} = \text{constant}$$

If P is to be constant, then

$$(V_v \cdot L_c)^{2/3} \cdot \phi \cdot n^{1/3}$$

is constant, and, therefore

$$h_c \propto \phi^{-3/2}$$

If $\phi \propto 1.305$, say (i.e., an increase from 230 degrees to 300 degrees), then $L_c \propto 0.67$. If both B and S are scaled down similarly

$$B^2 \cdot S \propto \phi^{-3/2}; \qquad B \propto S \propto \phi^{-1/2} \propto 0.875$$

Since $d \cdot h \cdot \phi$ is constant

$$d \cdot h \propto 1/\phi; \qquad d \propto h \propto \phi^{-1/2} \propto 0.875$$

The whole cylinder – B, S, d, and h – is scaled down in proportion

$$N \propto 1/L_c \propto 1.50$$

d/B is constant, but

$$V_p \propto \frac{d \cdot h \cdot \phi}{B^2} \propto \phi$$

These formulae define how the cylinder should be designed, but not what its performance will be, so obviously they should be used with discretion. Inlet tracts, carburettors, air silencers, and exhaust systems must all be scaled in area in proportion to the product $d \cdot h \cdot \phi$. Even then airflows should be checked on the final design. What these changes would do to intake and exhaust noise, carburation, starting, and slow running cannot be predicted quite so easily.

It is usual to start the design of an engine with the bore and stroke, but it is just as easy to start from a known performance by specifying the valve gear, and fitting the bore and stroke and number of cylinders to the performance of that gear. In fact, by the system explained here, piston engines can be designed as continuous-flow engines in exactly the same way as are turbines. The valves can be regarded as fixed orifices whose area is defined by the mechanical performance of the valve gear.

STROKE/BORE RATIO

There is no direct connexion between the stroke/bore ratio of an engine and the horsepower of specific output. Modern engines demonstrate that, with a given valve gear, piston area can be exchanged for piston speed without affecting power output.

Of the possible combinations of piston area and speed, two cases will be considered here. In the first, the revolutions per minute of the engine are constant, the bore being increased and the stroke reduced, at constant displacement, to reduce piston speed. The second case is that of constant piston speed: rotational speed is increased inversely as the stroke is reduced, again at constant displacement, to increase power and power per litre.

Constant revolutions per minute
Since the cylinder displacement is unchanged

$$L_c \propto B^2 \cdot S \propto B^3 \cdot \frac{S}{B} = \text{constant}$$

therefore

$$B^3 \propto \frac{B}{S} \quad \text{or} \quad B \propto \left(\frac{B}{S}\right)^{1/3}$$

If the piston, rod, and crankshaft are scaled with the bore, the inertia force per cylinder

$$\begin{aligned} \text{IF} &\propto M \cdot S \cdot N^2 \propto B^3 \cdot S \cdot N^2 \propto B^3 \cdot S \\ &\propto B/S \cdot S \propto B \propto (B/S)^{1/3} \\ V_p &\propto S \propto 1/B^2 \propto (S/B)^{2/3} \end{aligned}$$

Therefore

$$\text{IF} \propto B \propto (B/S)^{1/3} \propto V_p^{-1/2}$$

In spite of the reduced piston speed, the piston inertia forces increase directly with the bore. Assuming the crankshaft is scaled with the bore in diameters and lengths, its deflections under these inertia forces, which may affect smoothness of running, will vary as follows

$$\text{deflection} \propto \frac{W \cdot l^3}{D^4} \propto \frac{B \cdot B^3}{B^4} = \text{constant}$$

Deflections under gas loads – which vary as the square of the bore – will increase in proportion to the bore. Inertia stresses in rod and piston, and bearing unit loadings, will be reduced as

$$\text{Stress} \propto \frac{M \cdot S \cdot N^2}{A} \propto \frac{B}{B^2} \propto \frac{1}{B} \propto \frac{B^{-1/3}}{S} \propto V_p^{1/2}$$

To achieve this, bearings must be scaled up in diameter and length as the bore, and rubbing speeds will increase in proportion.

Therefore, let us suppose that a designer decides to change the S/B ratio of his engine from 1.2:1 to 0.6:1, on a 500 cm^3 engine.

The piston inertia forces would increase by up to 26 per cent, while stresses and bearing unit loadings would be reduced inversely as this, to 79 per cent of their previous values. Bearing rubbing speeds would be up by 26 per cent. The reduced stresses and unit loadings give a clue to possible further reductions in weight and inertia forces through the design of a lighter rod and big-end, not scaled up as much as the bore. However, these gains will not be significantly large.

Summing up for the constant revolutions per minute situation, since displacement and rotational speed are unchanged the pumping losses at the same power output will be substantially the same. Change in piston-friction losses will be negligible, since the reduced distance the pistons travel will be compensated by their greater weight, and by greater loads and side thrusts. Consequently, there is no fundamental reason for expecting lower piston speeds to effect appreciable improvements in engine smoothness, mechanical efficiency, fuel consumption, or life. There seems little point, too, in choosing a larger bore than will accommodate valves adequate for the required power output, unless future developments in valving and valve gear are planned.

When the low-piston-speed V-eight engines were first introduced in America, they replaced in-line sixes or eights which had two cranks per span between main bearings. The V-eights had five main bearings, and some had particularly stiff crankshafts and crankcases to permit possible higher compression ratios later. Their improved smoothness probably resulted from these features, and the low piston speeds disappeared subsequently in the 'horsepower race'.

Constant piston speed

Suppose that the designer of a new engine takes as his objective an increase in S/B ratio, still keeping the same displacement, but also attempting to keep the piston speed constant. He must provide for the revolutions per minute to increase inversely as the stroke. Let us first consider the required valving and valve-gear changes, without which there can be no increase in performance. The valve area must increase at least in proportion to the required rise in revolutions per minute and power or, for constant gas velocity, in proportion to the piston area. Then

$$\text{Power} \propto N \propto B^2 \propto \frac{1}{S} \propto \left(\frac{B}{S}\right)^{2/3}$$

If no change were made to the valve gear design, and existing valves were just scaled up in diameter and lift in proportion to the bore – and at the same time were required to run faster – the stresses would rise prohibitively. The valve-gear inertia forces would be as follows

$$\text{IF} \propto M \cdot h \cdot N^2 \propto \frac{B^3 \cdot B}{S^2} \propto B^8$$

$$\text{since } h \propto B \quad \text{and} \quad N \propto 1/S \propto B^2$$

$$\text{Stresses} \propto \frac{\text{IF}}{\text{area}} \propto \frac{B^8}{B^2} \propto B^6 \propto P^3 \propto \left(\frac{B}{S}\right)^2$$

Increasing the duration of valve opening ϕ, without changing valve diameter or lift, does not necessitate an alteration in S/B ratio. The revolutions per minute and piston speed of the engine can be increased in proportion without change in S/B or any penalty in valve-gear stress.

Alternatively an arbitrary change in S/B ratio can be made if desired. Suppose the designer wished to raise the specific output by increasing ϕ while holding the piston speed and displacement constant. Then

$$N \propto \phi; \qquad V_p \propto S \cdot N$$

If the displacement is unchanged

$$B^2 \propto 1/S \propto \phi$$
$$B/S \propto \phi\sqrt{\phi} \propto \phi^{3/2}$$

An increase in ϕ from 230 to 300 degrees – or by 1.305 times – allows a rise in S/B ratio to 1.49 times the original figure

$$\text{Ratio } \frac{d \cdot h \cdot \phi}{B^2} = \frac{\phi}{\phi} = \text{constant}$$

No change is made in d and h, so the value of $(d/B)^2$ is less for the same piston speed.

However, an increase of ϕ cannot be used as a stress relief to permit scaling up the valves and gear with the bore. At constant displacement

$$N \propto d \cdot h \cdot \phi; \quad V_v \propto \frac{h \cdot N}{\phi} \propto d \cdot h^2$$

It follows that stresses would rise rapidly unless the design were changed to permit operation at a higher valve speed, or to the use of multiple valves.

Multiple valves

Let us deal first with the latter possibility. Suppose the designer decides to increase the number of inlet valves from one to, say, two of the same design, but smaller, and capable of running at higher revolutions per minute at the same valve speed and stress. What size can the valves be and what will be the possible change in S/B ratio?

At the same gas velocity, mean effective pressure and piston speed, it was shown in a previous section that

$$p \propto N \propto n^{1/3} \propto B^2 \propto (B/S)^{2/3}$$

therefore

$$B/S \propto \sqrt{n}$$

Also

$$p \propto N \propto A_v \propto n \cdot d^2 \propto n^{1/3}$$

therefore

$$d \propto h^{-1/2}$$

By changing from one inlet valve to two of the same design, the S/B ratio can be changed from, say, 1.25 to $1.25/\sqrt{2} = 0.886$. The valve-head diameter and lift will be $1/2^{1/4} = 0.794$ times original figures, while the potential revolutions per minute and power increase will be as $2 \times (0.794)^2 = 1.26$ times.

Higher valve speed

Suppose ϕ, n, L_c, A_v/A_p, and V_p are all to be maintained as before, but a change of operating mechanism makes possible a higher valve speed

$$P \propto B^2 \propto (B/S)^{2/3} \propto V_v^{2/3}$$

therefore

$$B/S \propto V_v$$

This comes about as follows: due to the higher permissible valve speed, the valve diameter and lift can be increased until, at the new higher lift and revolutions per minute, the engine performs satisfactorily at the new designed valve speed.

Then

$$\text{r/min} \propto P \propto d \cdot h$$

$$V_v \propto h \cdot N \propto d \cdot h^2 \propto \frac{d}{h} \cdot h^3$$

It is assumed that $d \propto h$ and that d/h is unchanged

$$\text{r/min} \propto P \propto d \cdot h \propto \frac{d}{h} \cdot h^2 \propto V_v^{2/3}$$

Crankshaft torsional vibrations

On a six-cylinder engine a reduction of S/B ratio that might lead to higher revolutions per minute need not bring the destructive three/rev top torsional period into the running range. Torsional periods follow piston speed, and are not affected directly by change of S/B ratio. Assuming that the crankshaft is scaled up in diameters and lengths as the bore, and neglecting the effect of stroke on torsional stiffness, proof of the preceding statement is

$$\text{stiffness} \propto \frac{D^4}{L} \propto \frac{B^4}{B} \propto B^3$$

$$\text{inertia} \propto M \cdot r^2 \propto B^3 \cdot S^2$$

$$\text{frequency} \propto \sqrt{\left(\frac{\text{stiffness}}{\text{inertia}}\right)} \propto \sqrt{\left(\frac{B^3}{B^3 S^2}\right)} \propto \frac{1}{S}$$

That is, critical revolutions per minute, $N_c \propto$ frequency $\propto 1/S$. Critical piston speed, the product $N_c \cdot S$, is constant. The rule is the same with scaled similar engines. Therefore it is quite possible to design a short-stroke six-cylinder crankshaft to meet the situation satisfactorily.

Number of cylinders

If the stroke is reduced and the piston speed maintained constant, then the revolutions per minute increase and the engine should be smoother, since it fires more frequently at the same piston speed. Comparing engines of different S/B ratios, revolutions per minute, and numbers of cylinders, but of the same piston speed, total piston area, and power output, and the same frequency of firing

$n \cdot N =$ constant, therefore $n \propto 1/n$

If V_p is constant, $S \propto 1/N \propto n$

If the total piston area is constant, $L \propto S \propto n$

It is possible to design a short-stroke four-cylinder 2 litre engine to give the same power at the same mean effective pressure and piston speed, and to fire with the same frequency, as a long-stroke six-cylinder 3 litre.

Firing frequency in two engines of the same power output is proportional to the product $n \cdot N$. If the revolutions per minute and specific power output of one are greater, then for the same frequency $n \propto 1/N$.

Bearing inertia loads

It was shown in the section on piston and connecting rod stresses that, at constant displacement and piston speed – and assuming piston, rod, and crankshaft scaled with the bore (except in respect of rod length and stroke) – the following proportionality applies for rod stresses and bearing loads

$$\propto \frac{M \cdot r \cdot \omega^2}{A} \propto \frac{B^3 \cdot S \cdot N^2}{B^2} \propto \frac{B}{S} \cdot V_v^2 \propto \frac{B}{S}$$

since $S \cdot N = V_p =$ constant.

ENGINE MECHANISMS

Big-end bearings

In most attempts to uprate engines these components come off badly. The ratio A_b/A_p – or the big-end area as a proportion of the piston area – varies from about 30 per cent in touring cars to 55 per cent in racing car units. To meet higher inertia loads the obvious course is to increase bearing size, but this is, in fact, little help because the weight of a satisfactory big-end goes up faster than the area.

The clue to designing larger big-ends to carry greater loads comes from engine similarity. Bigger engines have higher inertia loads than small ones, and there is no reason to suppose there is any limit here if engines are strictly similar: the unit loading and the rubbing speed will be the same at the same piston speed. However, if the engines are strictly similar another condition will also exist: the comparative rigidities of the big-ends will be the same – that is, their deflections per inch of diameter will be the same. Therefore, in scaling up a big-end to carry higher load, the conditions are not all complied with unless the rigidity of the larger bearing is increased correspondingly. This is done if the big-end is strictly scaled up. If weight-saving is attempted in scale-up, it must not be at the expense of rigidity of rod or cap. More big-ends fail because rods are too light than too heavy.

Valve-gear comparison

Four typical valve mechanisms are illustrated in Fig. 1.1.

Pushrod and rocker

To appreciate the mechanical handicap under which this round-the-corner system has to work, the weight of material that has to be moved

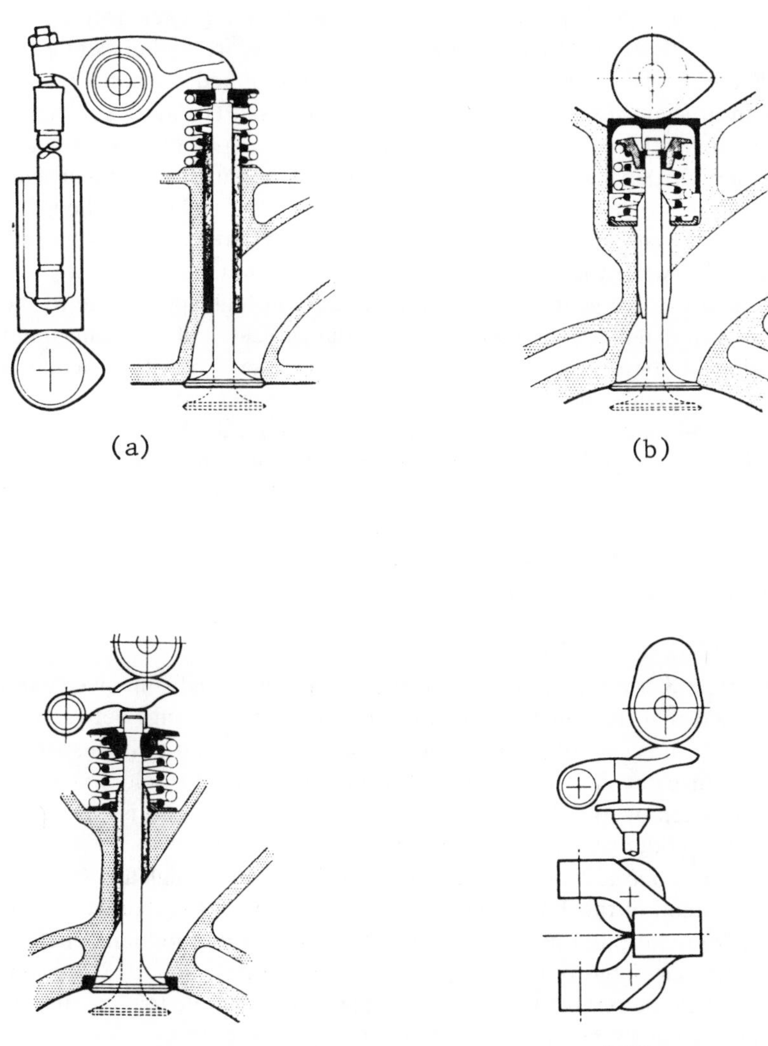

Fig. 1.1 Typical valve mechanisms compared. (a) Rocker, (b) cup-tappet, (c) finger lever, small ratio, (d) finger lever, large ratio

should be compared with the weight of the valve head, the object of it all. To help the spring to close this weight, a shorter rocker arm on the pushrod side is customary. Cam loads are therefore higher, while the deflections on the cam side are larger and are multiplied at the valve. With the insufficient support commonly used for camshafts and rocker shafts, these deflections are enough at high revolutions per minute to cause measurable departures from correct motion of the valve, and cam profiles have to be 'corrected' accordingly. The load on the rocker spindle is the sum of the loads on both sides – about three times that required to operate the valve – while that on the camshaft is even greater (Fig. 1.1(a)).

Pushrod valve gear tends to become noisy at high speeds due to the masses and deflections involved, and the difficulty of ensuring quiet seating of the valve. Also, it is usually necessary for the calculated bouncing speed to be about 35 per cent above the speed at which bouncing is required, because of the discrepancies just mentioned.

Pushrod valve mechanisms are best suited to long-stroke engines where the revolutions per minute are low relative to piston speed. The example illustrated is typical of many, but much better pushrod systems can be designed. They would be lighter, with equal-length rocker arms, return springs on the tappet side, scaled-up cams and stiff shafts.

Ohc with cup tappet

This type can be made quiet. Owing to the low weight involved, and provided that the camshaft is stiffly supported by bearings between all cylinders, it is capable of maintaining that quietness up to high speeds. Experience has shown that valve clearance adjustment is seldom necessary between overhauls, and need not be provided for; this is one of the system's hidden gains over the pushrod layout (Fig. 1.1(b)).

Another ohc advantage is that values of $R = \theta_a/\theta_d$ as high as 1/6 can be used instead of the 1/4 or 1/5 common with pushrods. Effective area is slightly greater, and bouncing speeds benefit. Another hidden benefit is that the calculated bouncing speed needs to be only about 12 per cent above the required bouncing speed – a big enough margin for error in cam profiles, deflections, spring surge, and the usual tolerances.

Racing type ohc; finger follower

Reciprocating weight is at a minimum with this layout. To realize an advantage over the type shown in Fig. 1.1(b), there must be an asymmetrical cam as shown in Fig. 1.1(d). By allowing for the change of leverage as the point of contact travels over the pad of the finger, this form produces a symmetrical lift curve. For smooth operation and long life, the rockers should trail on the cam as illustrated.

It was noted in the section on valve gear similarity that, in similar engines, inertia loads are constant. This means that large cams in large engines carry large loads. Obviously, therefore, there is advantage in scaling-up the cam relative to the valve by finger leverage as shown in Fig. 1.1(c). Higher cam loads can be carried for the same surface stress, with still further gain from the leverage ratio. If both cam width and diameter are increased, permissible valve accelerations should rise as the cube of the leverage ratio, allowing values of R even higher than the 1/8 used in ohc racing practice.

Twin-valve ohc layout

The S/B ratio can be reduced with this valve arrangement, since there is an increase in valve area together with higher possible revolutions per minute owing to the smaller and lighter valves. Figure 1.1(d) illustrates the change in leverage ratio desirable with multiple valves.

As explained previously, multiple valves will run at the same valve speed and stress as the single valve they replace when $d \propto h \propto n^{-1/3}$, and $N \propto n^{1/3}$, where n is the number of valves.

$$\text{Inertia force per valve} \propto m \cdot r \cdot \omega^2 \propto d^3 \cdot h \cdot N^2 \propto d^2 \propto n^{-2/3}$$

$$\text{Then if } r = \text{finger radius to the valve}$$

$$\text{and } R = \text{finger radius to the cam}$$

$$\text{Inertia force at the cam} \propto (r/R) \cdot n \cdot n^{-2/3} \propto (r/R) \cdot n^{1/3}$$

For this to be the same as with a single valve

$$r/R \propto n^{-1/3} \propto d \propto h$$

that is, the same cam should be used as with the original single valve, with the same finger radius to the cam, but the finger radius to the valve should be reduced as the diameter of the valve. With the scaled-down valves, this will give the correct lift, spring stresses, and cam stresses at the higher revolutions per minute.

In all valve gears, to avoid cam and tappet wear at very high loads, it is not enough merely to look after the surface loading in direct stress terms. It is necessary also to have flood or jet lubrication, surface hardness above Rockwell C-62 in extreme cases, and superfinish on both cam and tappet (or maybe an anti-scuff treatment).

Compression ratio and valving

It is a common delusion that large-bore engines will not accept high compression ratios. The ratio is not affected directly by *S/B*; it depends on the combustion-chamber volume and the swept volume, and not on the shape of the swept volume. A ratio of 12:1 has proved satisfactory on pump fuel, in an engine with an *S/B* ratio of 0.73:1.

If power is the objective, valve area should not be restricted to get a small combustion chamber and so reach a high compression ratio. More power is obtained by designing for valve area than for compression ratio. For example, if a combustion chamber and valves designed for 15:1 compression ratio are scaled-up by reducing the compressiuon ratio to 12:1, the volume increases by 27 per cent and the valve area by 17 per cent. The power lost by the reduced compression ratio, based on the air standard efficiencies, would be about 5 per cent, which, of course, gives a potential overall gain of 12 per cent.

It would be worthwhile to adjust the *S/B* ratio so as to accommodate the larger combustion chamber. As usual, the limit to this is how quickly the valves can operate, since they get larger and have to work faster at the same time. High compression ratios are difficult to achieve because of large valves and the clearance space required, not because of large bores.

CHAPTER TWO

Performance and combustion

This chapter comprises two sections, one on the limits set by existing physical, chemical, environmental, and operating conditions, and one on obtaining increased engine efficiency and output. After basic thermodynamic limitations comes a thorough description of the chemistry of combustion, flame propagaton, and cyclic dispersion effects; geometric limitations on performance by the valving; torque control for part-load economy; exhaust pollution control and designing inlet manifolds for the low-emission engine running; computer-modelling of carburettor fuel-evaporation, and the combustion process itself. In the second section an introduction is given to the filling and emptying technique used in performance prediction; next the parameters for enhancing power, economy, and pollution control are discussed with sections on combustion chamber design, lean burn and stratified charge. High compression ratio engines and variable valve timing are considered next, followed by case studies on engine development potential revealed by authors at Daimler Benz and BMW. The analysis of heat flow in engine components and an interesting study of cyclic dispersion effects explore two possible areas of future progress – to complete the chapter.

PERFORMANCE LIMITS SET BY PHYSICAL, CHEMICAL, ENVIRONMENTAL AND OPERATING CONDITIONS

Basic thermodynamic limitations

By examining the variations in the basic thermodynamic parameters – even throughout an individual engine cycle – W. T. Lyn[2] throws new light on the fundamental potentialities and limitations of the various combustion systems which he compares. He summarizes the solution to the overall problem of achieving high efficiency and output as follows:

> To achieve high efficiency, a high compression ratio has to be used, and the ignition will therefore be spontaneous. A premixed mixture is necessary to avoid carbon formation and hence increased air utilization, but must be prepared in small doses and consumed immediately so that high rate of pressure rise and detonation may be avoided. Such is the ideal combustion requirement which has so far defied, yet continues to challenge, the ingenuity of designers.

Available torque or brake mean effective pressure
It can be shown from simple energy considerations that

$$\text{bmep} = k1 \cdot AU \cdot \eta_v \cdot \eta_{br} \tag{2.1}$$

where

k = total energy of stoichiometric mixture in cylinder divided by swept volume (ideal maximum bmep)
AU = air utilization (per cent)
η_v = volumetric efficiency (per cent swept volume at ambient condition)
η_{br} = brake thermal efficiency (per cent)

Equation (2.1) can be written as

$$\text{bmep} = (k2 \times AU \times \eta_v)/\text{sfc} \tag{2.2}$$

where

sfc = specific fuel consumption

It is interesting to compare, on the basis of the above equation, the brake output of several current diesel and petrol engines – see Table 2.1. All figures are either directly measured or quoted from references, except those in brackets which are calculated from equation (2.2). As can be seen, the petrol engine derives its high output mainly from the high air utilization or fuel/air ratio, which is about 1.1 times the stoichiometric ratio for maximum power. This advantage, however, is partly offset by the higher specific fuel consumption as compared with the diesel. The

Table 2.1 Comparative performance of engine types

	Petrol, ordinary car		*Petrol, high-performance*	*Diesel, direct-injection*		*Diesel, indirect-injection*	
	1	2		1	2	1	2
a	1.1	1.1	1.1	0.64	0.64	(0.85)	(0.75)
b	(0.825)	(0.65)	(1.16)	(0.89)	(0.83)	0.77	0.75
c	298	304	283	228	231	469	456
d	931	717	1386	772	703	772	655
e	2500	5000	5000	1800	2600	2000	3500

a = air utilization; b = volumetric efficiency; c = specific fuel consumption (g/kWh); d = bmep (kN/m^2); e = speed (r/min).

bmep of the high-performance racing engine is, of course, due to the high volumetric efficiency created by 'ramming'.

Although the low-speed bmep of the direct-injection (d.i.) diesel engine is similar to that of the smaller indirect-injection (i.i.) unit, the two are obtained by different means: the d.i. engine has the better specific consumption but lower air utilization, whereas the reverse is the case for the i.i. engine. At high speed (3500 r/min) the air utilization of the d.i. engine, in spite of its better efficiency, is so reduced that the bmep cannot keep pace with that of the i.i. unit. It is suggested that the simple relationship shown in equation (2.2) serves as a useful comparison between different engines, and points the way in which improvements in efficiency are likely to be made.

Brake thermal efficiency

Although brake thermal efficiency is obviously important in its own right, it also contributes to the high bmep, as can be seen from equation (2.1). The brake efficiency can be written as a product of four component efficiencies

$$\eta_{br} = \eta_c \cdot \eta_l \cdot \eta_i \cdot \eta_m \tag{2.3}$$

where

η_c = combustion efficiency

$$\eta_l = \frac{\text{heat liberated} - \text{heat loss}}{\text{heat liberated}}$$

η_i = indicated or cycle efficiency

and

η_m = mechanical efficiency

Some general remarks on each of these efficiencies for petrol and diesel engines may be of interest.

Combustion efficiency. In the case of a petrol engine, the main factor causing incomplete combustion under high load conditions is the fuel/air ratio. For maximum power, where the mixture ratio is about 10 per cent rich, the combustion efficiency is bound to be low (the theoretical value is only about 90 per cent). Even for stoichiometric or lean mixture, the combustion efficiency in a single-cylinder engine is not 100 per cent. One reason for this is the small-scale inhomogeneity of the mixture within the cylinder. In the case of multi-cylinder engines, uneven distribution between the cylinders contributes a further loss and is essentially controlled by the design of the carburettor and intake manifold. An obvious way of improving mixture distribution is, of course, to use petrol injection.

Another factor affecting combustion is ignition timing, which greatly influences the cycle efficiency. It should be noted that late timing also causes late, and hence incomplete, burning. Excessive throttling, such as occurs under deceleration conditions, is seldom considered but is important in that the back flow of the burnt gas may interfere with the orderly consumption of the mixture by flame propagation throughout the chamber. Since combustion in petrol engines is essentially a premixed, near-homogeneous type, the unburnt products contain little carbon, but mostly carbon monoxide and broken-down, partially oxidized fractions of the original fuel. Table 2.2 shows typical exhaust products under various operating conditions.

With diesel engines the main cause of incomplete combustion is poor mixing. The mixing process has to be completed within a very short time (about 2 ms at 4000 r/min) and before the expansion stroke reaches such a stage that the gas temperature becomes low enough to quench the chemical reactions. Moreover, since in diesel engines the fuel is introduced into the air at high temperature and density, cracking and polymerization take place so quickly that the exhaust product contains a considerable quantity of carbonaceous solid particles. At 80 per cent air utilization, even if only 1 per cent of the original fuel is transformed into smoke, the exhaust state will be completely unacceptable. This is the limitation that restricts the air utilization in a diesel engine and, by the same token, maintains the high combustion efficiency in comparison with the petrol engine. Table 2.3 shows the typical diesel exhaust under various operating conditions, and is to be compared with the petrol engine case in Table 2.2. Figure 2.1 illustrates the percentage by volume of the main constituents of the incompletely burnt products in the exhaust, expressed as a percentage of the original fuel input. Using this information and the exhaust composition in Tables 2.2 and 2.3, the combustion efficiency, defined as (total heat input $-$ heat in unburnt fuel $\div$ (total heat input), can be calculated. These results are also shown in the respective tables, and the difference between the two types of engine is remarkable.

Heat loss. The heat loss from the cylinder gases to the combustion chamber wall is proportional to the difference between the gas and wall temperatures, the wall area and the coefficient of heat transfer (dependent on the density and velocity of the gases). For optimum efficiency the maximum gas temperature generally occurs at about 20 degrees about top

Table 2.2 Typical petrol engine exhaust

Operating conditions	*Carbon monoxide* % by vol.	*Carbon monoxide* % heat input	*Hydrocarbons* % by vol.	*Hydrocarbons* % heat input	*Hydrogen* % by vol.	*Hydrogen* % heat input	*Mean combustion efficiency* %
Idling	5–14	14.5–40.6	0.05–1.0	1.8–36.0	1.7	4.1	59.5
Accelerating	2.2–4.7	6.4–13.6	0.05–0.3	1.8–10.8	1.2	2.9	80.8
Cruising	1–5	3.5–17.5	0.02–0.3	0.9–13.7	0.2	0.6	81.6
Decelerating	2.7–6.3	7.8–18.3	0.4–2.6	14.4–94.2	1.7	4.1	28.6

Table 2.3 Typical diesel engine exhaust

Operating conditions	*Carbon monoxide*	*Hydrocarbons* % by vol.	*Hydrocarbons* % heat input	*Hydrogen*	*Smoke* mg C/m^2	*Smoke* % heat input	*Mean combustion efficiency* %
Idling	negligible	0.017–0.055	3.5–11.3	negligible	43	2.1	90.5
Accelerating	negligible	0.018–0.021	0.9–1.1	negligible	493	0.6	98.4
Cruising	negligible	0–0.015	0–0.8	negligible	279	0.4	98.8
Decelerating	negligible	0–0.061	0–3.1	negligible	0	0	98.4

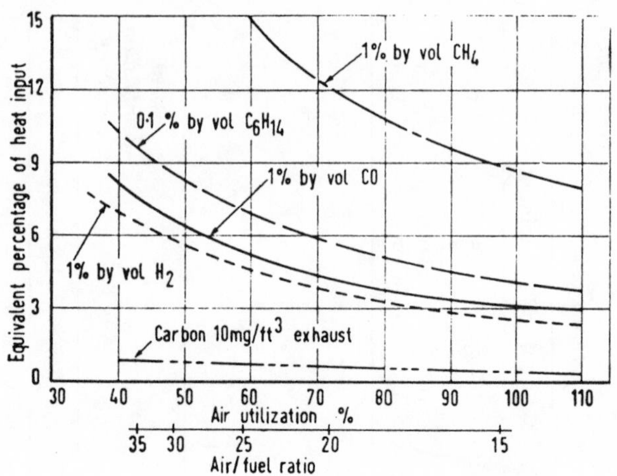

Fig. 2.1 Constituents of exhaust expressed as percentage of original fuel input

dead centre. The coefficient of heat transfer, for the type of chamber in which the swirl is not combustion-induced, as in petrol engines and direct-injection diesels, has its maximum somewhere around top dead centre. The chamber wall area, of course, is minimal at top dead centre and increases with the crank angle.

Several methods of calculating the heat transfer have been proposed. Although there are considerable variations in the absolute quantity, because each proposal is based on a particular type of engine, the distribution is remarkably similar. Figure 2.2 depicts heat flux distribution proposed or measured by different authors. The small differences in the peak position are due mainly to variations in the cyclic gas temperature profile. Only part of the heat loss, of course, is available for conversion into mechanical work. Any heat lost towards the end of the expansion stroke does not contribute significantly to the loss of efficiency because the availability or efficiency is zero at the end of the stroke. Figure 2.3 shows the indicated constant-volume cycle efficiency plotted against crank angle for engines of 20:1 and 10:1 compression ratio. With a realistic heat loss distribution as illustrated in Fig. 2.2, it can be shown that only about a third of the total heat loss can be converted into mechanical work for a 20:1 compression ratio, and a quarter for the 10:1 ratio.

There are two ways in which heat loss can be reduced. First, the

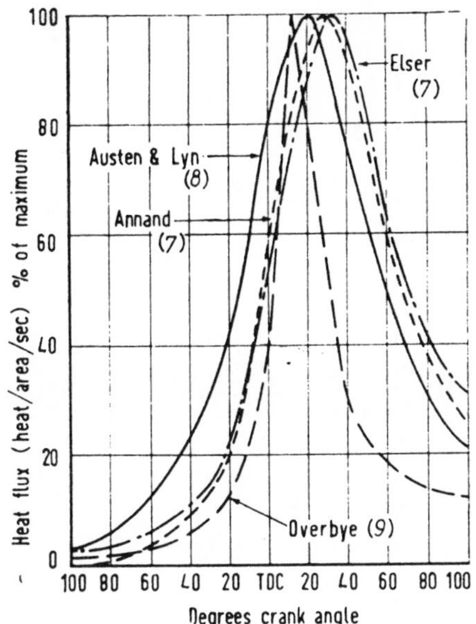

Fig. 2.2 Heat flux distribution within an individual engine cycle proposed (or measured) by different researchers

temperature difference between gas and wall can be reduced by increasing the wall temperature. Gas temperature at that point where the heat flux and the utilization factor are high is of the order of 2000°C, while wall temperature – except for the small insulated part of an indirect chamber engine – is never much more than 400°C, and generally much less. The temperature difference is, therefore, in the region of 1600°C. In order to reduce the heat loss by, say, 50 per cent, the wall temperature would have to be increased by some 800°C. This would present difficulty not only in the choice of material but also in piston lubrication.

The other way of reducing heat loss is to reduce the coefficient of heat transfer by lowering the air velocity. Here again, compromise has to be made because low air velocity means a low mixing rate and, hence, a low rate of heat release and air utilization. This can be compensated for, to some extent, by increasing the fuel velocity and improving distribution. There is, however, a limit in this direction, particularly at high engine speeds.

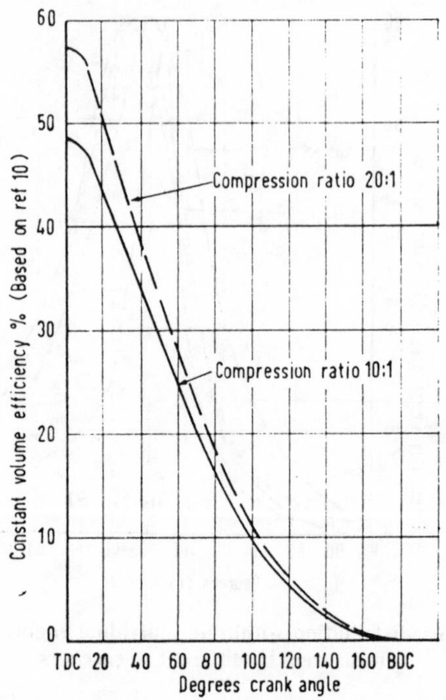

Fig. 2.3 Variation in constant-volume efficiency with widely different compression ratios

Some indicated work can be derived from the heat that is not lost as either unburnt mixture or flow to the wall. The efficiency with which the released heat can be converted into indicated work depends, of course, solely on the effective compression ratio – that is, on the crank angle at which the heat is liberated. Figure 2.3 already shows how this efficiency changes with crank angle and emphasizes the importance of completing the major part of combustion within 40 degrees about top dead centre. Incidentally, from the standpoint of indicated work, the efficiency would be the same whether the heat is released before or after top dead centre, so long as the crank angle is the same. In reality, though, any heat liberated before top dead centre increases the heat loss and peak pressure, the latter also decreasing the mechanical efficiency.

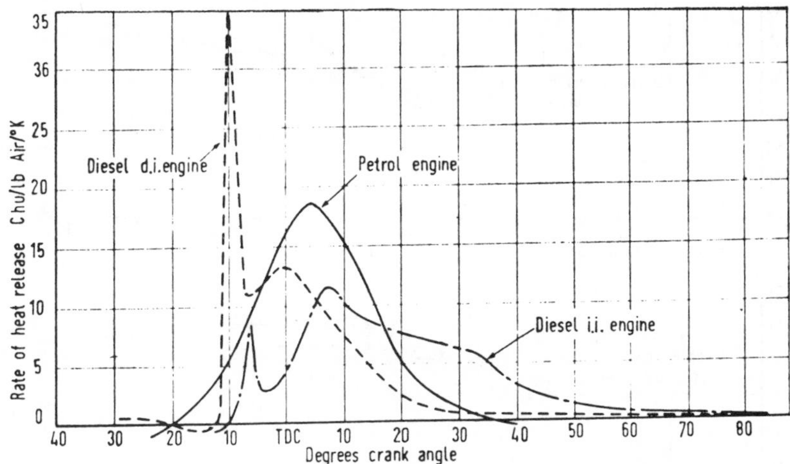

Fig. 2.4 Rate of heat release curves for both petrol and diesel engines

Cycle efficiency. Apart from its physical impossibility, instantaneous heat release at top dead centre is not desirable because it leads to excessive peak pressure. Normally, in a well-developed engine, be it petrol or diesel, the major heat release period covers about 40 degrees of crank angle (except in the case of the large, slow, marine diesel where a much longer release period has been recorded). Rate of heat release diagrams for a typical petrol engine, a direct-injection diesel, and an indirect-injection one, are shown in Fig. 2.4. Although the diagram shapes are quite different, the duration is very much the same in ease case.

A series of cycle calculations has been carried out for different shapes, duration, and timing of the heat release diagram, over a range of diesel compression ratios. The results can be summarized as in Fig. 2.5, in which the indicated cycle efficiency is plotted against peak cylinder pressure (which is varied by altering the timing). In this connection there are several noteworthy points. First, the indicated cycle efficiency depends, to a first approximation, solely on the peak cylinder pressure, and is independent of the shape of the heat release diagram. Secondly, for a given heat release period there is an optimum peak pressure beyond which no gain in efficiency can be obtained; this optimum pressure

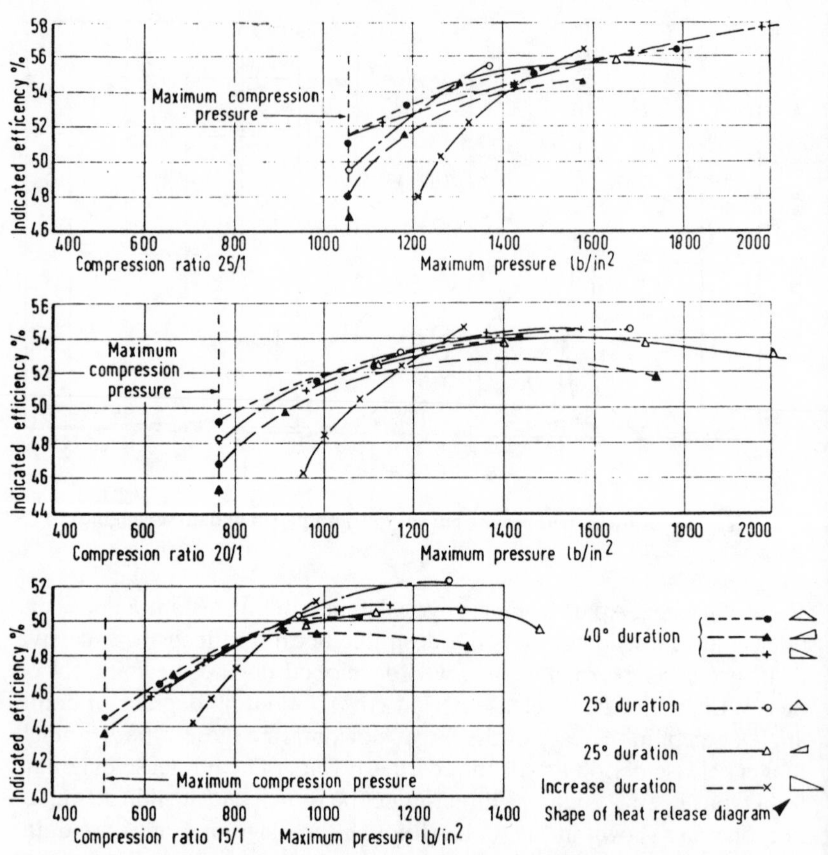

Fig. 2.5 A diesel engine example was used here to compare indicated thermal efficiencies for different heat release diagrams

increases with compression ratio. Thirdly, reducing the heat release period from 40 degrees to 25 degrees of crank angle gives only quite a small efficiency gain which is really not worthwhile since it can be achieved only with excessive increase in peak pressure. Experience has shown (Fig. 2.4) that 40 degrees crank angle duration is about the optimum. Although this investigation was carried out for the diesel cycle, the main conclusions are true also for the petrol cycle at a lower pressure range. In any case, the peak pressure in a modern high-compression

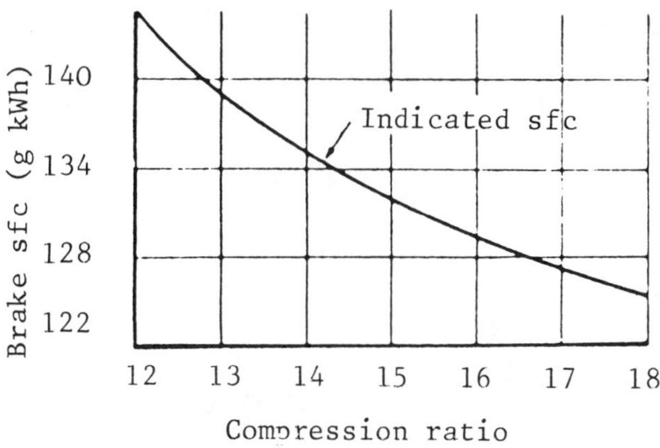

Fig. 2.6 Effect of compression ratio on engine efficiency

petrol engine is getting quite close to that of the diesel engine, as is the efficiency at full load.

Mechanical efficiency. It is clear that the mechanical efficiency directly influences the brake thermal efficiency and brake output. However, the subject is quite beyond the scope of the present review. It is proposed to mention here only how frictional losses affect the peak cylinder pressure and compression ratio. For instance, the optimum peak cylinder pressure for a given compression ratio would be lower than those indicated in Fig. 2.5 if brake thermal efficiency is considered. Again, the increase in maximum efficiency with rise in compression ratio will be less than is shown in the same figure, because this efficiency occurs at progressively higher peak pressures. One authority has assumed several rates of increase of frictional effective pressure with peak cylinder pressure, and has investigated the gain in brake specific fuel consumption as compression ratio rises, Fig. 2.6. It should be noted that the higher the frictional loss for a given peak pressure, the lower the optimum compression ratio.

Problems and prospects for the petrol engine

The foregoing points to the basic difference in the factors limiting the performance of petrol and diesel engines. Whereas the petrol engine has a high air utilization, leading to a high load or torque, its specific consumption at full load suffers because of lower combustion and cycle

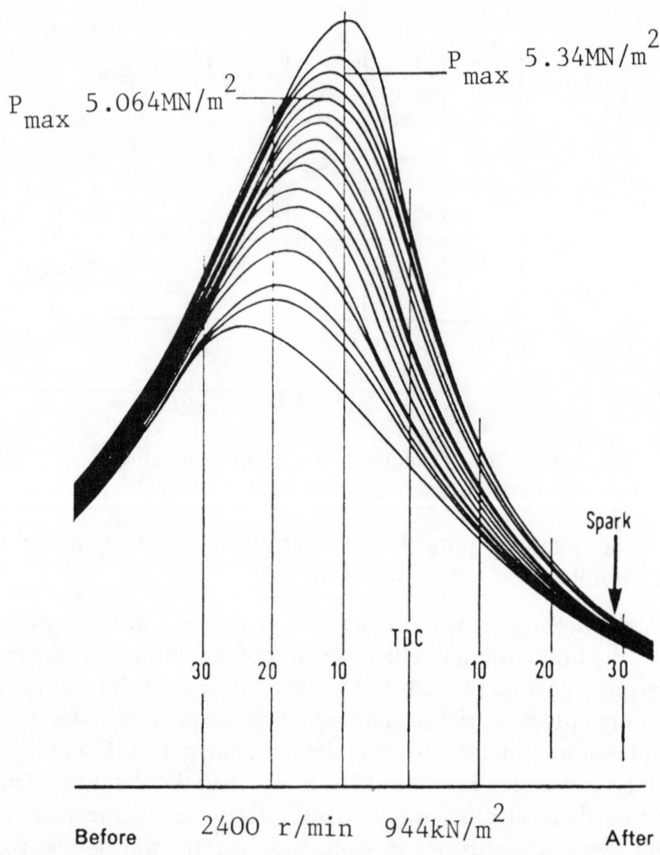

Fig. 2.7 Variation in petrol engine peak pressure over 17 cycles tested; these can cause up to 10 per cent loss in efficiency

efficiency (the compression ratio being limited by detonation or pre-ignition), and becomes worse still at part-load due to throttling. However, recent developments in fuel technology have greatly narrowed the difference in full-load performance, but that in part-load economy still remains. One attempt to solve the part-load problem has been the development of 'stratified charge' engines, having petrol injection.

Another source of efficiency loss in a petrol engine is cyclic irregularity in the pressure development. Figure 2.7 nicely shows this in the superimposed cylinder pressure diagrams from a petrol engine. It is estimated that, if all the cycles could be stabilized at the most efficient condition, there should be an improvement in indicated efficiency of 5–10 per cent. This irregularity is reasonably well established to be due to local variations in the turbulent velocity in relation to the flame speed. A multiple ignition source has been shown by one researcher to improve regularity.

Chemistry of combustion for SI engines

E. R. Norster[3] suggests that improvements in internal engine combustion can only follow from a better knowledge of the chemical kinetics of flame reactions. Such flames are defined as being the luminous gases associated with the liberation of large quantities of chemical energy, following rapid high-temperature oxidation of the hydrocarbon fuels – the fundamental chemical mechanism of combustion. Over the operating range of the conventional spark-ignition engine, pressures between 800 and 1400 kN/m^2 and temperatures between 400 and 500°C are the normal environmental conditions for the start of the combustion process. Whereas the amount of heat released and the equilibrium temperature of the products of reaction are not significantly affected over this range, the rate at which heat is released is particularly sensitive to the prevailing mixture conditions. Under the conditions mentioned, although the oxidation of the fuel would proceed without the aid of an ignition source, the process is forced to begin by the rapid production, from the spark, of active radicals that considerably reduce the induction period for the reaction. As the reaction proceeds through the combustion space, the pressure and temperature are raised, and therefore the rate of reaction increases according to the normal kinetic laws. A typical physical picture of the process in terms of the chamber pressure is shown in Fig. 2.8. The timing of the ignition is seen to be a vital factor in controlling the rate of rise and peak value of the pressure in the chamber.

If the ignition advance is considerable, this leads to high rates of pressure rise which result in the combination phenomenon known as knock or detonation, with its associated overheating and risk of damage to the engine. The same result can be obtained if the initial pressure and temperature of the mixture are raised by increasing the compression ratio. In such circumstances, the induction period of a local reaction may be sufficiently reduced at some point in the chamber for the reaction to proceed without assistance from the normal flame front. As the normal flame passes through the combustible mixture, the pressure and tempera-

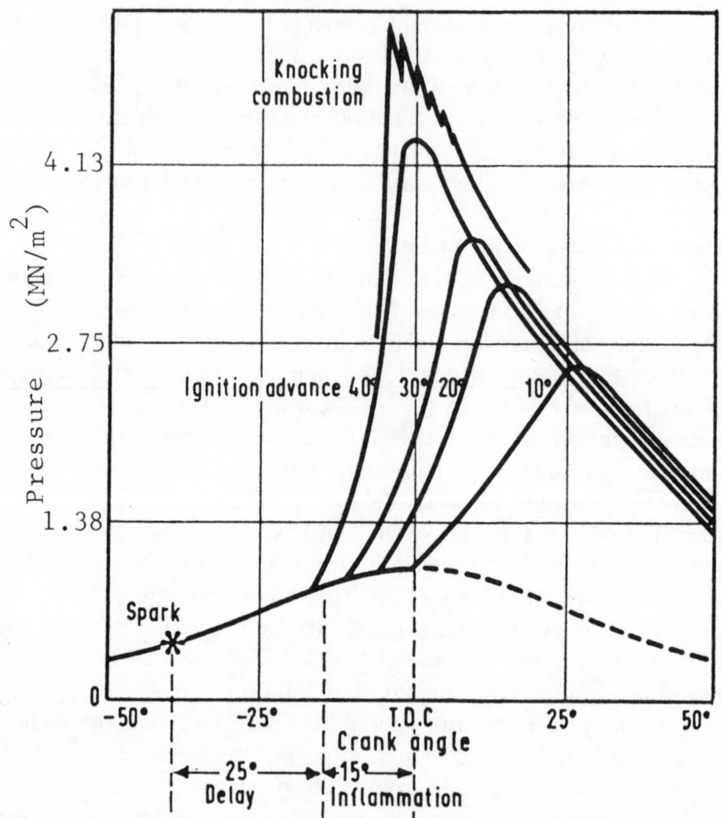

Fig. 2.8 Pressure variations (imperial units) during a combustion cycle showing the effect of ignition timing

ture of the local reaction are raised, and so is its rate. A situation can arise in which local branched chain reactions can proceed sufficiently quickly to cause an accelerated reaction rate and consequent detonation. The resulting pressure rise and normal flame propagation rate become very high and erratic, and a characteristic knocking noise is produced. Generally speaking, weak mixtures are much more favourable for detonation than rich ones, apparently because chain branching reactions are more prevalent.

Flame propagation

The ultimate goal of fundamental studies of flame propagation is a complete knowledge of the chemical kinetics of flame reactions. With this knowledge, the effect of variables on flame propagation rate could be predicted and the problems of combustion chamber design approached with greater confidence. Unfortunately, this knowledge is not complete; consequently, studies of the combustion process are necessarily conducted in terms of laminar flame velocity. Theoretical interrelations that are in accord with experimental results indicate this velocity to be a fundamental property of the mixture. For laminar flames, it is possible to define the flame velocity that, within reasonable limits, is independent of the surroundings. It would be desirable to define a turbulent flame propagation rate that would be independent of the surroundings and depend only on the fuel–air mixture and some easily defined property of the flow. Such is not the case, however, and values of turbulent flame velocity depend not only on the surroundings but also on the degree of turbulence.

Unlike the smooth and clear front of the laminar flame, the turbulent flame front is distorted, and as a result its surface is increased. Combustion products at high temperature are mixed with the unburnt mixture, igniting it at different points. In this case the mass of combustibles consumed in unit time and unit volume is far greater than for the laminar case. The turbulent flame speed has been found to depend on the laminar flame speed and, to varying extents, on the flow conditions defined by the Reynolds number.

Detonation and steady burning

The most obvious distinction between detonation and steady burning is in the speed of the respective wave of chemical reaction consuming the mixture. There is also, as a consequence, a great difference in the pressures produced by the reaction. Any burning wave has an associated compression wave ahead of it. Detonation is a particular case of burning in which the two waves travel at the same speed, which turns out to be the speed of sound in the products of combustion. This is the Chapman–Jouguet hypothesis, which enables the detonation velocity to be determined from the thermodynamic properties of the combustion products. From the fact that detonation occurs when the speeds of the reaction front and compression shock are equal, it can be inferred that the compression shock is responsible for the initiation of the chemical reaction, but there is no indication of how it does so. The most generally accepted reason for this phenomenon, however, is the heating effect attributable to the shock.

In the case of steady burning, the speed of the reaction wave is not determined solely by the products of the reaction; it also depends on the chemical structure of the reacting system in a way in which the chemical reaction rate is an important factor. Moreover, the same chemical system may give very different speeds in physical conditions that affect the transport of heat and radicals. Thus the speed in the combustible mixture increases when the flow changes from smooth to turbulent. The propagation process can be seen as one in which heat flowing from hot products causes the chemical rearrangement of the reagents, the rate of reaction increasing rapidly with temperature according to the normal kinetic laws. Where normal flame propagation occurs, the speed is limited by the rate of diffusion of heat from products to reagents. In gases, large-scale turbulence can progressively increase the speed of flame propagation until it approaches the steady detonation speed, and the shock ahead of the flame may become sufficiently intense to cause transition to the detonation state.

The burning of liquid droplets differs from that of a gas or vapour only in that the material has to be vaporized before it can react.

Engine factors in spark-ignition units

In a gasoline engine, after the charge has been ignited, the resulting flame takes a finite time to traverse the chamber. This time interval has an important bearing on the behaviour of the engine. In practice, too rapid burning or flame traverse causes excessive combustion pressures and high heat transfer rates. Burning at too slow a rate is uneconomical due to waste of heat during the expansion process. It is, therefore, important that the flame propagation rate should be correct for each chamber design and operating condition.

As indicated earlier, the flame propagation rate depends considerably upon turbulence. The greater the turbulence, the higher the propagation rate. An increase in engine speed increases the intake velocity and, hence, the velocity gradients in the chamber on the intake stroke. Although there is a considerable decrease in the turbulence level on the compression stroke, some sources have indicated that the velocity fluctuations in the chamber increase significantly with engine speed. In addition, considerable evidence has been presented by others to show that the mean flame travel time decreases linearly with the time of rotation. High-turbulence heads have a small clearance between part of the head and the piston when at top dead centre. When the piston approaches top dead centre, some of the combustibles in the clearance space are projected violently into the main part of the chamber. This high turbulence increases the rate of reaction and tends to prevent any stagnation of the combustibles, but it also serves to increase the heat transfer rates to the walls of the chamber.

The recent general increases in compression ratio might have been expected to introduce special problems for those responsible for combustion chamber design. This has not happened, though, and chamber shapes used today are generally refined versions of earlier types. Two main forms are in common use, the choice between them depending on whether high or normal levels of performance are required.

High-output chambers usually have intake valves in one line and exhaust valves in another (Fig. 2.9(a) and (b)). The chamber is sometimes part-spherical, and the spark plug is located in the top of the dome which makes a compact chamber. With this type of chamber, the flame front area increases rapidly and the length of flame travel is short. For lower levels of performance, however, the inverted bathrup or wedge chamber is currently favoured. The typical modern design shown in Fig. 2.9(c) bears considerable resemblance to earlier Chevrolet chambers (1939). Such chambers are compact and, with all valves in line, result in a neat cylinder head. Good idling and high-speed performance can be obtained from this type of chamber if the spark plug is situated near the intake valve.

The Otto cycle effectively fires a homogeneous mixture of fuel and air, and must operate within a narrow range of mixture strengths. When the mixture is set appreciably leaner than stoichiometric, misfiring begins and the combustion efficiency decreases. For maximum power it is necessary to have a relatively rich mixture, which also tends to reduce detonation.

Fuel factor

The structure of the hydrocarbon molecules forming the fuel of a spark-ignition engine, as well as the molecular size, can have a great influence upon the tendency to knock. The knock resistance of hydrocarbon types generally follows this descending order – aromatics, alcohols, napththenes, and paraffins. Molecules with a ring structure have greater resistance to knock than do unsaturated molecules, and it has also been

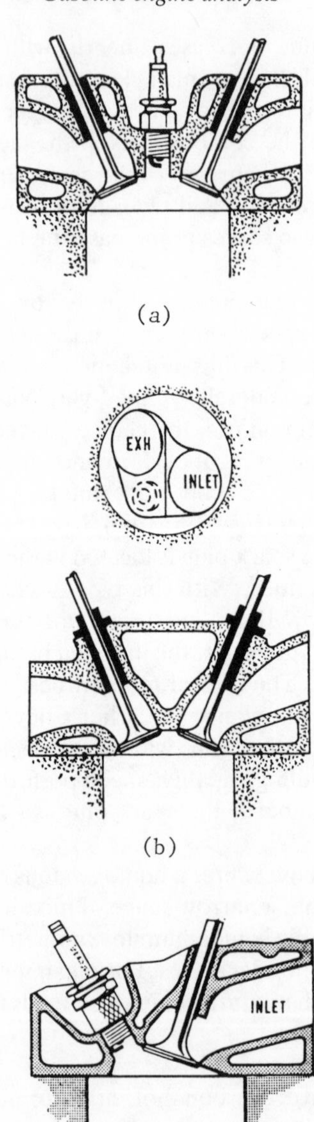

Fig. 2.9 Part spherical combustion chamber at (a), compared with pent-roof (b), and wedge types (c)

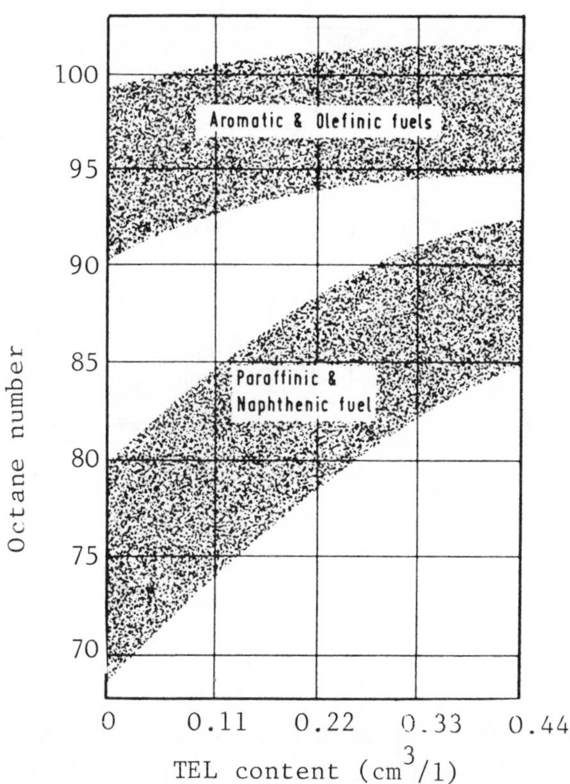

Fig. 2.10 Effect of lead additive on different fuels

found that the addition of methyl side chains raises the freedom from knock. An indication of these properties is given in Fig. 2.10, which shows the characteristic octane number for the different hydrocarbon types, together with the effect of adding tetraethyl lead.

The ever increasing demand for higher performance and improved economy is shown by the trend to higher compression ratios. Associated with this increased compression ratio is the increased octane number requirement of present-day chambers. To some extent these more stringent fuel quality demands can be offset by improvements and refinements in the combustion chamber design. The fuel factors that affect the turbulent flame propagation rate are of direct importance here and, being

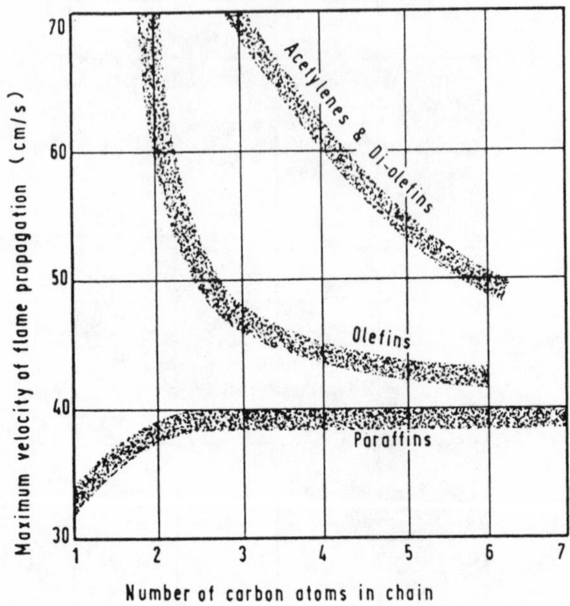

Fig. 2.11 Effect of fuel molecular structure on knock resistance

directly related to the constituents of the fuel, follow the trends of laminar flame speed. As a generality, propagation rates are higher with unsaturated hydrocarbons than with saturated ones of the same number of carbon atoms. The general effect of molecular structure is shown in Fig. 2.11.

Other high-compression-ratio combustion phenomena – which appear to limit compression ratios to about 10:1 – are known as 'wild ping' and 'rumble'. At high compression ratios, chamber deposits may glow during the compression stroke and thus initiate surface reactions in the mixture. This surface ignition can have the same effect as over-advancing the spark and can further raise the octane requirement of the engine. The intermittent knock arising from this type of ignition is called wild ping, and the tendency for it to occur increases with the lead content of the fuel because of the associated deposits. Rumble, on the other hand, is due to the simultaneous ignition of mixture at a number of different points in the chamber, causing very rapid burning and pressure rise which produces an unpleasant rumbling noise. These high-pressure phenomena are not yet fully understood, nor has a chemical cure yet been found.

Flame propagation and cyclic dispersion

G. G. Lucas and E. H. James[4] consider that one of the main factors influencing the development of the conventional spark-ignition engine in recent years has been the flame propagation process across the combustion chamber. There are many reasons for this, but perhaps its direct relevance to the exhaust emission problem is the most important. Of almost equal significance is the necessity to obtain the maximum power output from a given engine configuration without incurring knock. Another consideration which has now become of major importance is the desirability of predicting the performance of a spark-ignition engine at the design stage from computer simulations of the combustion process. (*A simulation by this research group is presented at the end of this section – Ed.*) Such a facility assists in the optimization of the engine design and operating parameters and can also be made to reveal more detailed information about the flame propagation process than can normally be obtained from measurements on an engine test bed. Essential to this ideal is the requirement of a detailed knowledge of the fundamentals of turbulent flame propagation in engine combustion chambers. Finally, the phenomenon of cyclic dispersion (the variation in the combustion development from cycle to cycle) must be investigated, since this is a direct function of the flame propagation process. One researcher has suggested that if the flame propagation rate under a given set of operating conditions could be maintained at a uniform constant maximum level, power increases from 2 to 10 per cent could be achieved. This would be accompanied by subsequent savings in fuel and emissions.

Flame propagation parameters for combustion simulation

The small flame kernel, which is generated at ignition, propagates outwards from the sparking plug through the premixed charge so that at any instant of time two distinct regions exist in the combustion chamber, viz, a region of burnt gas and a region of unburnt gas. These two regions have been found to be separated by a distinctly luminous flame front. The flame does not travel at a constant speed but, for a normal combustion, moves very slowly at first for 1 or 2 milliseconds, then gathers speed through the main part of the combustion chamber and finally slows down at the end of its travel. The period of time when the flame travels very slowly after ignition is often called the 'delay period' although, strictly speaking, there is no delay in spark-ignition engine combustion at all.

Treating this process in more detail, it is found that the spark provides the energy (in the form of heat and active particles) required to bring about the chemical reactions necessary to achieve a self-propagating flame. This only ensues when the release of heat and the diffusion of radicals from the spark is great enough to achieve a combustion reaction in the adjoining portion of the mixture. An explanation for the generally slow flame travel during the 'delay period' can be found in the relatively low temperatures of the burnt and unburnt fractions of charge at this time and in the visualization of a boundary layer covering the entire combustion chamber walls and separating them from the turbulent mean mass of charge. In this layer, the gas is practically stagnant and it is invariably in this that the sparking plug is positioned. Thus, until the flame has propagated out of it into the more turbulent mass, the burning velocity relative to the unburnt charge is predominantly laminar in nature. From correlations of computed and experimental pressure-time diagrams, the thickness of this layer has been estimated to be approximately 3.5 mm. On emerging from this boundary layer, the turbulent eddies of the main mass of charge act on the flame front, thereby enabling the turbulent burning velocity to 'take over' from the laminar burning velocity.

The main period of combustion follows the 'delay period'. This is the phase of the combustion process during which the flame speed and heat release rates are greatest since the temperature, pressure, and density of the unburnt gas is being continuously increased owing to its compression by both the expanding burnt gases and the movement of the piston when on the upstroke.

In normal combustion, the final period of the flame travel is characterized by the slowing down of the flame speeds and by a reduction in the heat release rates. One reason for this is that the piston is usually on its downstroke when this phase of the combustion process is reached. This has the effect of expanding the final unburnt charge volume with consequent reductions in its temperature and density. Additionally, lower levels of turbulence exist near the combustion chamber wall at this stage of the engine cycle.

It should be mentioned that the unburnt gas is subject to a further source of heat supply during flame propagation in addition to those from compression by the burnt gases and the piston motion. This is caused by some chemical energy release in the unburnt gas itself.

There is some controversy at present over the relationship between flame propagation and pressure development during the combustion process. On the one hand, there is the school of thought which disputes the premise that the complete chemical reaction and heat release occurs

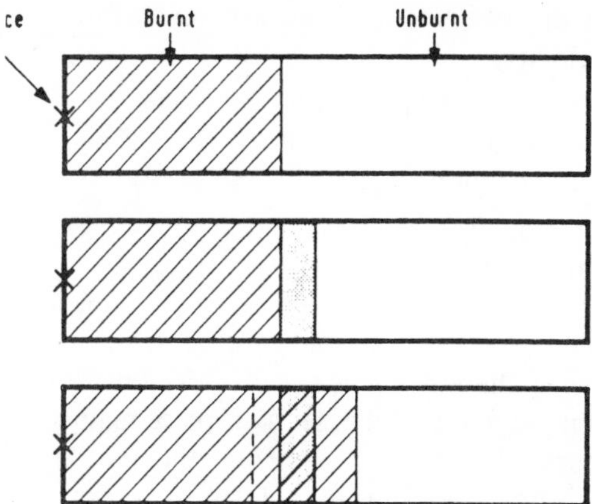

Fig. 2.12 Flame propagation at any instant (top), after burning of a small incremental volume (centre), and after expansion of this volume (bottom), into burnt and unburnt portions of the charge

in the flame front. In other words, the peak pressure is thought to be developed after the entire charge has been consumed by the flame. Others oppose these contentions, however, since their investigations have revealed that the heat release process is effectively completed at the flame front. This latter hypothesis is probably untrue of the processes which actually occur, since the chemical reactions take place at a finite rate and equilibrium is not instantaneously established. The extent to which the reactions are completed at the flame front is thought by some to be a function of engine speed.

Yet others have proposed that the flame propagation process can effectively be considered to consist of a continuous series of stages each of which comprises the combustion of a small volume increment and the subsequent expansion of this increment into the already burnt and the still unburnt portions of the mixture until pressure equilibrium is attained (Fig. 2.12). It is this process which generates the temperature gradient in the combustion chamber whereby the temperature of the burnt gases at the spark plug is higher than at the flame front. From Fig. 2.12 also it is clear that the flame speeds, which are measured or observed in engine combustion chambers relative to the cylinder head, are the vector sums of two fundamental quantities:

The linear velocity of the flame front with respect to the unburnt gas – the burning velocity.

The velocity of the unburnt gas itself in moving away from the flame front. This is due to the expansion of the burning gases and it is normally called the 'unburnt gas velocity'.

Thus, during the delay period, the 'unburnt gas velocity' accounts for the greater part of the flame speed since the expansion of the burnt gases takes place almost entirely in the direction of the unburnt portion of the charge. Towards the end of the flame progression in the combustion chamber, however (when the flame speeds slow down), the expansion of the newly burnt increments of charge takes place mainly in the direction of the already burnt gases (Fig. 2.12). Thus, the main component of the flame speed at this time is the burning velocity. These effects are shown in Fig. 2.13 taken from a computer simulation of combustion in the Renault

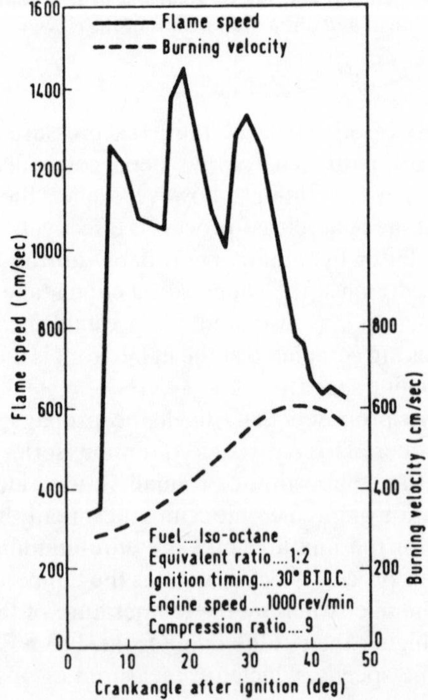

Fig. 2.13 Effect of burning velocity on flame speed measured on the Renault VCR engine

Variable Compression Ratio Engine. The unsteady flame speed values are a direct function of the technique used to obtain the fractions of charge burnt.

Factors in flame development for prediction modelling

It is sometimes assumed that the flame which propagates outwards from the spark plug is spherical in shape with the centre of the sphere at the sparking plug. Some works indicate, however, that the flame does not propagate spherically in any type of combustion chamber used in piston engines. Spherical combustion is only approached when a quiescent combustible mixture in a spherical constant volume bomb is ignited at the centre. Numerous factors are known to exert some influence on the manner in which a flame propagates through a combustible mixture in a combustion chamber. These include the following.

Combustion chamber design – this is obviously one of the main influences. Some researchers have proposed that the flame pattern development from the sparking plug is distorted in relation to the combustion chamber wall contour and the volumes ahead and behind the flame front.

Swirl – the main effect of a swirl is to deviate the flame pattern from a symmetrical path about the sparking plug and, in so doing, to increase the flame front area.

Piston movement – this affects the flame propagation pattern because combustion chamber shape is continuously changing. In addition, due to mass movements of the charge, the piston movement can change the position and contour of a burnt volume of charge especially just after ignition when it is small.

Charge homogeneity – inhomogeneity of the charge during combustion can lead to a condition where the air/fuel ratio at different points in the combustion chamber varies from very rich to very weak. During flame propagation, there will be an acceleration of the flame front in the richer areas and a deceleration in the weaker areas. Thus, the flame pattern development can be grossly distorted at the flame front.

Locally generated turbulence – experimental evidence indicates that whenever a flame front approaches a sudden change in the combustion chamber contour, there is an increase in the burning velocity. Such effects can be attributed to the generation of local turbulence. Its influence on the flame pattern is similar to the rich areas in an inhomogeneous charge.

Combustion chamber surface temperature variations – in all combustion chambers, considerable differences in the temperatures of the surfaces comprising the chamber are apparent. For example, some suggest that typical values for the cylinder wall, cylinder head, piston and exhaust valve are 395°K, 420°K, 520°K, and 610°K respectively. Experiments indicate that the flame moves faster over the hotter surfaces and more slowly over the relatively cooler surfaces.

Compression ratio – the influence of compression ratio is largely confined to its effect on the degree of flame curvature between the cylinder head and the piston since a spherical burning pattern is a good approximation in this plane.

Flame speed – this assumes its greatest influence on the flame pattern development whenever a large scale swirl is present, since engine operation under conditions which tend to increase the flame speeds result in a minimization of the effects of the swirl on the flame pattern.

Cyclic dispersion

The physical manifestation of the cyclic variation in flame propagation rates across a combustion chamber is a variation in the pressure development from cycle to cycle in the cylinder. Thus, basic research work in this field has attempted to minimize the extent of the cyclic dispersion and to try and trace its root cause. The following is a list of some operating conditions under which an engine must run in order to achieve minimum dispersion (for the flame to propagate uniformly and at a constant rate from cycle to cycle).

Good charge preparation – both in terms of mixture distribution between cylinders and in terms of air/fuel ratio and mixture motion homogeneity within individual cylinders from cycle to cycle.

Good cylinder scavenging – since the exhaust residuals left over from a previous cycle can be visualized as mixing in an inhomogeneous and random fashion with the fresh charge in the combustion chamber from cycle to cycle.

Engine operating conditions capable of sustaining the highest flame propagation rates without, of course, incurring knock.

No cyclic scatter in spark timing under a given set of operating conditions.

Even when such conditions are adhered to, however, cylinder pressure or flame speed measurements still indicate the presence of a certain amount of dispersion. Numerous factors, which are thought to influence the flame propagation process, have been investigated to try and reduce this even further and, if possible, to determine its fundamental cause. Such factors include variations in spark-ignition energy, type of spark, spark plug electrode shape, spark plug positioning, deposit accumulations, and compression ratio. Although, in some instances, slight improvements have been reported (by increasing the compression ratio, or when deposits are present in the combustion chamber), no one such effect has presented itself as the root cause.

This situation has resulted in the hypothesis being proposed that it is the cyclic variations in the gas flow velocities at a particular point in the combustion chamber which are fundamentally to blame. Earlier work appears to confirm this since it was noted that variations in the gas flow velocities at the same point in the combustion chamber of a single cylinder engine over four successive cycles took place as in Fig. 2.14. Measurements were made with a hot wire anemometer. The only remaining

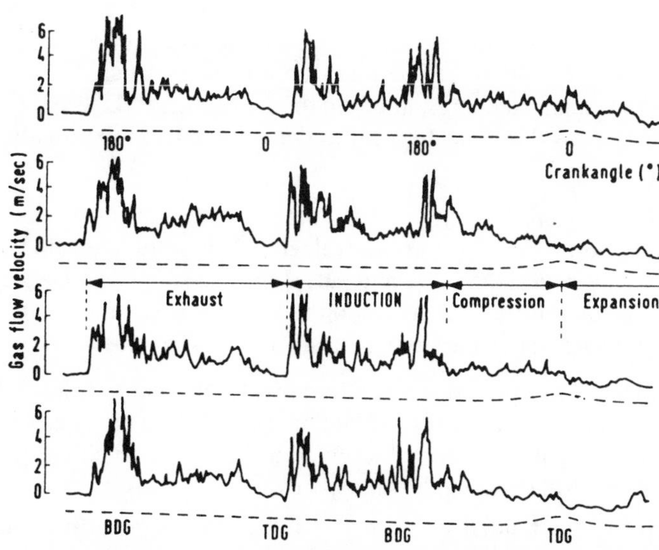

Fig. 2.14 Variations in gas flow velocities over four successive combustion cycles

doubt, therefore, appears to be centred on the following question – at what stage of the flame propagation process do these gas flow velocity variations exert their greatest influence in generating cyclic dispersion? In this connection, it has been suggested that it is primarily at the time of ignition and during the subsequent delay period that such effects are predominant. Alternatively, it is believed that the main influence is during the flame propagation through the bulk of the charge, since flame speed measurements have indicated variations in the speed of the fully developed flame.

Mechanisms of flame propagation

Since the flow in engine combustion chambers is predominantly turbulent in nature, except for the regions near the combustion chamber walls which has been shown to be mainly laminar, it is expedient to consider the mechanisms involved in laminar and turbulent flame propagation.

In laminar burning, the flame front is thin, continuous, and uniform and it progresses through the charge in a steady, eddy-free manner. Two limiting mechanisms of laminar flame propagation are conceived.

The thermal mechanism – in which heat conduction is taken to be the most important physical process involved in the transfer of the reaction zone from one layer of gas to the next adjacent to it. The temperature of the unburnt gas is considered to be raised to a point at which it ignites and releases the chemical energy stored in the fuel. The products of combustion then cool as they transfer heat to the yet unburnt mixture ahead of the flame front.

The diffusional mechanism – which maintains that diffusion processes principally determine the burning velocity. It is postulated that active particles and radicals (H, O, and OH), produced by dissociation and chain branching, diffuse rapidly into the unburnt gas in the local region directly ahead of the flame front, causing it to react explosively.

In turbulent flame propagation, alternatively, the flame front is much thicker and more complex than in laminar burning and the rate of burn is much greater. This is normally accepted as being due to either one or a combination of the three following processes.

The turbulent field distorts and wrinkles the flame front, thereby increasing the surface area across which combustion reactions occur. The normal component of the burning velocity, however, remains the laminar burning velocity.

The turbulence may increase the rate of transport of heat and active particles across the flame front thus increasing the actual burning velocity normal to the flame surface.

The turbulence may rapidly mix the burnt and unburnt gas in such a way that the flame becomes a homogeneous reaction the rate of which depends on the ratio of burnt to unburnt gas produced in the mixing process.

Turbulent burning theories, which are based on the assumption of increased surface area (first process, above) are given the name Surface or Wrinkled Flame Front theories. The remaining theories, which have their foundation in the second and third processes above, are defined as Three-dimensional or Volume Turbulent combustion theories. One researcher proposes that, in a given turbulent field, one of these groups of theories predominates. The criterion for the Wrinkled Flame Front mechanism to predominate is when

$$\frac{L}{V'} > \frac{\lambda}{U_{\mathrm{L}}}$$

where L is the scale of the turbulent eddies, V' is the root mean square turbulent velocity, λ is the width of the normal reaction zone, and U_{L} is the laminar burning velocity.

This is because the reaction in the flame front is faster at this time than the characteristic time for eddy fluctuation. The reverse case applies when the Volume Turbulent combustion theories predominate.

Effect of engine operation parameters on flame propagation rates

The difficulty in experimentally detecting the manner in which the flame speed is affected by a change in a particular operating variable is that other factors, which influence flame propagation, are altered at the same time. Thus, the unique effect of the variable is not ascertained. Nevertheless, the following general trends have been indicated from such experimental work.

(a) Increases in the temperatures of the burnt and unburnt fractions of charge result in corresponding gains in flame speeds (and vice versa).
(b) Increases in the degree of mixture motion give gains in flame speed (and vice versa).

These conclusions have been determined from observations showing that the flame speed increases with engine speed (Fig. 2.15(a)), attains a

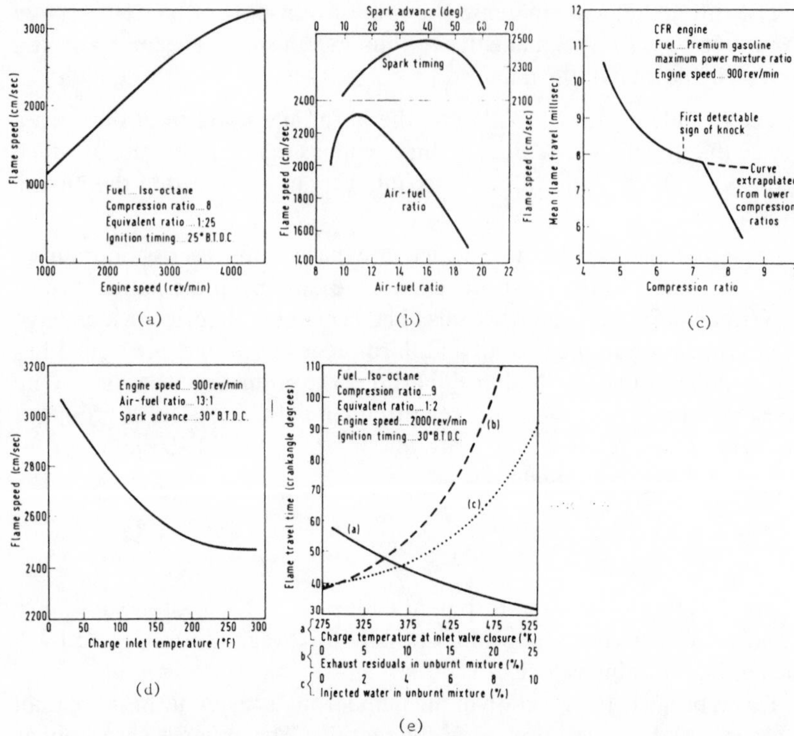

Fig. 2.15 Effect on flame speed and travel of (a) engine speed, (b) air–fuel ratio, (c) compression ratio, (d) charge inlet temperature, and (e) under the particular conditions stated below the diagram

peak value at relatively rich air/fuel mixtures (Fig. 2.15(b)), increases with increase in compression ratio (Fig. 2.15(c)), and has a pronounced peak value with spark timing variation (Fig. 2.15(b)).

However, some researchers found that the flame speed falls off markedly as the inlet charge temperatures increase which was most unexpected (Fig. 2.15(d)). It can, however, be attributed to corresponding increases in exhaust residual concentrations. This is an example of the unique effect of a particular variable not being obtained. The advantage of a computer simulation of the flame propagation process in obtaining such unique effects thus becomes evident. Such a simulation has been attempted in connection with the combustion in a Renault variable compression ratio research engine. For the case of inlet charge temperature rises. Fig. 2.15(e) was obtained under the stated conditions on the graph.

Geometric limitations on performance

According to Newlyn[5], benefits from the use of the pent-roof configuration of combustion chamber are evident in four areas: air consumption, thermal loading, valve gear dynamics, and thermal efficiency; the benefit of using two inlet valves as opposed to a single valve is in the potentially greater inlet flow area available. Considering dimensions given in three references by contemporary workers gives the relationships between inlet valve diameter D_I and the cylinder bore B in Table 2.4.

Table 2.4 Percentage increase in 4-valve curtain area over the 2-valve layout

D_I *(2-valve)*	D_I *(4-valve)*	*Increase* (%)
0.44B	0.33B	13
0.49B	0.36B	8.0
0.485B	0.39B	29.0

Figures in Table 2.4 assume that the valve lift/diameter ratio is kept constant. If the same lift is maintained, with doubled openings over the single valve, the figures become even more beneficial, being 50, 47, and 61 per cent, respectively. This comparison is purely based on the flow area available; added advantage occurs when the mean discharge coefficient is examined. The two-inlet-valve configuration exhibits an enhanced mean inlet discharge coefficient over the case of two isolated valves of the same design. Within a maximum lift set at L/D ratio = 0.25, the dual valve layout gives an improvement of 12 per cent over two isolated valves. If the L/D ratio is increased to 0.3 the benefit becomes 13 per cent. Thus the effective inlet valve area available is increased by this amount over the figures previously quoted, making the dual-inlet valve design even more attractive with regard to air consumption.

The 4-valve pent-roof chamber also benefits from the use of dual-exhaust valves, since smaller valves represent shorter heat flow paths. Cooling of the exhaust valves is further enhanced by the fact that the parameter available for heat transfer is increased in the same ratio as the basic flow area. As it is estimated that about 75 per cent of the heat loss from the valve is through the seat, this is particularly advantageous. The final benefit resulting from the use of dual-valves comes from the dynamics of the valve train. Whilst the use of dual-valves does not necessarily lead to a reduction in valve gear mass, it does reduce the valve gear characteristic speed (S_v). This is defined by one researcher as

$$S_v = V_p Q / n^{1/2}$$

where V_p = mean piston speed, Q = bore/stroke ratio, and n = number of inlet or exhaust valves per cylinder, whichever is less.

Hence, for the same bore/stroke ratio and piston speed, the valve characteristic speed of a 4-valve design is 0.7 that of a 2-valve. Alternatively, the piston speed can be increased without reaching dynamic limitations on the valve gear. Apart from the multiple valve aspect, the pent-roof chamber also has advantages for thermal efficiency. The research done by others on small displacement engines attempts to determine the 'actual combustion efficiency' of several designs of combustion chamber. Actual combustion efficiency η^*_{comb} is defined as the factor by which the air standard efficiency must be multiplied in order to obtain the actual indicated thermal efficiency. The data presented shows that the pent-roof chamber is comparable to the hemispherical chamber, having η^*_{comb} between 74.5 and 76.5 per cent. The wedge chamber is shown as 73–75 and the bath tub 70–71.5 per cent. This is almost certainly due to the excellent surface to volume ratio of the pent-roof and hemi-spherical chambers.

In addition to its importance with regard to heat loss, the surface to volume ratio has important implications with reference to exhaust emissions. The general opinion amongst research workers is that the production of hydrocarbons is due to flame quenching in the proximity of the combustion chamber surface and the crevice regions around the piston top land and any squish areas present. Some reduction of these hydrocarbons occurs due to oxidation in the exhaust process and some due to secondary reaction. The split between quench zone hydrocarbons and crevice region hydrocarbons depends on speed and load and combustion chamber geometry. However, the quench zone hydrocarbons present an area where very little can be done to effect a reduction per unit area. The fraction of emissions in the exhaust gas is, therefore, a function of the ratio of the volume of the quench zone to the volume of the combustion chamber (the surface/volume ratio of the chamber). The variation of non-dimensional surface to volume ratio for a typical pent roof chamber design with a flat topped piston with valve recesses is shown in Fig. 2.16(a). The non-dimensional surface to volume ratio is defined as: Chamber surface area at TDC/(volume at TDC)$^{2/3}$.

Engine geometry effects on surface/volume
Referring to Fig. 2.16(a) reinforces the conclusions of another researcher that to design for a low surface/volume (S/V) ratio requires a low compression ratio, a low bore/stroke ratio, and a large cylinder displacement. What is perhaps not so obvious is that the relative improvement in the reduction of S/V ratio is dependent on the bore/stroke ratio. For instance, increasing the cylinder size from 250 to 500 cm^3 gives a reduction of 21 per cent on S/V ratio at a bore/stroke ratio of 1.4, whereas at 0.8 the reduction is only 18 per cent. However, the higher percentage figure is the result of the S/V ratio falling by 0.1, whereas the lower percentage figure is a fall of only 0.058. Thus even though the percentage savings are of a similar order, the higher bore/stroke ratio saving represents a bigger improvement. These results are indicated centre. Changing the cylinder size without reference to total engine capacity is not truly representative of the actual situation and this shows the benefits of reducing the number of cylinders and effecting an increase in cylinder size on a 2 litre capacity engine. The S/V ratio in this case, with a fixed compression ratio of 12:1, can be reduced by 20 per cent when changing from a 6 to a 4-cylinder engine. Another aspect of the changes in geometry on S/V ratio occurs when considering the increase in compression ratio. With the current trend towards high compression, high turbulence, lean burn engines for economy purposes, this could result in high S/V ratios and, hence, high hydrocarbon (HC) emissions. However it can be seen that, with this configuration, a 2-litre 4-cylinder engine, at a compression ratio of 10:1, would have a valve included angle of 26 degrees and a S/V ratio of 0.298 l/mm at a bore/stroke ratio of 1:4.

For the design of an engine of the same configuration and capacity, but with an increased compression ratio of 12:1, and without detrimental effects of the S/V ratio, this could be achieved by decreasing the bore/stroke ratio to 1.03 and increasing the included angle of 24 degrees. It is true that this would result in a reduction of valve curtain area from 8200 to 6800 mm^2 or 17 per cent. But at this reduced valve area the pent-roof cylinder head valve area is only slightly below that for a hemi-spherical head at a bore/stroke ratio at the original 1.4 valve (right). This illustrates

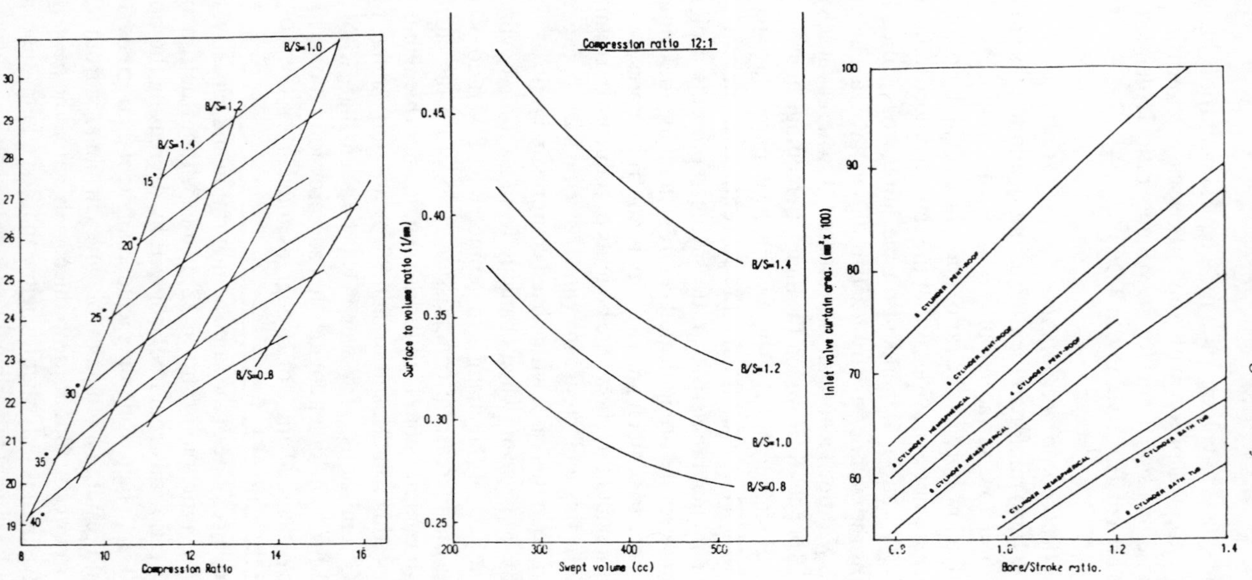

Fig. 2.16(a) (*Left*) shows surface/volume ratio against compression ratio for different valve included angles. In (*centre*) this data is plotted against swept volume In the (*right*) view, inlet valve curtain area is related to bore/stroke for a 2 litre engine

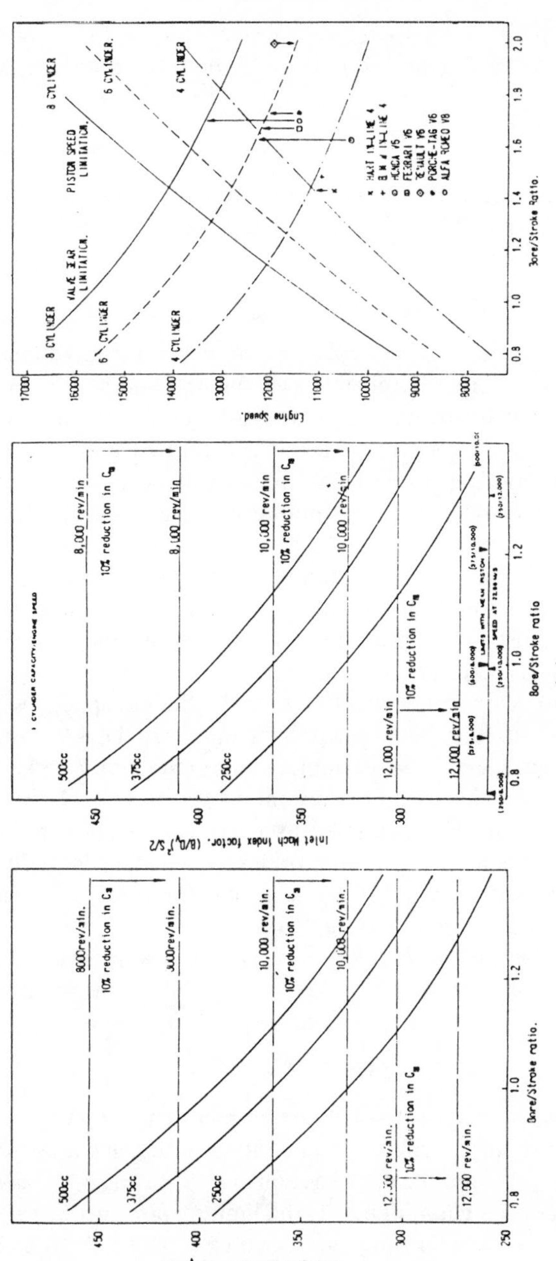

Fig. 2.16(b) ***(Left)*** **and** ***(centre)*** **shows inlet Mach index related to bore/stroke ratio for various swept volumes with 40 degree valve included angle, then 12:1 compression ratio, respectively.** ***(Right)*** **plots engine speed against bore/stroke to show limitations of valve gear characteristic-speed and piston speed**

the ability of the pent-roof 4-valve configuration to provide a compact chamber with a low S/V ratio without the penalty of poor breathing performance.

Effects of engine geometry on breathing

A measure of an engine's ability to consume air can be obtained from a study of the 'Mean Inlet Mach Index' or Gulp factor, Z, as defined by another worker as

$$Z = (B/D_T)^2 NS/(n C_m a_0)$$

where B = cylinder bore; S = stroke; D_T = valve throat diameter; C_m = mean valve discharge coefficient based on throat area; N = engine speed; a_0 = sonic velocity in intake duct, and n = number of inlet valves.

If the geometric contents are extracted and termed the inlet Mach factor this can be plotted against bore/stroke ratio for a constant compression ratio to indicate specific design limitations (Fig. 2.16(b)). Engine speeds can be plotted for a limiting inlet Mach index of 0.5, above which there is considerable reduction in volumetric efficiency. This limitation relates to engines with small valve overlap and with no dynamic effects in the intake or exhaust. The above results use a mean inlet discharge coefficient of 0.381 as determined from tests.

If it is desired to design an engine of 500 cm^3 cylinder capacity with as compact a combustion chamber as possible, to minimize the S/V ratio and to run at a maximum speed of 8000 r/min, the minimum bore/stroke ratio is 0.8. At 10 000 r/min, this minimum becomes 1.13, and for 12 000 r/min about 1.5. However, if the mean inlet discharge coefficient is reduced, perhaps as the result of inducing more turbulence, a 10 per cent reduction changes these bore/stroke ratios to 0.94, 1.32, and in excess of 2.0. Similar limitations can be determined for other cylinder capacities.

Another limitation on bore/stroke ratio is mean piston speed. If 22.9 m/s is taken as a maximum limitation on a production engine, the bore/stroke ratio at this limiting piston speed can be expressed as

$$Q = 6.28 \times 10^{-8} VN^{3/2}$$

where Q = bore/stroke ratio; V = cylinder capacity cm^3 and N = engine speed (r/min); hence, limits of bore/stroke ratio can be calculated. These are shown on the bottom of the centre graphs.

With a discharge coefficient of 0.381, the limiting bore/stroke ratio on volumetric efficiency at 8000 r/min for a 500 cm^3 cylinder is 0.8. With regard to piston speed it is 1.0. So piston speed is the limiting factor on the design bore/stroke ratio. However, if the C_m is reduced by more than 14 per cent, the volumetric efficiency becomes the limit on the selection of bore/stroke ratio. This change over between the two limiting conditions can also be seen with the changing cylinder volume. For example, at 10 000 r/min the cylinder of 500 cm^3 capacity requires a reduction of 13.5 per cent in C_m, the 375 cm^3 capacity needs 11.6 per cent reduction and the 250 cm^3 only 8.1 per cent, so the smaller the capacity the more likely it is to be limited by volumetric efficiency. With the increasing use of turbocharging the need for a low-loss induction system to provide adequate cylinder filling is no longer so critical. However, adequate valve flow capacity is still required for satisfactory scavenge. All the current Formula 1 turbocharged engines are of the 4-valve/cylinder design and the right-hand part of Fig. 2.16(b) illustrates how they relate with regard to piston speed and valve gear characteristic speed.

Torque control for part-load fuel economy

An examination of the thermodynamics of a method of improving the part-load specific fuel consumption of petrol engines has been made by S. F. Smith[6]. He explains that in a petrol engine, the exhaust temperature at the end of the power stroke is typically around 1500°K, so that the theoretical pressure immediately before the exhaust valve opens will be of the order of five times the inlet pressure. At full throttle the absolute value of the exhaust pressure in a real engine still would be well above atmospheric, and if the gas could be further expanded there would be a gain in power and specific fuel consumption (SFC). The volume in the cylinder at the end of the power stroke effectively determines the physical size of the engine and its weight, so an increase of engine capacity solely to provide increased expansion volume would give an engine whose specific weight, specific volume, and specific cost were all worse; in spite of a better SFC, this would be unacceptable for most motor car applications.

The normal method of reducing the power at a given engine speed to a level below full power is to throttle the air inlet and reduce the pressure, and hence the density of the air/fuel mixture entering the engine. To a first approximation, the power is reduced proportional to the reduction in the absolute pressure in the inlet manifold, although in practice the reduction will be greater than this due to a loss of thermal efficiency. A fundamental loss of thermal efficiency at part-load is caused by the pressure on the induction stroke being well below atmospheric pressure and the exhaust stroke pressure being above atmospheric. There will also be a loss, which is probably much more important, due to the friction and auxiliary losses remaining substantially constant at a given engine speed and, hence, being a greater proportion of the power output at part-load.

This section examines the thermodynamic advantages of an alternative method of reducing engine power that increases the part-load expansion ratio with benefit to the thermal efficiency. The proposal is that the reduced power output at a given speed and the lower mass of mixture required in the cylinder will be produced by delaying the closing of the inlet valve until some point in the 'compression' stroke, and by simultaneously reducing the combustion volume to maintain the same compression ratio. For example, at about half-power the inlet valve would close when the piston was halfway up the compression stroke and the combustion volume would be halved and, hence, the expansion ratio would be doubled, which improves the thermal efficiency. In this proposal the intake would not be throttled but, since some of the mixture drawn into the cylinder on the inlet stroke is rejected back to the inlet manifold, it can be applied only to engines which have four or more cylinders sharing a common manifold.

Thermodynamic considerations

In a petrol engine, the volumetric compression ratio is limited by considerations of pre-ignition and detonation to about 10:1 and, for the purpose of this section, it is proposed to consider the cycle of an engine with this compression ratio, as in Fig. 2.17.

In the classic Otto cycle the work done in compression is

$$W_c = \frac{P_1 V_1}{\gamma - 1} \{(V_1/V_2)^{\gamma-1} - 1\}$$

where $\gamma = 1.4$ is the ratio of specific heats for 'cold' air; for the chosen cycle

$$V_1/V_2 = 10 \quad \text{and} \quad W_c = 3.780 P_1 V_1$$

For the expansion work

$$W_e = \frac{P_3 V_3}{\gamma - 1} \{1 - (V_3/V_4)^{\gamma-1}\}$$

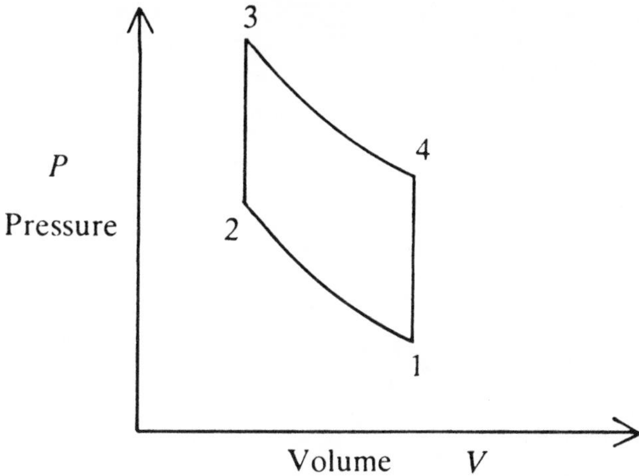

Fig. 2.17 Theroretical Otto cycle for a gasoline engine

Taking $\gamma = 1.3$ to allow for the effect of the combustion products, and high temperature, on the ratio of specific heats

$$W_e = 1.663 P_3 V_3$$

For an ambient temperature of 288°K the temperature at the end of the compression stroke will be 723°K for an adiabatic compression of volumetric ratio 10.

Taking the calorific value of the fuel as 10 800 CHU/lb, an air/fuel ratio of 16, and an average value of the specific heat at constant volume as $C_v\gamma = 0.2175$ to allow for the high temperatures involved, then the theoretical combustion temperature rise is 3103°K and T_3 is 3826°K.

$$\text{With no losses } \frac{P_3}{P_1} = \frac{3826}{288} \times 10 = 134.3$$

In a real engine, however, combustion will not be complete and there will be some loss of heat to the piston and cylinder walls, some leakage past the piston rings, and combustion will not be instantaneous. These will all have the effect of lowering the effective pressure on the expansion stroke, so to make the calculations more realistic P_3/P_1 is taken as 80 per cent of the theoretical; that is $P_3/P_1 = 107.44$. With this value

$$W_e = 17.85P_1V_1$$

The gross work output is then

$$W_e - W_c = 14.08P_1V_1$$

In a real engine some 20 per cent of the power will be dissipated as friction losses and to drive essential auxiliaries, so that the net output work

$$W_n = 11.26P_1V_1$$

In terms of P_1V_1 the fuel energy input is

$$\frac{10\,800}{16 \times 19.74} = 34.19P_1V_1$$

so that the thermal efficiency at full power output is

$$\frac{11.26}{34.19} = 32.9 \text{ per cent}$$

which is sufficiently representative of a real engine for these figures to be used for the subsequent calculations.

Part-load performance with intake throttle

The basic cycle is unchanged under conditions of intake throttling (Fig. 2.18(a)) but the pressure on the exhaust stroke is theoretically the

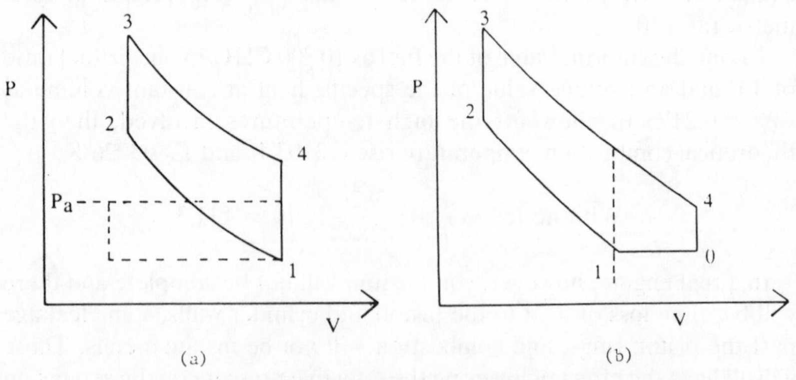

Fig. 2.18(a) Intake throttling expressed on an Otto cycle

Fig. 2.18(b) Otto cycle modified as part load by intake throttling and variable combustion chamber volume

atmospheric pressure P_a, while the intake pressure P_1 is less than P_a, so that there is a loss of net cycle work. Hence

$$(V_1 - V_2)(P_a - P_1) = P_a V_1\left(1 - \frac{V_2}{V_1}\right)\left(1 - \frac{P_1}{P_a}\right)$$

Taking a number of values of P_1/P_a the cycle calculations of part-load thermal efficiency at a constant engine speed then become

$\frac{P_1}{P_a}$	=	1.0	0.75	0.5	0.4	0.3	0.25
$\left(1 - \frac{V_2}{V_1}\right)\left(1 - \frac{P_1}{P_a}\right)$	=	0	0.225	0.45	0.54	0.63	0.675
$\frac{W_g}{P_a V_1}$	=	14.08	10.56	7.04	5.632	4.224	3.52
Friction losses $\frac{W_f}{P_a V_1}$	=	2.81	2.81	2.81	2.81	2.81	2.81
$\frac{W_n}{P_a V_1}$	=	11.27	7.525	3.78	2.282	0.784	0.035
% power	=	100	66.7	33.5	20.0	6.96	0.31
Fuel energy $\frac{F}{P_a V_1}$	=	34.19	25.64	17.09	13.67	10.26	8.54
% fuel	=	100	75	50	40	30	25
T%	=	33	29.3	22.1	16.7	7.6	0

Part-load performance with intake throttling and combustion chamber volume control

In this case the basic cycle is altered at part load (Fig. 2.18(b)). From 0 to 1 the mixture is rejected back to the inlet manifold at atmospheric pressure and the combustion volume is reduced to keep $V_1/V_2 = 10$, which here has been taken as the limited compression ratio for detonation. The expansion ratio V_4/V_3 is increased in the same ratio as V_0/V_1 and the cycle calculations are then

$\dfrac{V_0}{V_1}$	= 1.0	2.0	3.0	3.5	4.0
$\dfrac{W_{c12}}{P_0V_0}$	= 3.78	1.89	1.26	1.08	0.945
$\dfrac{W_{c01}}{P_0V_0}\left(1-\dfrac{V_1}{V_0}\right)$	= 0	0.5	0.667	0.714	0.75
$\dfrac{V_4}{V_3}$	= 10	20	30	35	40
$\dfrac{W_e}{P_3V_3}$	= 1.663	1.976	2.132	2.186	2.23
$\dfrac{P_3}{P_0}$	= 107.44				
$\dfrac{W_e}{P_0V_0}$	= 17.86	10.61	7.645	6.71	5.988
$\dfrac{P_3}{P_4}$	= 19.95	49.1	83.22	101.67	
$\dfrac{W_f}{P_0V_0}$	= 2.81	2.81	2.81	2.81	2.81
$\dfrac{W_n}{P_0V_0}$	= 11.27	5.41	2.898	2.106	1.4828
% power	= 100	48	25.7	18.7	13.1
$\dfrac{F}{P_0V_0}$	= 34.19	17.09	11.39	9.768	8.547
T%	= 33.0	31.6	25.4	21.56	17.3
% fuel	= 100	49.1	32.9	28.4	24.9

The figures show the fuel flow at various power levels at constant engine speed, both for control of torque by the conventional intake throttle and for the method proposed in this article. It will be noted that from 0 to 50 per cent power the fuel flow for the latter is reduced by about a quarter of the intake throttling flow for the same time. A modern car of fairly high performance with a top speed of over 160 km/h (100 mile/h) would rarely use full throttle in normal driving and would usually be operated at 50 per cent full engine torque or less. This means that, if such a car had an overall fuel comsumption of 11.3 litre/100 km (25 mile/gal), an improvement of this order would reduce comsumption to about 8.8 litre/100 km (32 mile/gal).

Exhaust pollution control: effect of engine design parameters

C. G. Lucas, E. H. James, and R. Chrast[7] of Loughborough University explain that present and impending legislation suggests that the most important pollutants in the exhaust of a spark ignition engine are carbon monoxide, hydrocarbons, the oxides of nitrogen and lead compounds.

Carbon monoxide is formed in the bulk gases within the cylinder. If the engine is run with a rich mixture, the measured concentration of CO in the exhaust is higher than it should be according to equilibrium theory and, in fact, corresponds more to the equilibrium concentration of CO at the peak cycle temperature. As the mixture is weakened to stoichiometric or leaner conditions, the exhaust concentration of CO more nearly approaches the concentration that would be expected if equilibrium were to exist throughout the expansion process. The carbon monoxide concentration in the exhaust gas is primarily a function of air–fuel ratio and the quality of pre-mixing and distribution of the mixture. It is impossible to explain the presence of unburned hydrocarbons in the exhaust gases by equilibrium or non-equilibrium considerations. Except during engine idling and deceleration, it has been shown that the principal mechanism producing unburned hydrocarbons in the exhaust is flame quenching at the relatively cold walls of the combustion chamber. During idling and deceleration, incomplete combustion is the main cause of the high hydrocarbon concentration. Thus, any factor which influences the quantity of fuel at the wall, the temperature of the wall, or the area of the quench zone will affect the quantity of hydrocarbons emitted.

A significant amount of hydrocarbons may be destroyed in the exhaust system if the temperature in the exhaust system is sufficiently high. Approximately 99 per cent of the oxides of nitrogen emissions from the engine exhaust are initially nitric oxide (NO) which may then oxidize to form nitrogen (NO_2). The NO is formed in the bulk gases and its concentration in the exhaust is primarily a function of peak combustion temperature and the amount of available oxygen. The amount of nitric oxide in the exhaust is greater than would be expected from equilibrium considerations. The considerable quantities of nitric oxide which are formed at the peak pressure and temperature should, assuming equilibrium, reduce to almost zero in the low temperature exhaust. The fact that this is not so indicates that nitric oxide 'freezes' at some relatively high temperature. In general, anything which increases the oxygen concentration in a cylinder, when the engine operates near or richer than stoichiometric, or which increases the peak cycle temperature, will increase also the NO concentration in the exhaust gas. Cylinder pressure has a significant effect on the NO concentration at rich mixtures, but very little effect at lean mixtures. For rich mixtures, an increase in pressure decreases the amount of NO formed under equilibrium conditions.

Lead compounds are the most significant fractions of particulate emissions from a spark-ignition engine exhaust and results from use of lead compounds as petrol additives for improving the anti-knock characteristics of a fuel. It is estimated that approximately 70 per cent of the lead in the burnt fuel is exhausted and the remaining 30 per cent is retained within the engine, either in the form of deposits on the walls of the combustion chamber, valves, ports, and in the exhaust system, or is scavenged into the engine lubricating oil. The amount of lead exhausted from the engine for any particular condition is proportional to its concentration in the burnt fuel.

Considerable effort has been devoted to establishing the effects of operating and design variables on exhaust emissions from spark-ignition engines. The most important of these are mentioned below.

Air–fuel ratio

As a rule of thumb guide, the CO concentration in the exhaust is proportional to the amount of excess fuel in the cylinder. An increase in air–fuel ratio (when the mixture is rich) substantially reduces the concentration of CO, but the concentration remains relatively constant at a low level with air–fuel ratios above 15:1 (Fig. 2.19). A similar trend had been shown for hydrocarbons. The decrease in hydrocarbon level is rapid if the air–fuel ratio is increased from a rich mixture. Further weakening of the mixture beyond stoichiometric does not help (Fig. 2.20). The weakening of the mixture beyond an air–fuel ratio of 18:1 may, in some cases, increase the concentration of hydrocarbons because the excess air may cause cooling and increase the quenching effect at the wall. In weakening the mixture, one must be aware of the fact that any misfiring increases the exhaust emissions considerably.

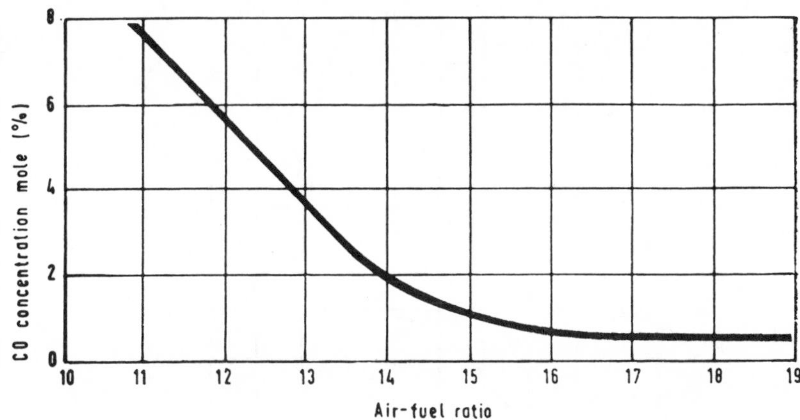

Fig. 2.19 Air–fuel ratio effect on CO concentration in exhaust gas

The concentration of nitric oxide is low for very rich operation and increases with air–fuel ratio to peak at a ratio slightly leaner than stoichiometric. This occurs because the combination of high cylinder gas temperatures and sufficient oxygen concentrations is achieved with the mixture strength at about 15–16:1. A further weakening of the mixture results in a reduction of NO concentration in the exhaust gas (Fig. 2.21).

Spark timing

Retarding the spark timing has no significant effect on the CO concentration, but it decreases the concentration of unburned hydrocarbons in the exhaust gases considerably. This decrease is greater for lean mixtures. Retarding the spark timing increases the exhaust temperature and the

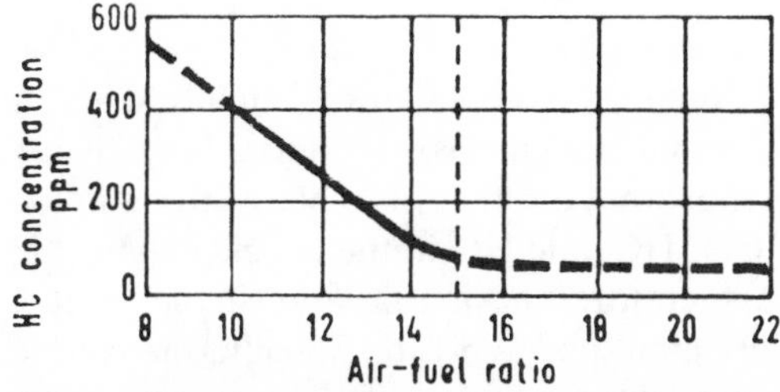

Fig. 2.20 Air–fuel ratio effect on HC concentration in exhaust gas

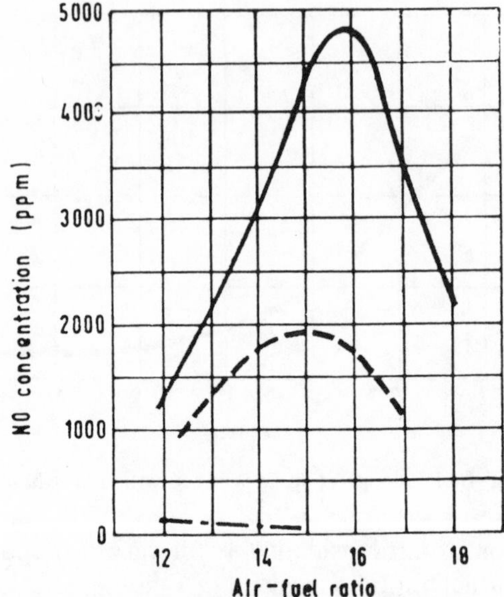

Fig. 2.21 Air–fuel ratio effect on NO concentration in exhaust for speed of 2500 (full-line), 1500 (dotted) and 550 (chain) r/min, with corresponding spark timings at 40, 30, and 5 degrees btdc

rate of hydrocarbon oxidation in the exhaust. As the retarded spark timing tends also to slow down the combustion process and thus lowers the peak combustion temperature, it decreases substantially the exhaust emission of NO. The reduction is greater at air–fuel ratios near stoichiometric or leaner. Unfortunately, such a setting of spark timing results in lower thermal efficiency and in increased fuel costs per unit of power.

Preparation of mixture

A well pre-mixed, homogeneous mixture, equally distributed to the cylinders of a multi-cylinder engine, is essential for low exhaust emissions. Figures 2.19–2.21 show that air–fuel ratio has a strong influence on the concentrations of CO, HC, and NO in the engine exhaust. Maldistribution of the mixture, therefore, causing variation in the air–fuel ratio from cylinder to cylinder, should be avoided. Equally, the charge should be uniform within a particular cylinder, if local effects of air–fuel ratio variation are to be avoided.

Compression ratio

Increasing the compression ratio slightly increases the CO exhaust levels when expressed on a percentage by volume basis. This trend is much more severe in the case of hydrocarbons since there is a general increase in the concentration of unburnt hydrocarbons with compression ratio, regardless of the mixture strength, spark advance, or engine loading. The main reasons for this are the decrease in exhaust temperature and the better scavenging as a result of the increase in compression ratio. As the maximum cycle temperature rises with compression ratio the NO concentration tends to increase. The compression ratio is known to affect the power output and the thermal efficiency of an engine also. It should not be concluded from the above that, in order to reduce emissions, one should lower the compression ratio since, if by doing so a much larger engine is required for a given power requirement, this may result in greater emissions on a mass basis.

Surface–volume ratio

While the CO concentrations are unaffected by changes in the surface–volume ratio of a cylinder, the combustion chamber walls surface area is the primary source of the hydrocarbons found in the exhaust. Thus a decrease in the surface–volume ratio (which suggests an increase in stroke–bore ratio and greater displacement per cylinder with fewer cylinders) results in a decrease of the exhaust concentration of hydrocarbons. It is assumed that the NO concentration should be less if the surface–volume ratio of the engine's cylinder increases. This is because the heat transfer increases and, hence, the maximum cycle temperature decreases.

Valve timing

When the valve overlap is not excessive, the valve timing seems to have very little effect on exhaust emissions. However, it is reported that a decrease in valve overlap reduces considerably the hydrocarbon concentrations during idle and deceleration if the air–fuel ratio is 15 : 1 or leaner.

Intake manifold vacuum

The CO concentration shows essentially no variation with intake manifold vacuum. However, the concentration of HC rises with vacuum and the rate of emission rise becomes much more rapid as the vacuum increases beyond approximately 15 in Hg. Conversely, a decrease in intake manifold vacuum increases the NO concentration in the exhaust and the increase is greater for leaner air–fuel ratios.

Inlet air temperature

Increasing the inlet air temperature has no significant effect on CO concentration and only slightly (perhaps due to the improvement in mixture vaporization) reduces the HC concentration in the exhaust. It is reported that the concentration of NO increases with the inlet air temperature, reaches a peak at about 205°F for almost the whole range of air–fuel ratios, and then decreases with further temperature rise.

Coolant temperature

CO concentration does not seem to be affected by changes in coolant temperature. However, an increase in coolant temperature decreases the HC concentration and this effect is greater at lower coolant temperatures. The concentration of NO rises with increase in coolant temperature.

Exhaust back pressure

The exhaust back pressure has no significant effect on CO and NO concentration, but the emission of hydrocarbons is slightly reduced with its increase.

Exhaust temperature

An increase in exhaust temperature slightly reduces the CO emission and decreases the emission of HC. The effect on NO emission is not significant.

Mode of engine operation

The levels of CO concentrations are high during the warming-up of the engine for all modes of operation. Once the engine is at operating temperature, the highest values of CO are during the idle period. If the air–fuel ratio of the mixture entering a cylinder is kept constant, the CO concentration in the exhaust is unaffected by changes in the engine power output and speed. The largest hydrocarbon concentrations in the exhaust are during idle and deceleration. As the engine speed increases, the HC emission decreases. For rich mixture operation, knock generally decreases the HC levels and the extent of the decrease is greater with the intensity of knock. The concentration of NO rises rapidly with increased engine loading. The effect of engine speed is slight and direction uncertain. The NO emission during idle is relatively very low.

Inlet manifold fuel evaporation modelling

Picken, Soliman, and Fox[8] point out that for many years designers have sought to supply sufficient heat to evaporate the liquid fuel in the manifold without drastically reducing the volumetric efficiency of the engine. Many methods have been tried to achieve these objectives, using engine cooling water or exhaust gases to supply the heat for evaporation, and the results have proved that complete evaporation has a great effect on mixture distribution, engine performance, and exhaust emission levels. Here a theoretical investigation into the manifold requirements to produce a given level of evaporation, to understand the further evaporation processes during the compression and combustion periods, and the effects of any remaining liquid fuel on the formation of unburned hydrocarbons and carbon monoxide is described. The investigation is limited to a single-cylinder engine (Table 2.5). It traces the relationship between the manifold length, temperature, and percentage evaporation of liquid fuel, and then traces the life history of the remaining liquid droplets. The theoretical results for the effects of these droplets on cylinder maximum pressure and the products of combustion indicate that for optimum results it is necessary to ensure complete evaporation before ignition, but not necessarily at the start of compression.

Manifold conditions

The evaporation of the liquid fuel in the manifold was investigated by considering the manifold to be a tube and that the heat required to produce any level of vaporization was supplied from its wall. Calculations were based on the following.

(1) The heat required for vaporization and heating the air is given by

$$Q = m_a C_{p\ air} \Delta T_{air} + \varepsilon m_f L \qquad (2.4)$$

Table 2.5 Data for engine simulation

Engine capacity (l)		0.450
Cylinder bore (mm)		88
Piston stroke (mm)		74
Compression ratio		6.0:1
Engine speed (r/min)		2400
Timing:		
Inlet valve	opens 20° BTDC	closes 240° ATDC
Exhaust valve	opens 240° BTDC	closes 20° ATDC
Ignition		20° BTDC

where

Q is the required heat flow rate (kJ/h)
m_a is the air flow rate (kg/h)
C_p is the specific heat of air (kJ/kg/K)
ΔT is the air temperature rise in the manifold (K)
ε is the percentage of vapour required at the end of the manifold
m_f is the liquid fuel flow rate (kg/h)
L is the latent heat of vaporization of fuel (kJ/kg)

(2) The heat transferred from the manifold wall to the mixture (based upon forced convective heat transfer through a cylindrical pipe)

$$Q_t = hAT_1 \qquad \text{(kJ/h)} \tag{2.5}$$

where

Q_t is the heat transferred (kJ/h)
h is the heat transfer coefficient (kJ/m^2 sec K)

$$h = 0.023(\lambda/d)Re^{0.8}(Pr)^{0.4}$$

where

Re is Reynold's number
Pr is the Prandtl number
A is the manifold surface area $= (\pi d)X$ (m^2)
d is the manifold diameter (m)
X is the manifold length (m)
ΔT_1 is the temperature difference $(T_s - T_g)$ (K)
T_s is the manifold surface temperature (K)
T_g is the mixture temperature (K)

Using these two equations, (2.4) and (2.5), the required manifold length to produce a given level of vaporization can be calculated from

$$X = \frac{m_a}{\pi d h \Delta T_1}\left(C_{p\ air}\Delta T_{air} + \varepsilon \frac{L}{A/F}\right) \qquad \text{(m)} \tag{2.6}$$

For the average condition the air was assumed to rise in temperature by 40 K, and using the following data:

C_p = 1.005 kJ/kg K,
L = 375 kJ/kg for gasoline,
m_a = 23.17 kg/h, at 2400 r/min 20°C, 1 bar,
d = 0.0254 m,
A/F = 12,

the required manifold length for a given temperature difference (ΔT_1) was calculated for complete vaporization and is presented in Fig. 2.22.

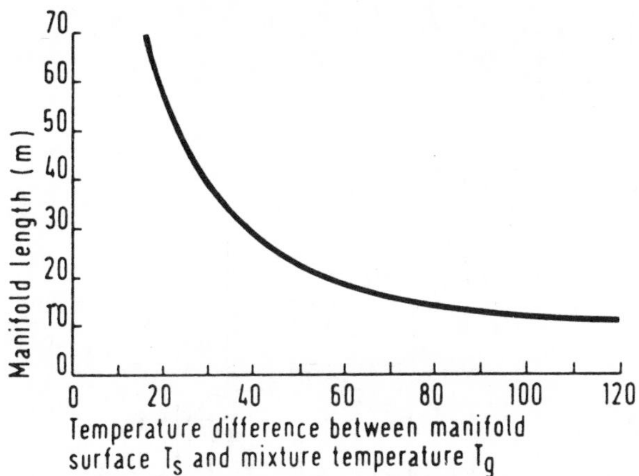

Fig. 2.22 Manifold length required in relation to temperature difference between surface and mixture for 100 per cent vaporization

The results of the calculation indicate that for normal operating conditions a manifold length of more than two metres is required for complete vaporization of the fuel. Thus, most practical engines are likely to be operating with at least some liquid fuel droplets entering the cylinder. This result makes it necessary to follow the life history of the liquid fuel droplets after they have entered the cylinder.

Cycle calculation

A cycle calculation was developed on the first law consideration and taking into account the rate of droplet vaporization, based on Spalding's equation for a single droplet vaporization, vapour/air ratio, changes in mixture internal energy, and heat transferred from the cylinder walls, together with a two-zone combustion model. It was considered that the following information would be useful in predicting the effects of the droplets on the combustion process: the indicator diagram, unburned hydrocarbons from the quench zone and the liquid fuel overtaken by the flame front, and the products of combustion.

Figure 2.23 shows the logic of this calculation and Fig. 2.24 shows the life history of a single droplet during the compression, combustion, and expansion processes, together with cylinder pressure and mean gas temperature for a typical cycle. It is considered significant that droplets in the cylinder at the start of compression remain as droplets throughout the burning period. These droplets will be overtaken by the flame front and

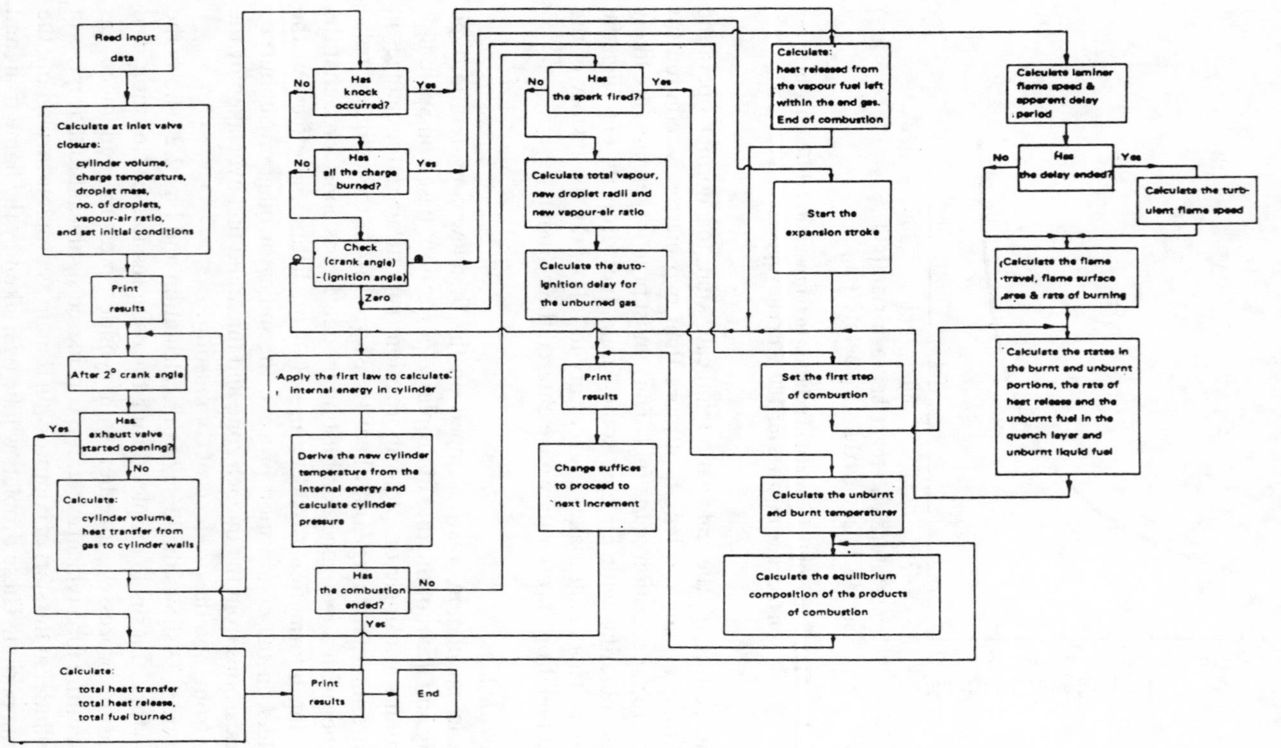

Fig. 2.23 Combustion model flow diagram

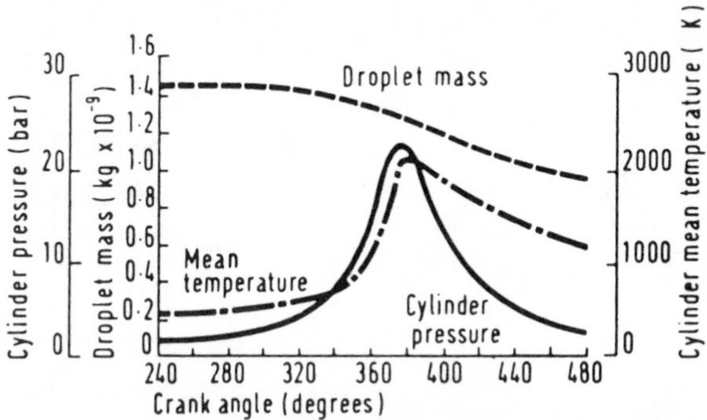

Fig. 2.24 Calculated cylinder pressure, mean temperature and droplet mass

further evaporation will result in fuel vapour in a hot, oxygen-free, atmosphere which can only break down the hydrocarbons into smaller and partially burned components.

Figure 2.25 shows the manifold length required for a given temperature difference between manifold surface and the mixture to produce various

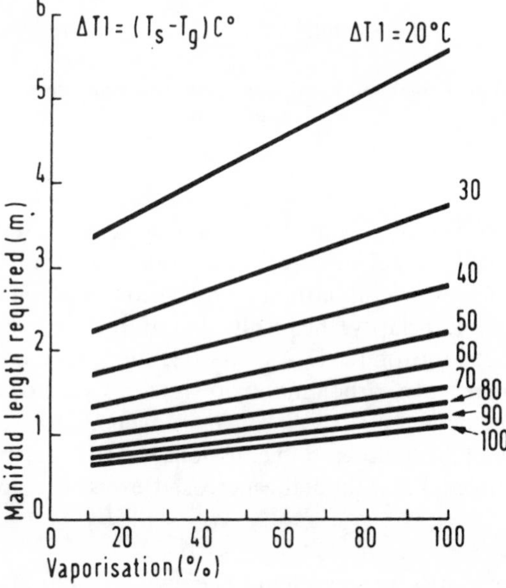

Fig. 2.25 Manifold length requirement in relation to percentage vaporization

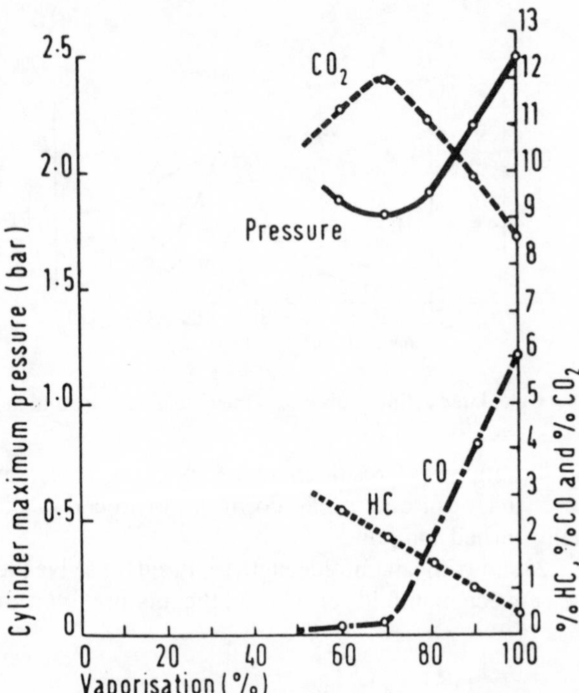

Fig. 2.26 Cylinder maximum pressure and emissions relative to percentage vaporization

degrees of vaporization. Figure 2.26 shows the effects of different degrees of evaporation of the fuel mixture droplets introduced into the cylinder on the maximum pressure, unburned hydrocarbons, carbon monoxide and carbon dioxide. For clarity, the results show only the curves relating to an overall air/fuel ratio of 12:1. As vaporization increases with this rich mixture an increase in carbon monoxide is to be expected as the unburnt hydrocarbons decrease. At closer to stoichiometric conditions the improvement in the hydrocarbon emissions will still take place as vaporization is improved, but without the increased levels of carbon monoxide.

Conclusion

For normal operation of spark ignition engines, manifold heating cannot vaporize the fuel droplets to any significant extent. Even for the case where a manifold length of 0.5 m is used a surface temperature of over 200°C would be required for complete vaporization, and a manifold length of about two metres would be required at more normal operating temperatures. Most fuel evaporation takes place inside the cylinder because of the heat transferred from the hot points of the cylinder bore and head and from the top surface of the piston, and also continuously during compression because of the increasing mixture temperature. Nevertheless, evaporation before combustion is critical in improving hydrocarbon emissions. Further, if fuel can be completely vaporized before ignition it is possible to operate at peak performance with leaner overall mixtures, thus keeping carbon monoxide emissions to a minimum.

Carburettor fuel evaporation modelling

The carburettor considered in this investigation by Finlay, Boam, and Bannell[9] is of the constant depression type as shown idealized and simplified in Fig. 2.27. Fuel entering at the jet is atomized and entrained by the air stream to form a two-phase mixture which is firstly expanded into a mixing section and then accelerated past the throttle plate where, at part load conditions, further atomization and mixing occur. The air flow in the carburettor can be strongly pulsating and highly turbulent, with liquid fuel present as rivulets flowing on the duct walls and as entrained droplets having a broad spectrum of size and velocity. A rigorous treatment of the problem of predicting rates of heat, mass, and momentum transfer under these conditions would clearly involve prohibitive computational effort.

Some simplifying assumptions are therefore necessary to reduce the problem to manageable proportions. These are: (a) liquid fuel is dispersed within the air as spherical droplets having uniform size at any given location; (b) initial distributions of temperature and concentration within each phase are uniform; (c) steady-state conditions prevail; (d) within the gas phase, gradients of temperature and concentration exist only in the general direction of flow; (e) within each droplet, temperature and composition are assumed to be uniform (The droplet temperature and composition will, however, vary in the direction of flow. The validity of this assumption has been discussed by earlier researchers, who concluded that with low-viscosity fuels, in situations where there is forced convection outside the droplet, liquid circulation within the droplet may be sufficiently strong to maintain spacially uniform composition and temperature, a point which is made later in this article.); (f) there is no heat flow to or from the walls of the carburettor.

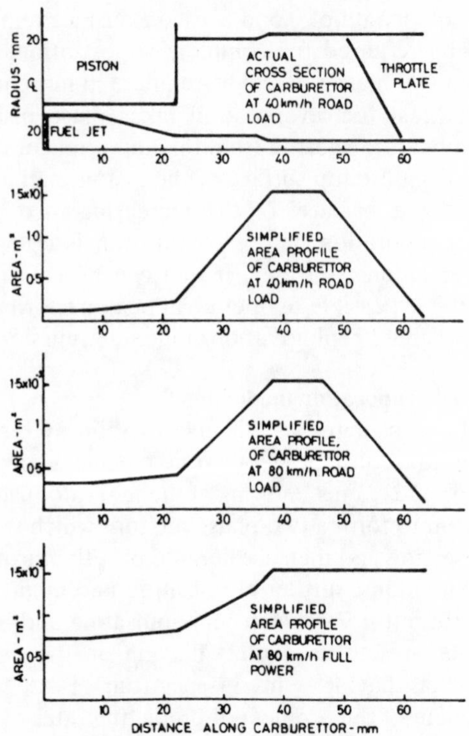

Fig. 2.27 Simplified carburettor geometries used in model

Table 2.6

Notation	
A_c	Cross-sectional area normal to general direction of flow
A_c	Surface area of droplets in length $\delta 1$
C	Specific heat capacity
C_D	Drag coefficient
D	Diffusion coefficient
d	Droplet diameter
F_D	Drag force on a droplet
f	Friction factor
h	Specific enthalpy
k	Thermal conductivity
L	Enthalpy of vaporization

Table 2.6 *continued*

Notation	
λ	Distance measured in direction of flow
M	Molecular weight
m	Number of components in liquid fuel
n	Number of droplets per second passing any given section
P	Pressure
Pr	Prandtl number, $c_g\mu_g/k_g$
p	Partial pressure
R	Specific gas constant
$\mathrm{Re_d}$	Droplet Reynolds number, $\varrho_g(V_g - V_d)d/\mu_g$
r_h	Hydraulic radius
Sc	Schmidt number, $\mu_g/\varrho_g D$
T	Temperature of liquid
t	Temperature of gas
V	Velocity
W	Mass flowrate
α	Heat-transfer coefficient
β	Mass-transfer coefficient
μ	Absolute viscosity
ν	Kinematic viscosity
ϱ	Density
σ	Surface tension
Subscripts	
a	Of air
c	At the critical point
d	Of droplets
f	Of film
g	Of gas
i	Of i-th component
L	Of liquid
m	Of mixture of air and fuel vapour
S	Saturated vapour
T	At temperature T
t	Total
v	Of vapour
Superscripts	
–	Indicates effective vapour pressure of component in mixture; see equation (2.17)
*	Denotes absolute temperature

By applying the laws of conservation of mass, energy, and momentum to both liquid and gaseous phases the following equations are obtained using the notation in Table 2.6.

Conservation of energy applied to air and fuel vapour

$$\frac{\mathrm{d}t_a}{\mathrm{d}l} = \frac{\Sigma_{i=1}^{m} (\mathrm{d}W_{vi}/\mathrm{d}l)C_{vi}(T - t_a) + \alpha(n\pi d^2/V_d)(T - t_a)}{(W_a C_a + \Sigma_{i=1}^{m} W_{Li} C_{Li})} \tag{2.7}$$

where

$$\alpha = \frac{k_a}{d}(2 + 0.6\mathrm{Re}_d^{0.5}\mathrm{Pr}_a^{0.33})$$

Conservation of energy applied to liquid fuel

$$\frac{\mathrm{d}T}{\mathrm{d}l} = \left\{\frac{n\pi d^2\alpha}{V_d}(t_a - T) + \sum_{i=1}^{m} L_i \frac{\mathrm{d}W_{Li}}{\mathrm{d}l}\right\} \Big/ \sum_{i=1}^{m} W_{Li}C_{Li} \tag{2.8}$$

Conservation of mass applied to air and fuel vapour

$$\varrho_a V_a A_c + \sum_{i=1}^{m} \varrho_{vi} V_v A_c = W_a + \sum_{i=1}^{m} W_{vi}$$

whence, if $V_a = V_v = V_d$

$$\frac{1}{V_a}\frac{\mathrm{d}V_g}{\mathrm{d}l} + \frac{1}{A_c}\frac{\mathrm{d}A_c}{\mathrm{d}\lambda} + \left(\varrho_a + \sum_{i=1}^{m} \varrho_{vi}\right)^{-1}\left(\frac{\mathrm{d}\varrho_a}{\mathrm{d}l} + \sum_{i=1}^{m}\frac{\mathrm{d}\varrho_{vi}}{\mathrm{d}l}\right)$$
$$= \left\{A_c V_g\left(\varrho_a + \sum_{i=1}^{m} \varrho_{vi}\right)\right\}^{-1}\frac{\mathrm{d}W_{vi}}{\mathrm{d}t} \tag{2.9}$$

Conservation of momentum applied to system

$$A_c\frac{\mathrm{d}P_t}{\mathrm{d}l} - \pi f \varrho_m V_g^2 r_h + \left(W_a + \sum_{i=1}^{m} W_{vi}\right)\frac{\mathrm{d}V_g}{\mathrm{d}l}$$
$$- (V_g V_d)\sum_{i=1}^{m}\frac{\mathrm{d}W_{vi}}{\mathrm{d}l} - W_L\frac{\mathrm{d}V_d}{\mathrm{d}l} = 0 \tag{2.10}$$

where from the gas laws

$$\frac{1}{p_a}\frac{\mathrm{d}p_a}{\mathrm{d}l} - \frac{1}{\varrho_a}\frac{\mathrm{d}\varrho_a}{\mathrm{d}l} - \frac{1}{t_a^*}\frac{\mathrm{d}t_a^*}{\mathrm{d}l} = 0 \tag{2.11}$$

and

$$\frac{\mathrm{d}p_{vi}}{\mathrm{d}l} = \frac{(\beta_i R_{vi}/W_a R_a)(n\pi d^2/V_d)(p_{SiT} - p_{vi})}{\{(P_t - \Sigma_{i=1}^{m} p_{vi}) + p_{vi}\}(P_t - \Sigma_{i=1}^{m} p_{vi})^2} \tag{2.12}$$

where

$$\beta_i = \frac{D_i}{dR_{vi}T_f^*}(2 + 0.6\mathrm{Re}_d^{0.5}\mathrm{Sc}_i^{0.33})$$

The diffusion coefficient for each component has been calculated as

$$D_i = \frac{1.277 \times 10^{-6}}{P_t}\left(\frac{M_i + M_a}{M_i M_a}\right)^{1/2}(P_{ci}P_{ca})^{1/3}(t_{ci}^* t_{ca}^*)^{-0.495} t_i^{*1.823}$$

This equation assumes that each component is diffusing into air alone, whereas, in practice, each component will be diffusing into a mixture of air and other hydrocarbon vapours. Using corrections from earlier researchers, these pure values were computed and found to be negligible. These corrections were, however, based on the maximum concentrations likely to be achieved under well mixed conditions and so did not take account of the higher concentrations that may occur close to the surfaces of the evaporating droplets.

From the law of partial pressures

$$\frac{\mathrm{d}P_t}{\mathrm{d}l} = \frac{\mathrm{d}P_a}{\mathrm{d}l} + \sum_{i=1}^{m}\frac{\mathrm{d}p_{vi}}{\mathrm{d}l} \tag{2.13}$$

Using a previously derived equation for the drag force on spherical particles

$$F_D = \frac{C_d}{2}\frac{\pi d^2}{4}\varrho_g(V_g - V_d)^2$$

and equating the drag force to the rate of change of momentum gives

$$\frac{\mathrm{d}V_d}{\mathrm{d}l} = \frac{0.75 C_D \varrho_g (V_g - V_d)^2}{d\varrho_L V_d} \tag{2.14}$$

Droplet diameter is normally calculated from the continuity of mass equation

$$\frac{\mathrm{d}d}{\mathrm{d}l} = \frac{2}{\pi d^2 \varrho_L n}\frac{\mathrm{d}W_L}{\mathrm{d}\lambda} \tag{2.15}$$

where

$$\frac{\mathrm{d}W_i}{\mathrm{d}l} = -\frac{\mathrm{d}W_v}{\mathrm{d}l} = -\sum_{i=1}^{m}\frac{\mathrm{d}W_{vi}}{\mathrm{d}l} = -\sum_{i=1}^{m}\frac{n\pi d^2\beta_i}{V_d}(\bar{p}_{SiT} - p_{vi}) \tag{2.16}$$

Simulation of petrol

A normal petrol can have in excess of 40 components. Many of these, however, particularly the heavier ends, are present in very low concentrations. The model used in this study accurately represented the light and medium fractions of the fuel, but simplification was achieved by omitting the low concentration, heavier fractions that, as will be shown later, would contribute negligibly small amounts to the overall quantity of fuel evaporated within the carburettor. Altogether 16 components were used.

In a multi-component liquid the saturation vapour pressure of any component is modified by the presence of the other components in proportion to its mole fraction

$$\frac{W_{\mathrm{Li}} M_{\mathrm{i}}}{\Sigma_{i=1}^{m} (W_{\mathrm{Li}}/M_{\mathrm{i}})} \quad \text{Thus } \bar{p}_{\mathrm{SiT}} = p_{\mathrm{SiT}} \left\{ \frac{W_{\mathrm{Li}} M_{\mathrm{i}}}{\Sigma_{i=1}^{m} (W_{\mathrm{Li}}/M_{\mathrm{i}})} \right\} \tag{2.17}$$

Since W_{Li} varies continuously and by different amounts for each component, the mole fraction and effective vapour pressures are continuously updated during a calculation.

Integration of the model

A fourth order Runge–Kutta procedure was used to integrate equations (2.7) to (2.16). Integration was started at the needle and proceeded along the bore of the carburettor to a location just beyond the throttle plate. In an air valve carburettor the internal flow area changes as the mean air mass flowrate changes, due to piston movement. A relationship between distance from the needle and the cross-sectional area for flow is therefore required for each engine condition studied. In this investigation three engine conditions were examined, namely 40 and 80 km/h road load and 80 km/h full power. Details of the carburettor, engine, and vehicle combinations assumed are given in Table 2.7, the geometries used for each condition being shown in Fig. 2.36.

Table 2.7 Details of assumed carburettor, engine, and vehicle combination

Carburettor	Constant depression type with 45 mm bore
Engine	4 cylinder overhead valve 1.8 litre four stroke
Body	Saloon type with 1090 kg inertial weight
Gearing	26.7 km/h per 1000 r/min

Since the integration procedure was unable to handle a step change in the internal flow area of the carburettor, the sudden increase in flow area that occurs behind the piston could not be reproduced exactly. As can be seen, the gradient $dA_c/d\lambda$ immediately following the piston is, therefore, made finite. Results obtained from various values of $dA_c/d\lambda$ at this location show that the predictions are largely insensitive to the sharpness of the transition, and so the error resulting from this simplification is small. A small initial positive value of the droplet velocity, V_d, is assumed to avoid a singularity in the initial droplet surface area. The value used in this study was 0.1 m/s. Entering a previously derived equation for petrol, namely

$$d = 10^{-6}\left\{\frac{3045}{(V_{\bar{a}}V_d)} + 30\left(1000\,\frac{W_L}{W_a}\frac{p_a}{\varrho_L}\right)^{1.5}\right\} \tag{2.18}$$

where V is in (m/s) and d in (m).

With this value of V_d and with the computed air velocity at the bridge of the carburettor, the initial droplet diameter is calculated. Integration of the model then begins and the changes in the values of the dependent variables are computed at various positions within the carburettor.

Factors influencing fuel evaporation

The effect that various parameters have on the amount of fuel evaporated within the carburettor was studied with the model. Firstly the effect of engine speed and load was examined. It will be seen from Fig. 2.28 that the three engine conditions selected, namely 40 and 80 km/h road load and 80 km/h full power, give very similar results in terms of percentage evaporation of fuel; indeed it is only as the mixture approaches the throttle plate that slight differences develop. In an air valve carburettor the bridge velocity remains largely constant over a wide range of operating conditions; since the bulk of the evaporation occurs while the mixture is in the vicinity of the bridge, the three cases considered must inevitably produce similar results, particularly when, as is the case here, wall effects are neglected.

At road load conditions the mixture is subjected to a large acceleration as it approaches the throttle plate. Since the fuel droplets cannot match the large increase in the velocity of the air and fuel vapour, a large velocity differential is built up which enhances heat and mass transfer between droplets and the air. This is shown in Fig. 2.28 by the turn up of the road load curves in the vicinity of the throttle plate.

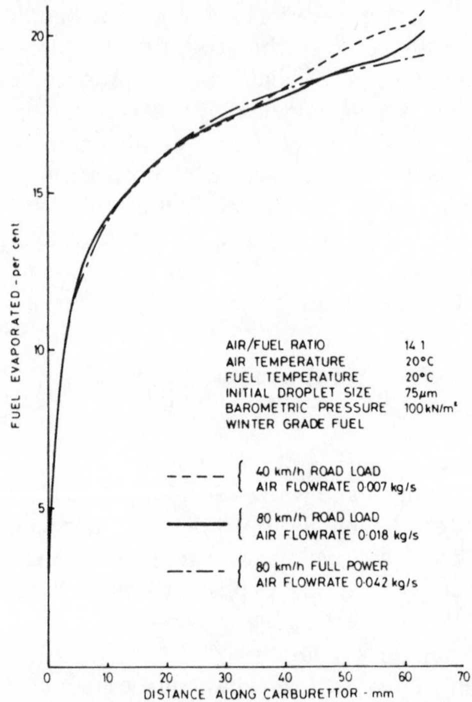

Fig. 2.28 Effect of engine speed and load on fuel evaporated

Combustion modelling in spark-ignition engines

Dr E. H. James[10] here describes the modelling techniques available for the simulation of engine combustion. Combined with the recent publication of CAD techniques for predicting the performance of other engine systems, the work indicates how such combustion modelling can assist in the computer design analysis of a complete engine (*other aspects of which are considered in the final chapter of this book – Ed.*).

Continuing efforts to improve the thermal efficiency of the spark ignition engine whilst simultaneously reducing its undesirable exhaust emissions has resulted in close attention being focussed on a total understanding of the combustion process. The initiation of combustion towards the end of the engine's compression stroke provides a flame kernel at the spark plug which develops and propagates out across the chamber through the predominantly pre-mixed fuel–air–exhaust gas residual charge. The linkage between the fluid dynamic and chemical reaction processes ensures a determining effect of the turbulent flow field on the rate of flame propagation.

Computer modelling of this process is extremely complex since there are present fluctuating flow velocities, temperatures, and multiple specie concentrations. Nevertheless, great progress has been made in the development of suitable models to fulfil a wide range of engineering needs. The early types were relatively simple and restricted but, as turbulent flame propagation was more closly studied, more realistic models became available. So much progress has been made that, at the present time, the most detailed and predictive of models is emerging. From this historical perspective, the following classification of model types can be constructed: 'zero-dimensional' (thermodynamic) models incorporating defined rate of burn (DRB) models, 'wrinkled' turbulent flame front (WTFF) models and fragmented combustion zone (or entrainment) models; one-dimensional (1-D) models, multi-dimensional (MDM) models, and finally, 'hybrid' models.

Zero-dimensional (thermodynamic) models

The term zero-dimensional was first ascribed to these models in 1974 because their mathematical basis is independent of any space coordinate. Thus, as time is the only independent variable, ordinary differential equations can be used in their formulation. They are sometimes called 'phenomenologial' models because they invoke sub-models to aid in the description of certain aspects of the combustion process (exhaust pollutant generation, turbulent flame propagation, heat transfer). Such models are non-predictive in nature and relate to the engine concept being modelled. They cannot be extrapolated confidently to other designs. Their formulation involves the conservation equations of mass and energy applied to a closed system process. The neglect of the momentum equation precludes any predictive coupling between the turbulent flow field and the chemical processes taking place – hence, their alternative title of 'thermodynamic' models. This constitutes a major disadvantage.

The mathematical solution of the equations involved in 'zero-dimensional' models can be approached by using either numerical integration techniques or iterative procedures. In both, calculations are made in a stepwise fashion through the combustion period in the engine. Differences between the three sub-groups within the 'zero-dimensional' model classification revolve around the way the mass burning rate is accounted for.

Defined rate of burn (DRB) models. Details of the combustion process are here fixed as an input to the calculation. From experimentally obtained curves of mass fraction burnt, one approximate functional relationship is

$$\alpha(\theta) = \tfrac{1}{2}[1 - \cos\{\pi(\theta - \theta_0/\Delta\theta_c\}] \qquad (2.19)$$

where θ_0 denotes the onset of combustion and $\Delta\theta_c$ is the combustion interval. The form of this equation is shown in Fig. 2.29(a).

'Wrinkled' turbulent flame front (WTFF) models. Attempts are made here to relate the turbulent burning velocity, U_T, to the intensity of the turbulent flow, u', and infrequently also to the integral length scale of turbulence, L. In these models, the large scale eddies in the flow are considered to wrinkle the flame front, thereby increasing its surface area, A_T, in comparison with that which would exist under laminar conditions, A_L. Thus

$$U_T/U_L \simeq A_T/A_L \simeq f(u', L) \qquad (2.20)$$

The mass fraction burnt during a specified time interval, dt, is then

$$d\alpha = \varrho_u U_T A_f \, dt/m \qquad (2.21)$$

where ϱ_u is the unburnt gas density and A_f is the flame front surface area.

Fragmented combustion zone (entrainment) models. Improvements in the conceptual understanding of the effects of turbulence on flame propagation led to doubts that WTFF models adequately depicted the detailed physics of the process. Such doubts were voiced by several investigators who found that turbulent flames do not have 'thin' reaction zones (as would ensue from WTFF theory visualizations) but intead have 'thick' ones of width between 10 and 15 mm. This concept was developed into a mechanism of burning in which the propagating flame engulfed or entrained 'lumps' of the reactant mixture which were subsequently burnt behind the advancing flame front (Fig. 2.29(b)). The entrained mass, m_e, within the turbulent reaction zone is given by

$$dm_e/dt = \varrho_u A_f(u' + U_L) \qquad (2.22)$$

The later burn-up of this mass takes place in eddies of size λ at a rate

$$dm_b/dt = (m_e - m_b)/\tau_b$$

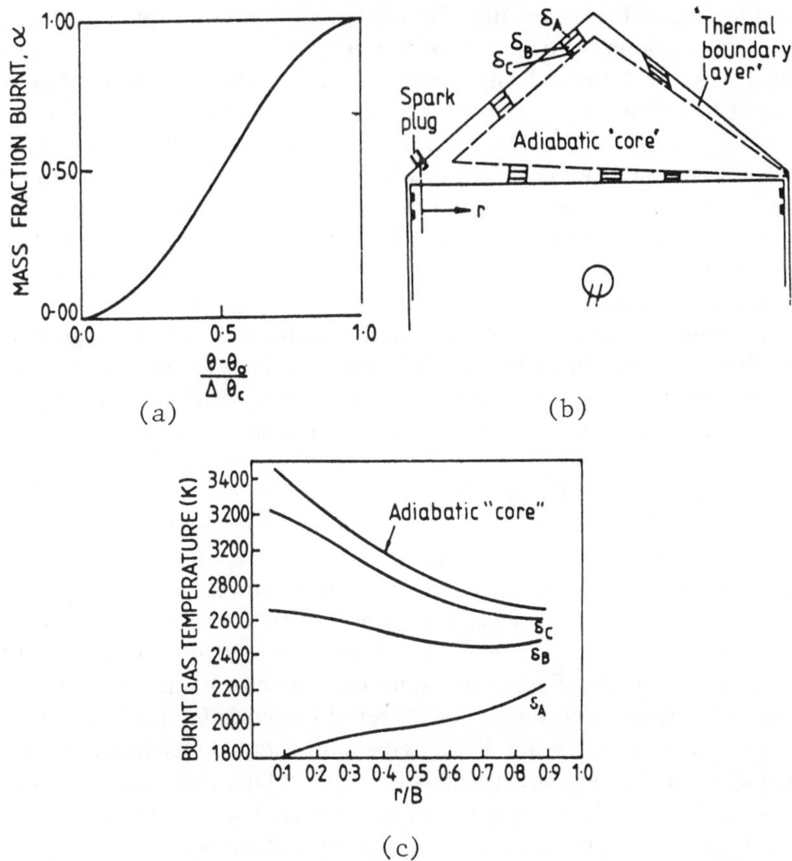

Fig. 2.29 Computer modelling of combustion: (a) defined mass burning rate, (b) thermal boundary layer after combustion, and (c) burnt gas temperature variations in the combustion chamber

where

$$\tau_b = \lambda / U_L \tag{2.23}$$

The WTFF and fragmented combustion zone models have sometimes been referred to as 'quasi-dimensional' since they attempt to introduce an element of dimensionality by linking the factors affecting the combustion process to the engine design under study and also to the operating conditions. This is often aided by the specification of a spherical flame pattern development from the spark plug.

N.B. More realism can be incorporated into zero-dimensional models by the inclusion of temperature gradients and a realistic thermal boundary layer development from the spark plug. This enables heat transfer and the nitric oxide and unburnt hydrocarbon exhaust emissions to be more accurately modelled. For the thermal boundary layer variation shown in Fig. 2.29(b), the corresponding temperature profiles in the individual zones of the burnt charge are given in Fig. 2.29(c) at one particular engine operating condition.

Zero-dimensional models are used particularly when parametric studies of the combustion process are required. They are relatively cheap to use and will probably continue to be pre-eminent in modelling engine performance and exhaust emissions for some time to come.

One-dimensional (1-D) models

These have been referred to as 'the first of a new generation of more detailed and informative combustion models'. Because both time and a dimension are now operative, the defining equations are of the partial differential kind and specify the principles of conservation of mass, momentum, energy, and chemical species. The form they assume is not denoted here to avoid repetition since they can be obtained by imposing the above restrictions on the MDM set of partial differential equations (PDEs) in the next section. Perhaps the most detailed such model to date employs finite difference methods to solve 9 PDEs including equations from an 'energy-dissipation' turbulence model. Validation of the model has been established from an extensive comparison with certain experimental quantities (flame speeds, exhaust emissions). Such models are worthy of a separate classification since, historically, they represented a step-forward in modelling development. Latterly, they have been largely superseded by MDM models.

Multi-dimensional (MDM) models

These models have the potential, in principle, to provide detailed information on the spatial and temporal distributions of the flow velocities, temperatures and species concentrations in the engine, enabling the turbulent burning rate and geometrical progression of the flame front to be accurately monitored. In such terms, they could be completely predictive, allowing the engine geometry to be incorporated into the simulation in a fundamental way. They would also appear to be well-suited for the accurate prediction of the exhaust pollutants. As will be shown, however, incomplete information, coupled with numerical and computational problems, preclude against this ideal at the present time.

The first stage in their development is the definition in general terms of the controlling set of coupled PDEs for mass, momentum, energy, and individual chemical species. One form this can take for a multicomponent reacting gas mixture is given below using Cartesian, tensor notation.

Mass: $\partial\varrho/\partial t + \partial(\varrho v_i)/\partial x_i = 0$ (2.24)

Momentum: $\varrho Dv_i/dt = -\partial p/\partial x_i + \partial p_{ij}/\partial x_j$ (2.25)

where D is the substantial derivative, v_i represents the three components of the velocity vector, $\boldsymbol{V}$, and the viscous stresses are given by

$$p_{ij} = -\mu'\,\partial v_k/\partial x_k \cdot \delta_{ij} + \mu(\partial v_i/\partial x_j + \partial v_j/\partial x_i)$$

Here, μ and μ' are coefficients of viscosity and δ_{ij} is a unit tensor.

Notation

B	Engine bore diameter
C_p	Constant pressure specific heat
e	Specific internal energy
E	Activation energy
h_0	Specific enthalpy
h_f	Heat of formation at reference temperature T_0
k_f, k_r	Rate constants for forward and backward reactions
k_b	Thermal conductivity
K_c	Concentration equilibrium constant
m	Mass of charge in engine
M	Molecular weight
N	Total number of species
p	Pressure
Pr_t	Turbulent Prandtl Number
R	Gas constant
t	Time
T	Temperature
u	Diffusional velocity
U_L	Laminar burning velocity
Y	Mass fraction
Z	Pre-exponential factor

Subscripts

b refers to burnt charge
n refers to nth specie
u refers to unburnt charge
s reaction number

Energy: $$\partial(\varrho e_t)/\partial t + \partial(\varrho v_i e_t)/\partial x_i + \partial(v_i p)/\partial x_i = -\partial\varepsilon_i/\partial x_i \quad (2.26)$$

neglecting energy dissipation where $e_t = e + V^2/2$

$$e = \sum_{n=1}^{N} h_n Y_n - p/\varrho = \sum_{n=1}^{N}\left(h^0_{f,n} + \int_{T_0}^{T} Cp_n \cdot dT\right)Y_n - p/\varrho$$

$$\varepsilon_i = -k_t\ \partial T/\partial x_i + \varrho \sum_{n=1}^{N} h_n Y_n u_n$$

Chemical species

$$\partial(Y_n\varrho)/\partial t + \partial\{Y_n\varrho(v_i + u_n)\}/\partial x_i = \dot{w}_n \quad (2.27)$$

where $\dot{w}_n$ is the rate of generation/destruction of the nth specie and can be determined by expressions of the form

$$\dot{w}_n = M_n \sum_{s=1}^{T} (v''_{n,s} - v'_{n,s}) k_{fs} \varrho^{m_s} \prod_{n=1}^{N} (Y_n/M_n)^{v'_{n,s}} B \quad (2.28)$$

where

$$B = \left(1 - (\varrho^{n_s}/K_{c,s}) \prod_{n=1}^{N} (Y_n/M_n)^{v''_{n,s}-v'_{n,s}}\right)$$

$$m_s = \sum_{n=1}^{N} v'_{n,s} \qquad n_s = \sum_{n=1}^{N} (v''_{n,s} - v'_{n,s})$$

T = number of elementary, stoichiometric reaction equations representing the chemical kinetics and of the form

$$\sum_{n=1}^{N} v'_{n,s} R_n \underset{k_{fs}}{\overset{k_{fs}}{\rightleftharpoons}} \sum_{n=1}^{N} v''_{n,s} P_n$$

in which R_n and P_n refer to reactant and product species.

A further equation required is one accounting for the diffusional velocity, u_n. The ideal gas equation of state is also needed.

The solution of these equations in the unsteady context of engine cylinders is evidently a most difficult assignment. Ideally, one would hope for detailed information becoming available on a small scale, highly localized basis, relating to turbulent fluctuations and chemical reaction rates. Practically, however, it is not possible to achieve anything approaching this. The reason for this is that finite difference, numerical methods have been used in their solution and this necessitates the specification of a grid or mesh to fully occupy the combustion chamber volume. Computer storage and execution time constraints determine the use of a fairly coarse grid size invariably in a 2-D coordinate frame. This dictates the need for sub-grid scale models for turbulence and chemical reactions.

The requirement for these scale models invokes the definition of the PDEs in their averaged form. Additional terms now become involved (turbulent viscosity, μ_t, and the thermal eddy diffusivity ($=\mu_t/\varrho \mathrm{Pr}_t$)).

Turbulence modelling to obtain μ_t. The evaluation of μ_t presents a major problem. The earliest attempts in this context involved the use of constant or time-dependent 'global' values. More recently, an 'energy-dissipation' model of turbulence has provided a more realistic approach. Local values of the kinetic energy of turbulence, k, and its dissipation rate, ε, are evaluated and used to calculate μ_t according to

$$\mu_t(x, t) \simeq \varrho k^2/\varepsilon \tag{2.29}$$

Although clearly representing an improvement in μ_t estimations, this type of sub-grid scale turbulence model has dubious validity in the engine context since its initial development was largely directed towards incompressible, boundary layer flows. Its unaltered extension to the unsteady, compressible flows in engines is suspect. Among the attempts to correct such deficiencies, some have modified the equation giving the dissipation rate of turbulence so that the k–ε model simulates their conceptual understanding of the behaviour of the turbulence length scale during the engine compression stroke.

Combustion chemistry modelling. Besides alleviating computer storage and execution time constraints, sub-grid scale modelling of the chemical kinetic processes also overcomes the lack of available knowledge on the detailed reaction mechanisms and rate constants, for fuels used in engines. As such, it also precludes against any detailed, small scale study of the formation and possible destruction of certain undesirable exhaust pollutants. Several such models have been recommended and/or used in this respect including: a two or three step oxidation mechanism describing the major exothermic reactions only, a single-step, 'global' chemical kinetic equation of the form

$$\dot{\omega}_{\text{fuel}} = -d[\text{fuel}]/dt = Z[\text{fuel}][\text{O}_2] \exp(-E/RT) \qquad (2.30)$$

and also a 'turbulence-decay-control' model of combustion in which turbulence energy dissipation controls the rate at which reactions occur. This has the great advantage of not requiring a prior knowledge of the detailed chemistry involved.

Numerical diffusion ('smearing'). This is a measure of the amplification in the extent of diffusion in the model compared with that which occurs in practice in the engine. In a reacting flow, it is particularly serious since the flame front can become 'smeared out' with the result that the burning velocities are over-estimated. The most apparent corrective measure is a reduction in grid size but this can outlandishly increase computing costs (Table 2.8). The use of an arbitrary Lagrangian–Eulerian grid system is one answer and another is the explicit monitoring of flame propagation to ensure correct flame travel times.

Computing times. Table 2.8 lists some estimates published for computing times for 'working' models over a complete engine cycle. Clearly, reacting flows with adequate grid resolution impose severe economic penalties and this prohibits against any extensive parametric studies. This is unfortunate since finer grids and smaller time-steps will be needed to

Table 2.8 Computation of grid sizes and times

Ref.	*Flow*	*Grids (number)*	*Grid sizes (cm)*	*Time step*	*Comp. time*
5	Reacting	250	0.37	0.05 msec	10–20 min CDC 7600
5	Reacting	1700–10 000	(<0.01–0.06)	0.05 msec	1–5 hr CDC 7600
9†	Non-reacting	900	0.25	3°CA	40 min IBM 360/195
13	Reacting	1024	0.24	* $(\text{r/min})^{-1}$	$1\frac{1}{2}$ hr UNIVAC 1100/40

* During comb. $10^{-2}\ (\text{r/min})^{-1}$.
† Gosman and co-workers.

completely predict the behaviour within boundary layers and reaction zones and to fully understand the wall quenching processes. Moreover, before 3-D computations become feasible, numerical solution methods will have to be much more efficient. Dr Gosman at Imperial College, London is at the forefront of such advances at the time of writing.

N.B. Validation of predicted results from MDM models has resulted in reasonable 'global' behaviour in non-reacting flows being achieved whilst, in reacting flows, research workers have obtained the correct trends in burning rate and pressure development when some engine operating conditions were changed. It would appear, therefore, that at the present time, MDM modelling is of the greatest benefit in enhancing the conceptual understanding of certain of the phenomena occurring in the engine. Indeed, there is no other way of getting the sort of detail that MDM models can in theory provide, since relevant experimental techniques are extremely difficult to apply in the engine combustion chamber environment. In this respect, even at this early stage in their development, MDM models contribute greatly to the fund of knowledge which the engine designer and 'phenomenological' modeller can exploit.

Hybrid models

This final classification has been proposed as a means of bridging the gap between MDM and 'quasi-dimensional' models. This would enable the turbulent flow field to be directly coupled to phenomenological sub-models of the turbulent burning process, heat transfer, and so on. Some progress to this end has already been made. Thus, research workers have linked a modified k–ε model for turbulent flow to the entrainment burning model – see equation (2.22). With the inclusion of dissipation in the kinetic energy evaluations, this equation now assumes the form

$$\mathrm{d}m_e/\mathrm{d}t = \varrho_u A_f(k^{1/2} + U_L) \qquad (2.31)$$

OBTAINING INCREASED EFFICIENCY AND OUTPUT

Computerized engine performance prediction

Careful studies[11] of the basic parameters of one of Ford's engine series have allowed the company's engineers to develop a performance prediction program whereby revised versions of the engine can be conceived 'on paper' by feeding appropriate data to the company's central computer installation at Dunton. A thermodynamic model of the engine is drawn up

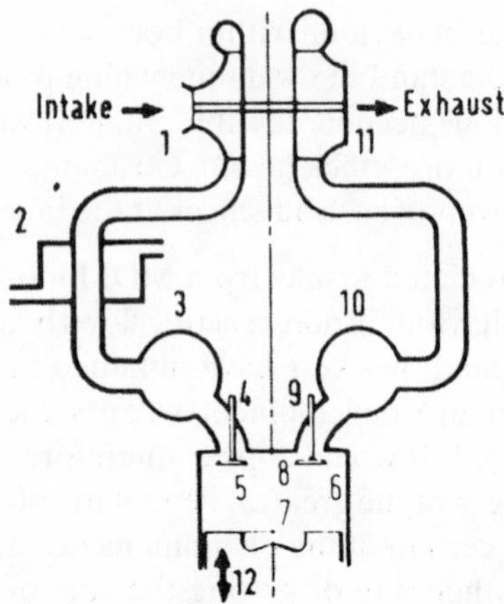

Fig. 2.30 Thermodynamic model for 'filling and emptying' technique

as shown in Fig. 2.30 and, using the so-called 'filling and emptying' technique, the performance of derived versions of the engine can be computed by feeding in changes to the basic design parameters.

Input data for the computation include engine geometry, valve effective areas, turbocharger characteristics, heat release rate, component temperature estimates, intake and exhaust conditions, speed and fuelling. The computation will then reveal indicated and brake mean effective pressure, power and fuel consumption, air flow, heat transfer, gas pressures and temperatures, and mass flow. Development of the program has taken place in conjunction with Imperial College, London.

The various volumes of the engine are as follows: (1) compressor, (2) charge cooler, (3) inlet manifold, (4) intake valve flow, (5) gas properties, (6) heat transfer, (7) combustion, (8) gas exchange, (9) exhaust valve flow, (10) exhaust manifold, (11) turbine, and (12) engine friction (see Fig. 2.30). Flow between the volumes can be calculated from the pressure/temperature characteristics in each volume and a knowledge of the interconnecting effective areas between the volumes. Heat released from combustion is the basis for calculations showing how much energy is used in powering the piston and how much is released to coolant and to the exhaust. Turbocharger characteristics are supplied by the component manufacturer and exhaust systems are allowed for in the calculations.

The computation can provide performance parameters at each degree of crank rotation for a given load and speed, so a full understanding of a

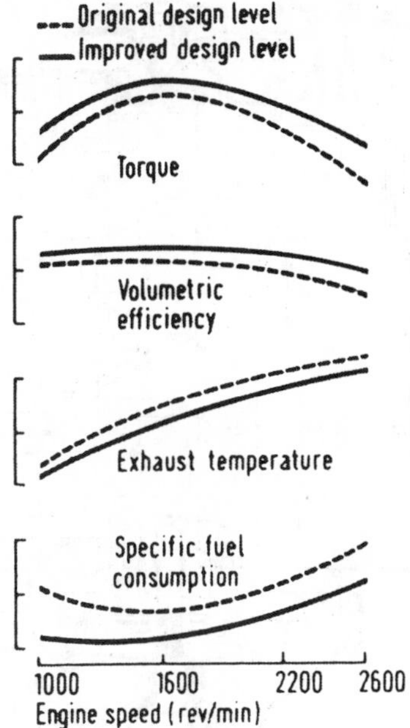

Fig. 2.31 Predicted performance of original and improved design-level engines

complete combustion cycle can be obtained. The pressure development throughout one cycle is, of course, crucial to the combustion characteristics, mechanical loads, and the noise emitted from the engine.

Effects of changes in valve timing, cam profile, and lift can be evaluated, and the prediction of the overall breathing characteristics (volumetric efficiency) of the engine saves considerable time otherwise spent with 'in the metal' development programmes. Turbocharger matching is another facility of the system that saves extensive development time and cost; even the effect of charge cooling can be predicted. Effects of changes in fuelling, injection pump can form, and inlet porting can also be studied.

Figure 2.31 shows characteristic gains in performance available over the base engine using the prediction program described and effecting appropriate changes to the design parameters. Generally, an order of magnitude up to 5 per cent applies to the level of improvements evaluated. These theoretical predictions have been verified in performance measurements on real engines modified in the same way.

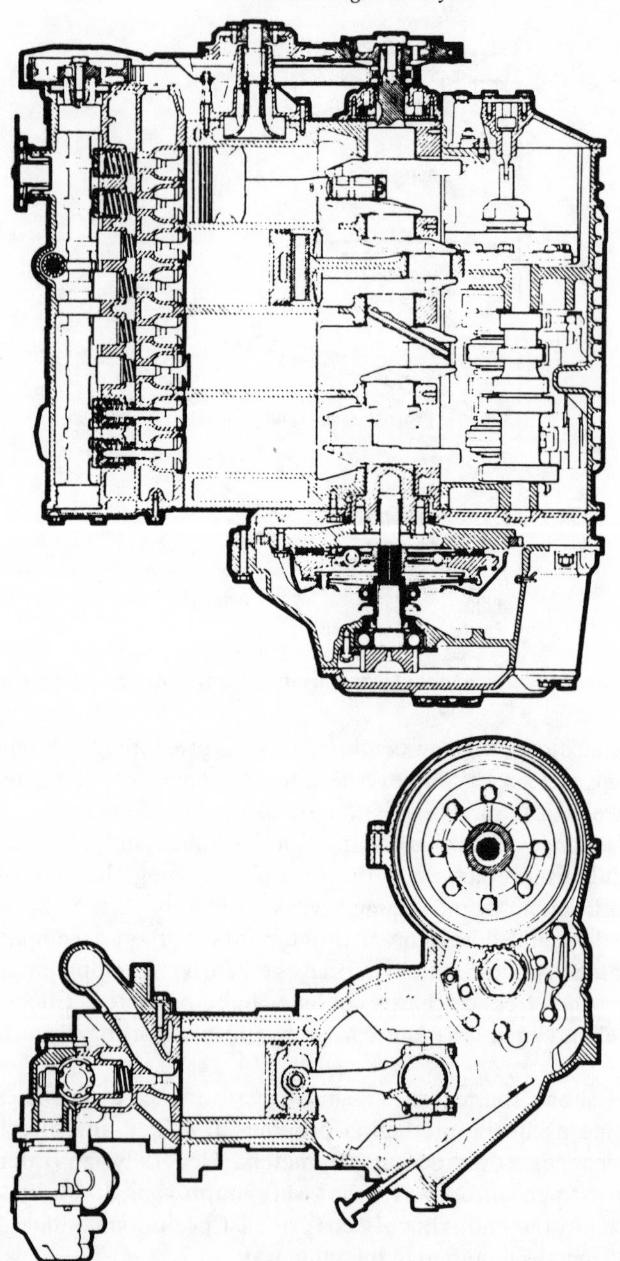

Fig. 2.32 Austin O-series engine of 1.7 l

Results from the engine performance prediction can also be fed into the Ford 'Vehicle Mission Simulation' program which is used for overall vehicle performance prediction. The latter system has been well documented and computes performance in relation to such parameters as transmission ratios and efficiencies, rotational mass factors, and tyre characteristics.

Enhancing performance: the principal parameters

Whether to increase power from a given cylinder capacity, improve fuel economy, or reduce exhaust pollutants, improvement in combustion efficiency has been approached from several directions. In what follows, short referenced extracts from *Automotive Engineer* are used both to demonstrate the (production) state-of-the-art in 1984 and to examine individual avenues of development. This is followed by sections on high compression ratio engines, and development potential revealed in concept engines by Daimler-Benz and BMW.

The 'big-four' UK motor manufacturers Austin/Rover and Ford/GM (the latter coupled with their mainland European counterparts) had the typical designs below in current production during 1984.

The Austin O-Series 4-cylinder engine (*Automotive Engineer*, 1978, **3**, (4), 41–43) is shown in Fig. 2.32 to have a piston-bowl combustion chamber. The 9.0 : 1 compression ratio 1.7 litre version developed 65 kW at 5200 r/min in its originally introduced form using a single SU HIF 6 horizontal carburettor with 44.4 mm diameter choke. Oil-cooled pistons are employed and an aluminium alloy cylinder head. Unit weight was reported to be 134 kg and for reduced overall length siamesed bores and a recessed water pump are used.

The Rover SD1 6-cylinder engine (*Automotive Engineer*, 1977, **2**, (5), 15–16, 20) also had a specific output of 38 kW/litre and is seen in Fig. 2.33 to use part-spherical combustion chambers (over dished-crown pistons) with inclined valves and an offset sparking-plug location. Twin SU HS6 carburettors are employed for both 2.3 or 2.6 litre versions, the former developing 90 kW at 5500 r/min when first introduced. Dry weight was reported to be 196 kg and compression ratio 9.25 : 1.

Ford's 4-cylinder CVH engine (*Automotive Engineer*, 1980, **5**, (5), 43–45) when introduced was for a manufacturing rate (in the UK at Bridgend and the US at Dearborn) of a million units per year. CVH stands for compound valve-angle hemi-spherical to describe the combustion chamber (Fig. 2.34) common to 1.1, 1.3, and 1.6 litre variants. The

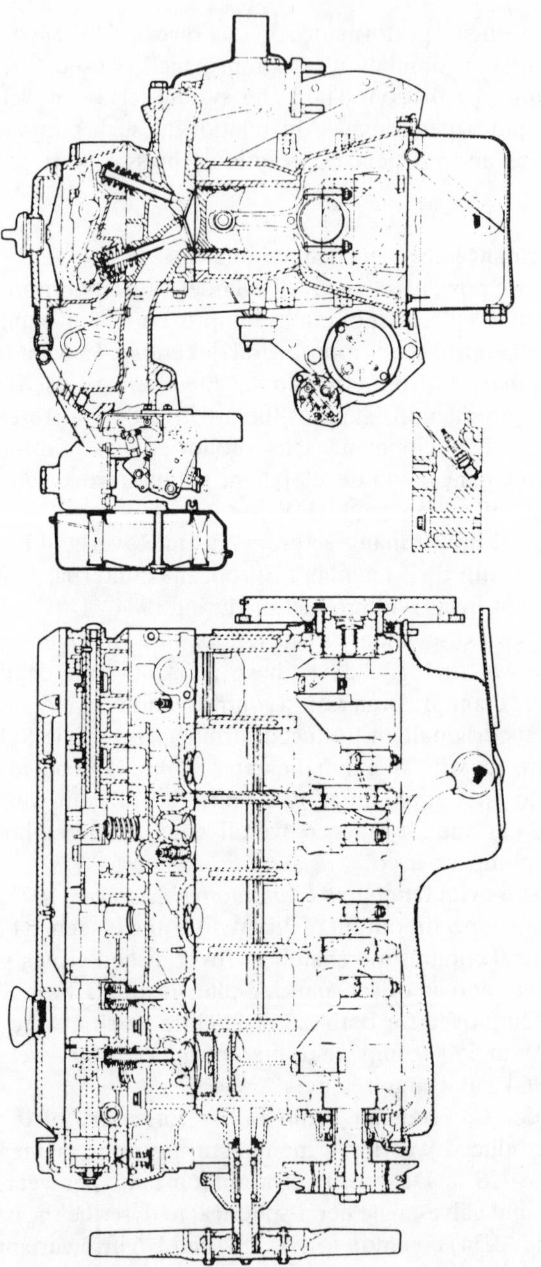

Fig. 2.33 Rover SD1 engine of 2.3 l

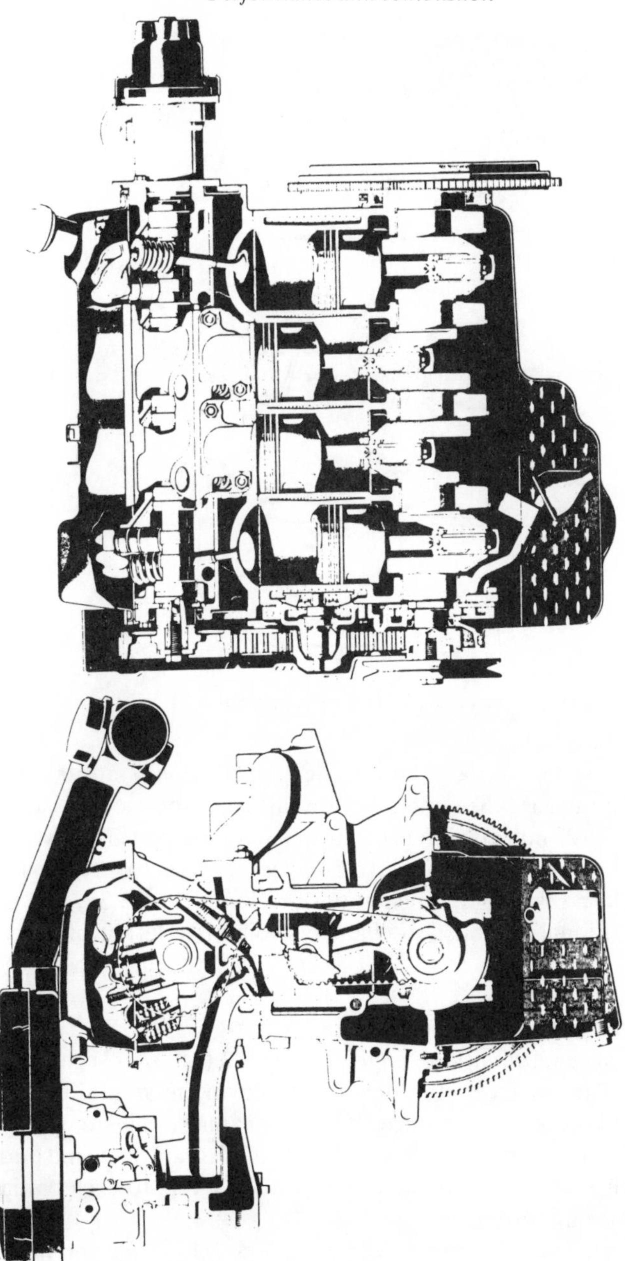

Fig. 2.34 Ford CVH engine of 1.3 l

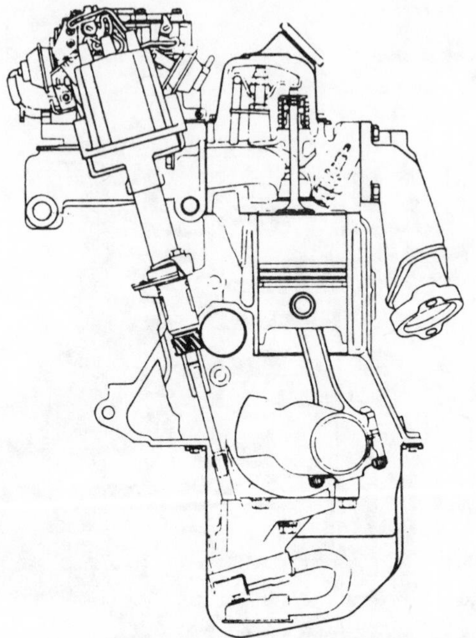

Fig. 2.35 GMs J-car engine of 1.8 l

1.3 litre unit weighed 106.2 kg and developed 51 kW at 6000 r/min. Compression ratio is 9.5 : 1 and a Motorcraft constant-vacuum carburettor is used, with 23 mm maximum choke diameter for the 1.3 litre version.

GM's 1.8 litre engine (*Automotive Engineer*, 1981, **6**, (6), 94–95) for the J series world car is a high camchaft push-rod unit (Fig. 2.35) and uses a more conventional combustion chamber which is tolerant of unleaded gasoline.

Stratified-charge and lean-burn engines

Stratified charge (*Automotive Engineer*, 1982, **2** (1), 12–17) combustion was intended to improve efficiency by employing moderate peak temperatures and pressures to reduce NO_x in the exhaust; high temperatures during the expansion phase to oxidize unburned HC and CO, and burning the fuel with excess air with similar objective. Cycle-to-cycle variation had also to be minimized.

In the experimental Porsche SKS modified 911 engine the auxiliary

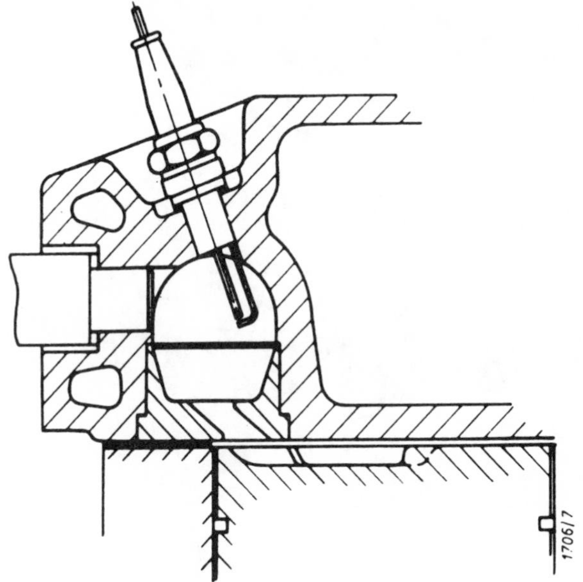

Fig. 2.36 Ricardo Comet mark V combustion chamber

chamber contains a high proportion of residual exhaust gas at the end of the inlet stroke plus all the fuel injected during that stroke. The main chamber is supplied in the normal way via the engine induction system with lean mixture of 1.5–3.0 equivalence ratio. During the compression stroke some of the mixture is forced into the auxiliary chamber reducing the value to 0.4–0.8; resultant overall air/fuel ratio is above 1.0. The engine operates on 86 octane fuel at 10:1 compression ratio. By contrast, the Ricardo Comet Mark V system (Fig. 2.36) has fuel injected near TDC into the rapidly swirling air within the pre-chamber.

With lean-burn engines (*Automotive Engineer*, 1979, **4**, (4), 55–58) operation is at air–fuel ratios up to 24:1. Dual ignition systems are common-place and typified by the Nissan NAPS Z engine operating at 21.8:1. The May Fireball system, put into production since by Jaguar, uses a very compact combustion chamber which is a pocket below the exhaust valve. The inlet valve seats in the head face, as with a Heron head; there is a guide channel from this valve to the combustion chamber. A 14.6:1 compression ratio is used.

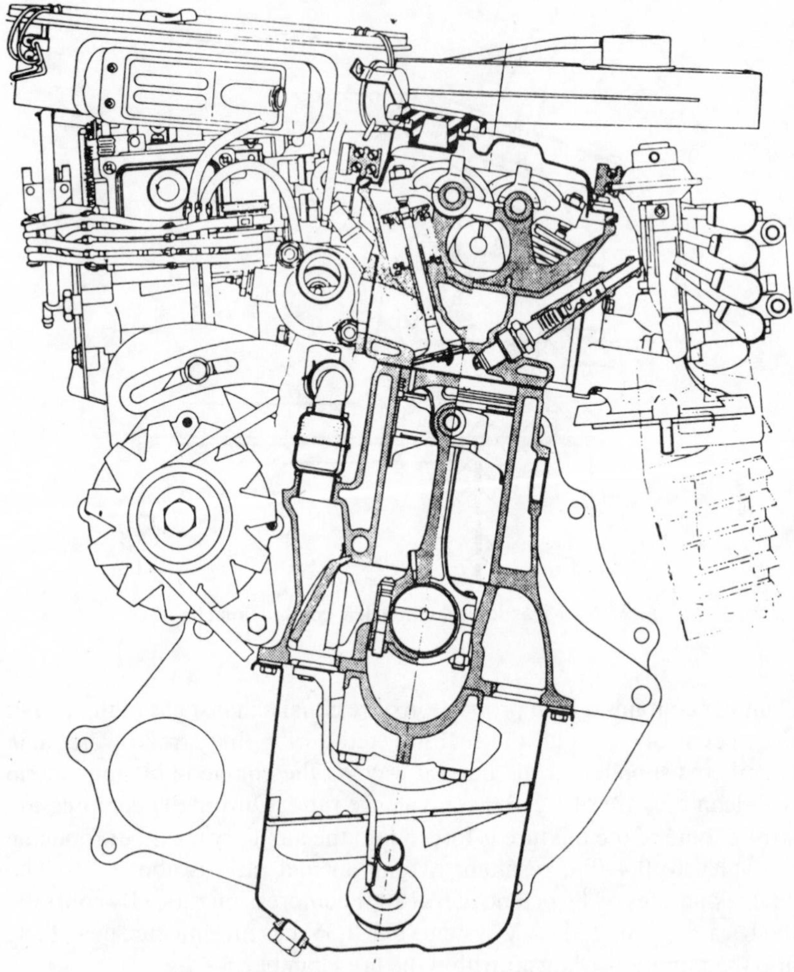

Fig. 2.37 Nissan NAPS Z lean-burn engine

Nissan's NAPS-Z system has since been productionized (Fig. 2.37) and featured on the company's 1.8 litre engine. Swirl ports have also been added (*Automotive Engineer*, 1981, **6** (6), 48–49). The Honda CVCC system (Fig. 2.38) is also in production on the *Civic* 1500 car. Five outlets between auxiliary and main chamber help to spread the combustion flame more quickly. The car uses only 3.57 l/100 km at 60 km/h.

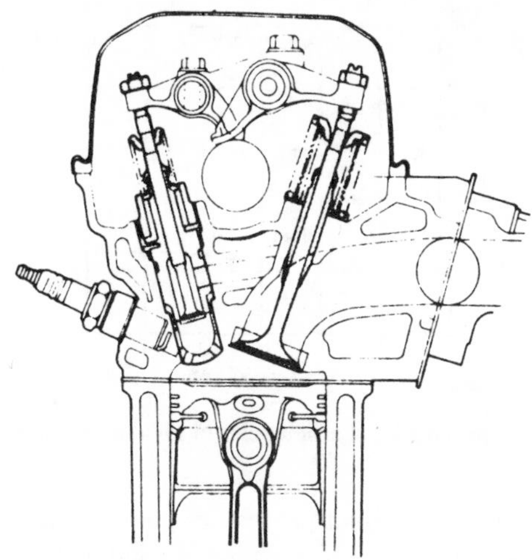

Fig. 2.38 Honda CVCC lean-burn engine

High compression ratio engines

M. T. Overington[12] points out that ideal air cycle efficiency increases continuously with compression ratio but, in a real engine, using imperfect fluids and with heat losses through the chamber walls, the indicated efficiency reaches a peak at a compression ratio of about 13:1 (Fig. 2.39). Since most European engines have compression ratios around 9:1, there is considerable scope for improvement by the use of higher compression ratios. Many researchers have investigated the effects of various design variables such as combustion chamber geometry and inlet air motion on the engine's fuel quality requirement, to eliminate detonation while raising compression ratio.

Experimental investigation at Ricardo has concentrated on combustion chamber designs which are compact by nature (where the largest dimension of the chamber is kept to a minimum). This family has been called High Ratio Compact Chamber (HRCC) designs. The design philosophy may be applied to combustion chambers situated both in the cylinder head or in the piston and typical designs are shown in Figs 2.40 and 2.41. In the first case, the combustion chamber is formed below a recessed exhaust valve. The recess has vertical sides and in plan is 'bean' shaped. The sparking plug breaks out into the top surface of this chamber. The remainder of the cylinder head face and the piston are flat. The inlet valve, however, is recessed to avoid interference with the piston crown

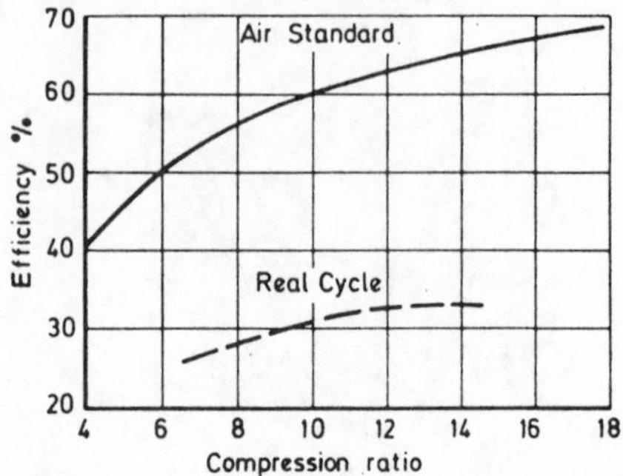

Fig. 2.39 Comparison of efficiencies of ideal and real engine cycles

near top dead centre. The inlet port is designed to give high flow with a relatively low level of swirl. A partially shrouded valve has, however, been used to determine the effects of higher levels of swirl and turbulence. The investigations have also evaluated a slot in the cylinder head between the inlet valve and the combustion chamber thus removing the thin metal section. The HRCC designs in the piston have been generally circular in plan with vertical sides located close to the sparking plug. The inlet port design is similar to that described above and use is also made of modified inlet valves to change the inlet gas motion. Both types of design allow for the production of conventional and HRCC designs with few changes to the production machinery since, in both, the line of the valves is parallel to the crankshaft axis and the axes of the valves are vertical. This is an important cost factor in the evolution of an engine family.

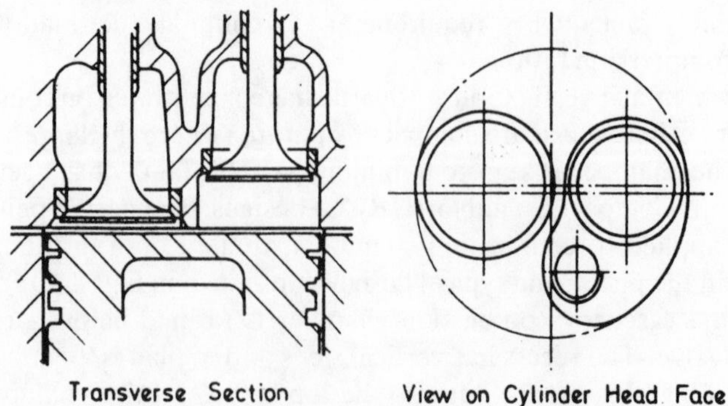

Fig. 2.40 HRCC chamber-in-head design

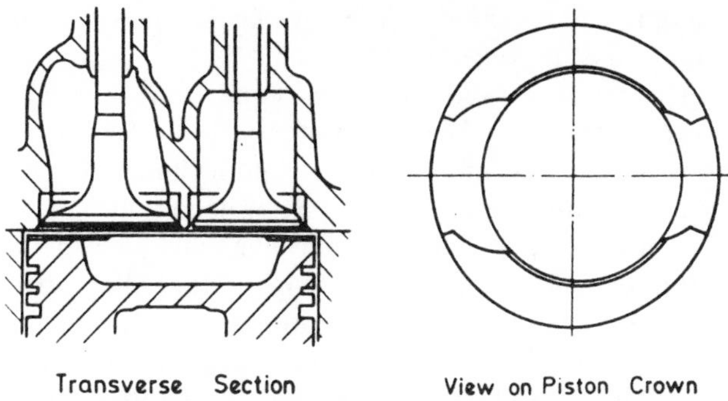

Fig. 2.41 HRCC chamber-in-piston design

Engine performance

Most high compression ratio engines exhibit a number of common features. They will run with much leaner air/fuel ratios than conventional designs and also show their best economy at leaner mixtures (Fig. 2.42). The limit on how far these attributes may be exploited is set by detonation at full load; here the design of the chamber and port is important. The aspect ratio of the bean chamber affects the performance (in our experiments a range from 2.7 to 4.5 was investigated), and may be chosen to give the best compromise of octane requirement and part load fuel economy. The intake port swirl ratio, produced either by port shape or by using a shrouded valve, has also been varied up to a value of 1.5 and influences

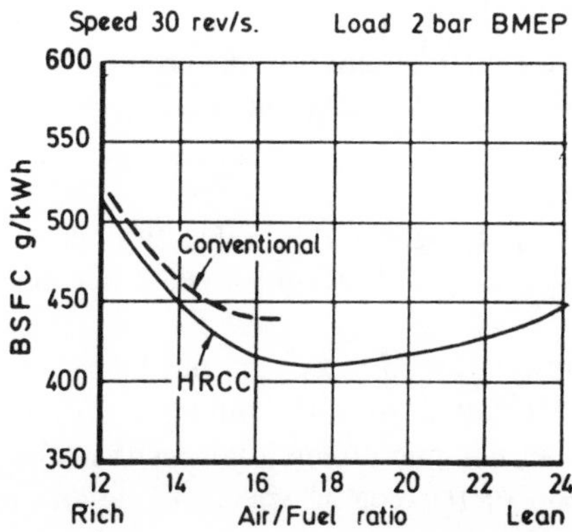

Fig. 2.42 Part load fuel economy comparison with and without HRCC

the fuel economy and octane requirement. Again the best compromise has to be found for each type of design.

The high compression ratio chambers require less ignition advance than conventional designs and have the potential for higher torque output through improved air utilization. Thus in a multicylinder engine application, both the carburation and ignition systems need returning to suit the new combustion chamber design, and the valve timing may be revised to produce a more desirable torque curve shape. Other alternatives would be to reduce the engine swept volume in order to maintain engine power output constant or to decrease the final drive ratio to reduce engine speeds while maintaining the same top gear acceleration. A computer drive cycle simulation program can be used to calculate vehicle fuel economy figures over a fixed drive cycle from a comprehensive set of test bed data. By using this technique the effect of the combustion system modifications alone can be evaluated. This type of analysis has been used to compare the fuel economy of conventional and HRCC designs over the US Federal Urban drive cycle for a typical European car. The results are shown in Fig. 2.42 for a 1135 kg car with conventional engine as 9.75 litre/100 km and for HRCC piston and head as 8.88 litre/100 km, some ten per cent improvement. Changes in swept volume or gearing can take this figure to nearer 15 per cent or more without loss of vehicle performance.

Variable valve timing for i.c. engines

C. R. Stone and E. K. M. Kwan[13] explain that variable valve timing is a well proven method of improving the performance i.c. engines, and results are summarized for improvements in power output, economy and emissions from and exhaustive study of the literature. Since no comparison has been made between the numerous different variable valve timing mechanisms, they maintain, a classification system is also developed here.

Spark-ignition engines

The main claims for applying variable valve timing to spark-ignition engines are: increased power output, improved efficiency, and reduced emissions.

Many authors claim increased power output by using variable valve timing on the grounds that fixed valve timing inevitably represents a compromise, since the dynamic effects that occur in the gas exchange process are a function of the engine speed. Thus an engine which is designed to produce maximum torque in the mid-speed range (3000

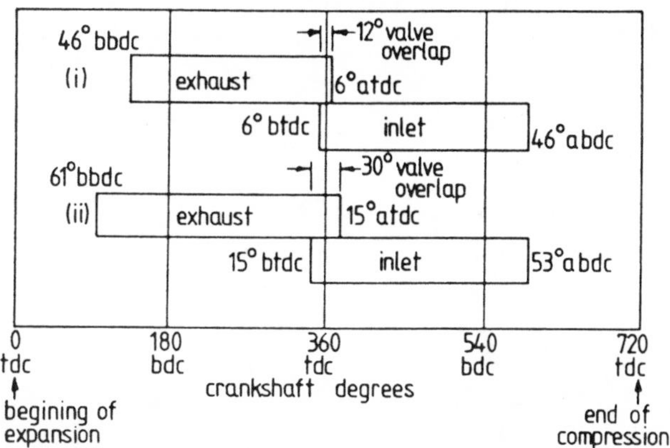

Fig. 2.43 Typical valve timing diagrams for maximum torque at (1) 3000 (2) 5000 r/min

r/min), has its valves open for a shorter period with a smaller valve overlap, than an engine designed to produce maximum torque at a high speed (5000 r/min) (Fig. 2.43). While the large valve overlap (30 degrees) enables full benefit to be obtained from the pulse effects at high speeds, it has a detrimental effect on the mid-speed torque. The fuel economy will also suffer at part load since the pressure in the induction manifold is sub-atmospheric whilst the exhaust pressure is greater than atmospheric, consequently the large valve overlap permits the undesirable mixing of the incoming mixture and exhaust products. The problem is most severe at idling, an important operating condition in urban fuel economy cycles.

It is thus not surprising that in order to show maximum benefits, variable valve timing has most often been applied to engines with a production camshaft designed for maximum torque at a high speed. The results from one worker are typical and are shown in Fig. 2.44. In general, variable valve timing enables the torque curve to be modified such that the new curve lies about half-way between the maximum torque and the original torque curve (Fig. 2.45). The torque curve is not completely flat since, whilst the valve timing may be optimized, the resonance effects in the induction and exhaust system will each have been optimized for particular speeds. This emphasizes the point that valve timing should not be considered in isolation from the design of the induction and exhaust tracts. Indeed, the torque curve can be modified profoundly by varying

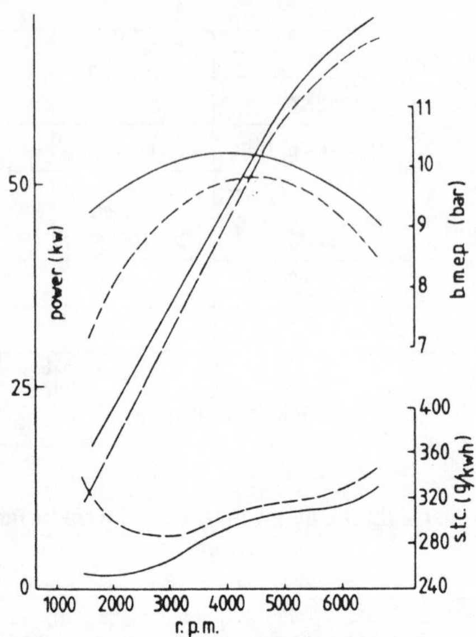

Fig. 2.44 Effect of valve timing on spark-ignition engine performance

the geometry of the induction and exhaust systems, without varying the valve timing; this is illustrated in Fig. 2.46 for an Austin Rover R Series engine. In this experiment the geometry of the induction system was controlled by varying the length of the induction pipe, upstream of the carburettor.

When considering valve timings the following points should be borne in mind: 40 degrees before bottom or top dead centre represents about 12 per cent of the engine stroke; 5 degrees after starting to open the valve may only be 1 per cent fully open, after 10 degrees only 5 per cent, and remain not fully open to half way through the valve period; valve timing is modified by valve clearances; theoretical valve motion is modified by the finite mass and elasticity of the valve train components; camshafts are subject to manufacturing tolerances – all important to remember.

Gains in efficiency through using variable valve timing can arise from three causes: elimination or reduction of throttling (or pumping) losses; control of valve overlap to reduce mixing between induction and exhaust processes; improvement in volumetric efficiency and better control over the combustion.

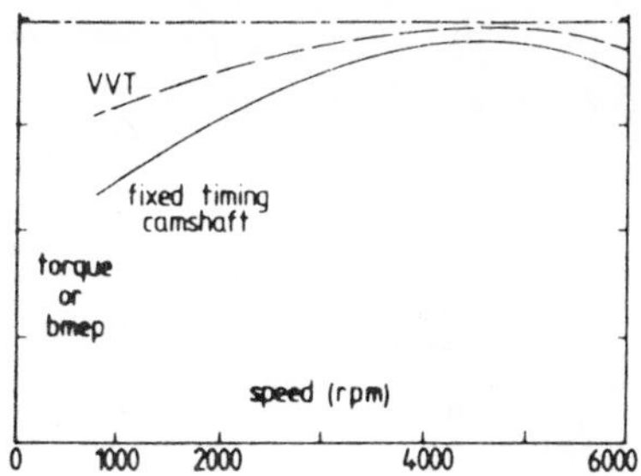

Fig. 2.45 Effect of valve timing on torque curve

A comprehensive set of experiments by another researcher investigated the effect of controlling the engine load by early closing of the intake valve. Spark-ignition engines require an approximately stoichiometric mixture of fuel and air. Thus, at part load the quantity of air has to be reduced as well as the quantity of fuel, and the reduction is usually achieved by throttling. The throttling process dissipates work done on the gas by the piston during the induction stroke. However, if the inlet valve is closed when the correct mass of air and fuel has been admitted, the piston then does work in expanding the gas to below atmospheric

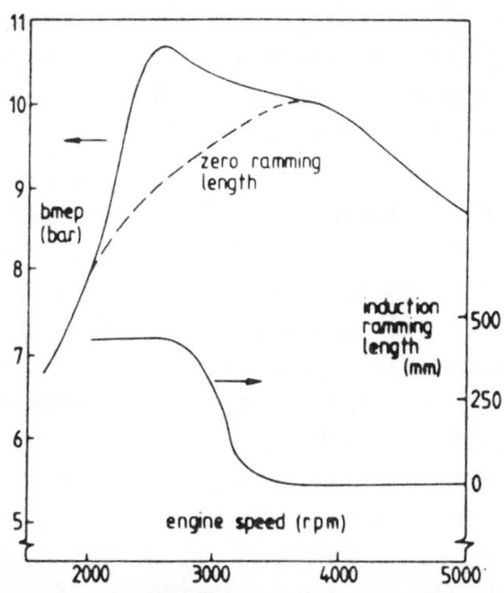

Fig. 2.46 Effect of ramming length upstream of the carburettor

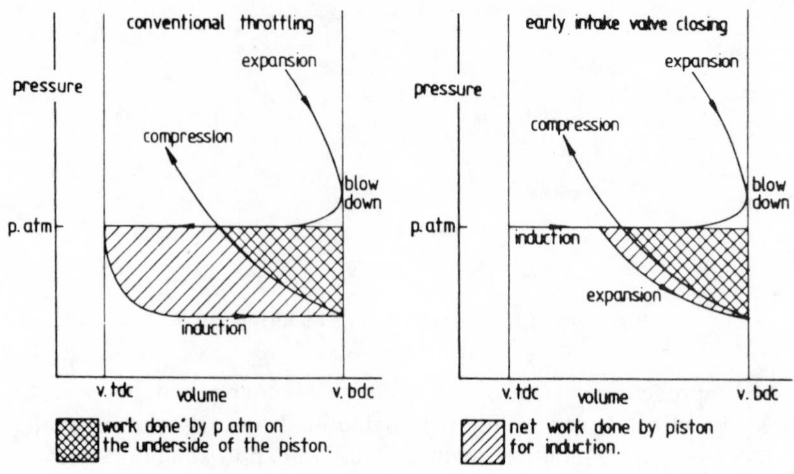

Fig. 2.47 Comparison of idealized PV diagrams at part load

pressure. If the engine is reversible then this work would be wholly recovered as work done by atmospheric pressure on the underside of the piston. This would occur at the beginning of the compression stroke as the cylinder pressure rises to atmospheric pressure. Figure 2.47 shows a comparison between idealized pressure volume diagrams, to illustrate the work lost in throttling. The researcher conducted experiments by using separate camshafts with different cam profiles to vary the engine output. Results are presented for pressure–volume diagrams at 1600 r/min that show the reduction in work; they also show a lower pressure at the start of compression. In addition, since the induction manifold remains close to atmospheric pressure, there are fewer exhaust gas residuals during induction. This leads to a lower charge temperature, and consequential lower temperatures and pressures throughout the cycle. Another researcher deduced that the reduced heat transfer more than offsets any reduction in cycle efficiency, and that this contributes about 20 per cent to the overall gain in efficiency at the lightest loads. The improvement in fuel economy as a function of load is shown in Fig. 2.48.

An alternative means of controlling the air flow without resort to throttling is to vary the geometry of the induction system. Just as the induction system can be tuned at a particular speed to maximize air flow,

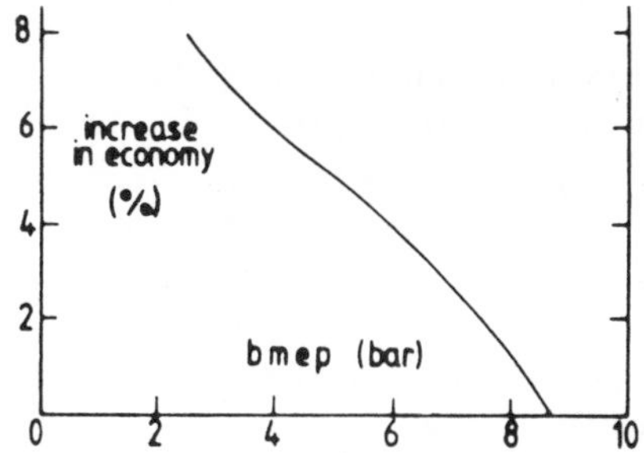

Fig. 2.48 Improvement in fuel economy using early inlet valve closing

it can also be detuned to reduce the air flow. Finally, the engine should not be considered in isolation from its application. In the case of vehicles, optimization of the powertrain has led to the use of wider span gear ratios, and these reduce the importance of part-load fuel economy. This trend will continue with the introduction of continuously variable transmissions. The second way that variable valve timing can improve the engine efficiency is by controlling the amount of valve overlap. If the valve overlap is restricted at part throttle operation the mixing of the incoming fuel and air mixture with the exhaust products will be reduced. Evidence that this assists fuel economy is only indirect; the section on engine emissions below reviews the use of increased valve overlap to control engine emissions, but this is at the expense of fuel economy. However, if early inlet valve closing is used to eliminate the need for throttling, then valve overlap would have a less serious effect on the exhaust gas dilution of the incoming charge.

The third way that variable valve timing can improve engine fuel economy is by improving the volumetric efficiency. The valve timing can be optimized to maximize the air flow rate at each speed, and this manifests itself as an increase in torque at each speed, and the reduced significance of mechanical losses leads to improvement in fuel economy at full throttle, Fig. 2.44. The larger trapped mass of air leads to improved cycle performance and an increased output. In addition, the mechanical losses do not rise in direct proportion to the indicated power; as the output increases the significance of the mechanical losses reduces, and this further improves the fuel economy.

Engine emissions

Many sources provide evidence of variable valve timing being used to control engine emissions, and comprehensive results have been presented. In general, engine emissions are reduced by increasing the valve overlap, since this allows mixing between the unburnt charge and the exhaust gas residuals. The exhaust gas is effectively inert and thus lowers the gas temperatures during combustion and expansion. Since the formation of nitrogen oxides (NO_x) is highly temperature-sensitive, a small reduction in combustion temperature leads to a substantial reduction in the formation of nitrogen oxides. Exhaust gas recirculation (EGR) is an alternative way of increasing the exhaust gas residuals in which a valve is used to control the flow of exhaust gases from the exhaust manifold to the induction manifold. The use of EGR to control exhaust gas residuals is less selective than controlling valve ovelap since the latter recirculates the gases from the end of the exhaust stroke. These gases contain the greatest concentration of unburnt hydrocarbons (HC), and thus the use of valve overlap enables a greater reduction in the emission of unburnt hydrocarbons. Unburnt hydrocarbons are formed by the combustion reaction being frozen or quenched in the thermal boundary layer (or quench zone). Since these gases are adjacent to the combustion chamber walls they are the last to leave during the exhaust stroke.

The data presented on emissions reduction is difficult to interpret since the emissions are dependent on engine operating conditions (load, speed, ignition timing, air/fuel ratio). In contrast, other researchers report a trade-off between reducing NO_x emissions, and reducing CO and HC emissions, as valve overlap is varied (Fig. 2.49); development of legislation along with the corresponding solutions is shown in Table 2.9. The US

Table 2.9 History of emission control solutions

Model year	*CO*	*HC*	*NO_x*	*Solution*
1966	87	8.8	3.6	Pre-control
1970	34	4.1	4.0	Retarded ignition, thermal reactors
1974	28	3.0	3.1	Exhaust gas recirculation (EGR)
1975	15	1.5	3.1	Oxidation catalyst
1977	15	1.5	2.0	Oxidation catalyst and improved EGR
1980	7	0.41	2.0	Improved oxidation and three-way catalysts
1981	7	0.41	1.0	Improved three-way catalyst and support material

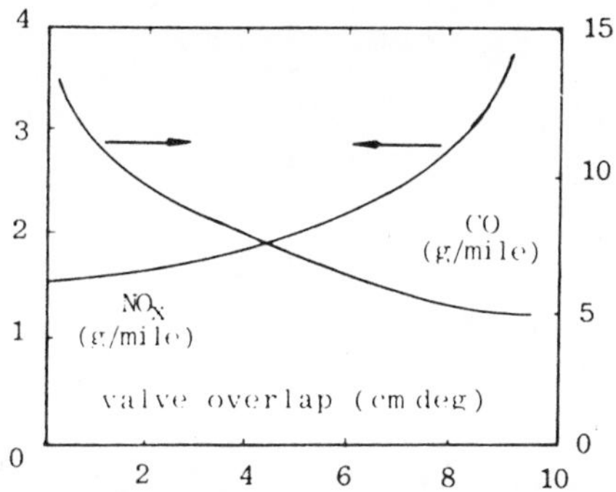

Fig. 2.49 Trade-off between emissions for varying valve overlap

Federal test is a simulation of urban driving from a cold start in heavy traffic. Vehicles are driven on a chassis dynamometer (rolling road), and the exhaust products are analysed using a constant volume sampling technique in which the exhaust is collected in plastic bags for subsequent analysis.

Catalyst systems are now used in conjunction with engine management systems (to accurately control the air/fuel ratio and ignition timing), in order to meet stringent emissions legislation without the fuel economy penalties associated with EGR and retarded ignition timing. Consequently, the need and scope for reducing engine emissions in the 1980s by variable valve timing is greatly reduced.

Variable valve timing mechanisms

Considerations in selecting a mechanism will include: cost, durability, reliability, complexity, the ownership of patents, and number and disposition of the cylinders spatial constraints, such as number of camshafts. A taxonomy for variable valve timing mechanisms has been con structed in Fig. 2.50. Two principal means of providing variable valve timing, each of which will be discussed in turn, are: a variable geometry cam follower system with fixed camshaft properties and variable camshaft properties with fixed geometry cam followers. Of the first, Figs 2.51 and 2.52 both show similar approach, by using a variable position element between cam and the follower on the valve stem. The first system uses a rocking cam driven by an eccentric and strap, with a variable position 'finger' type follower. The system built had potential for a 30 degree variation in valve opening, and a 40 degree variation in valve closing. The valve lift was also

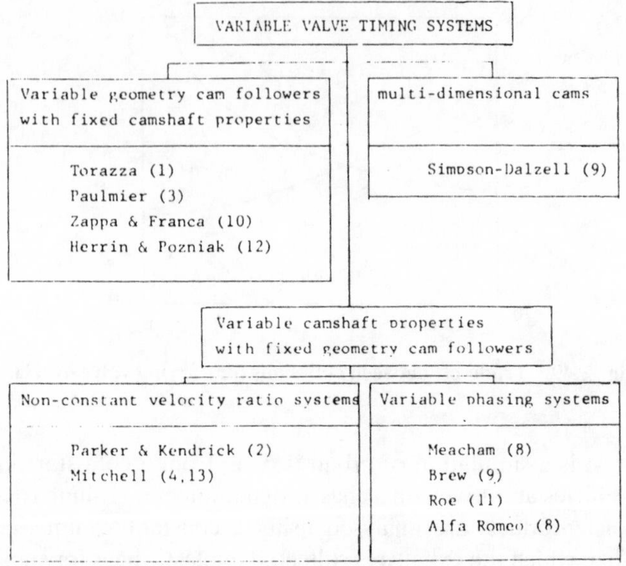

Fig. 2.50 Taxonomy of variable valve timing mechanisms

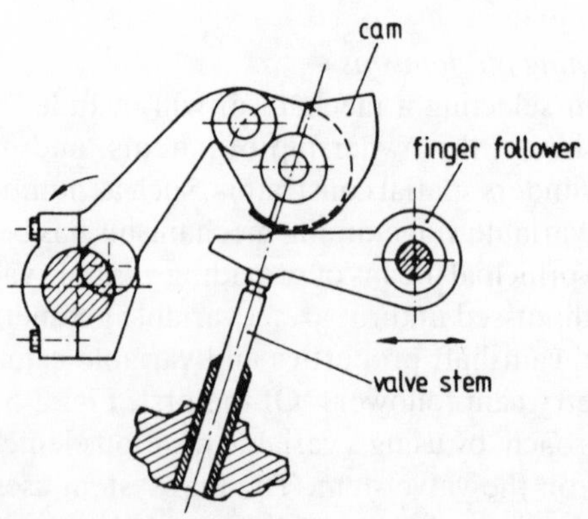

Fig. 2.51 Variable valve timing by cam-follower geometry

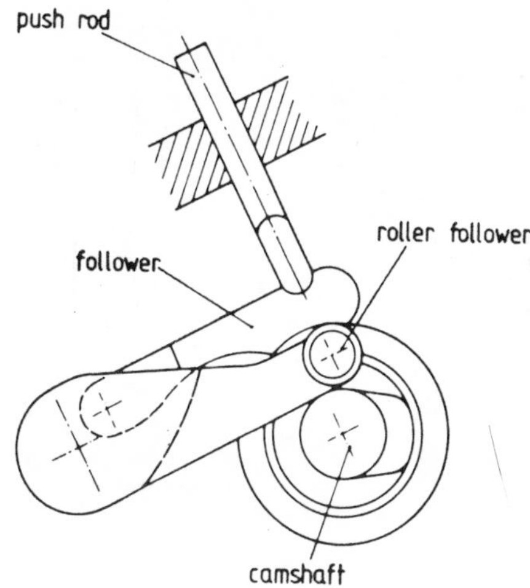

Fig. 2.52 Variable phasing by cam-follower geometry

reduced for the shorter valve period. Such parameters are controlled by the geometry of the cam and the finger follower, and how the finger follower is moved.

The second mechanism provides a variable phasing of 40 degrees; the valve period is constant since the valve opening and valve closing are affected equally. This particular application was for the inlet valve, there is no reason why the exhaust valve could not be controlled in the same way. Both these mechanisms have the advantage that they can give independent control of the inlet and exhaust valves of single overhead camshaft engines.

One researcher describes a 'ballistic follower' system that is shown in Fig. 2.53; this system is for an overhead valve engine and the push rod and rocker are not shown. The light spring causes the follower to lose contact with the cam; the higher the engine speed the earlier the loss of contact occurs, leading to a greater valve lift and a greater valve period. To cushion the impact of the valve on its seat there is a dashpot arrangement in the cam follower. A similar approach can be found in the lost-motion system shown in Fig. 2.54. The flow rate sensitive valve, that is incorporated into the hydraulic tappet, means that the cam motion is only transmitted to the valve, when the cam follower exceeds a certain speed.

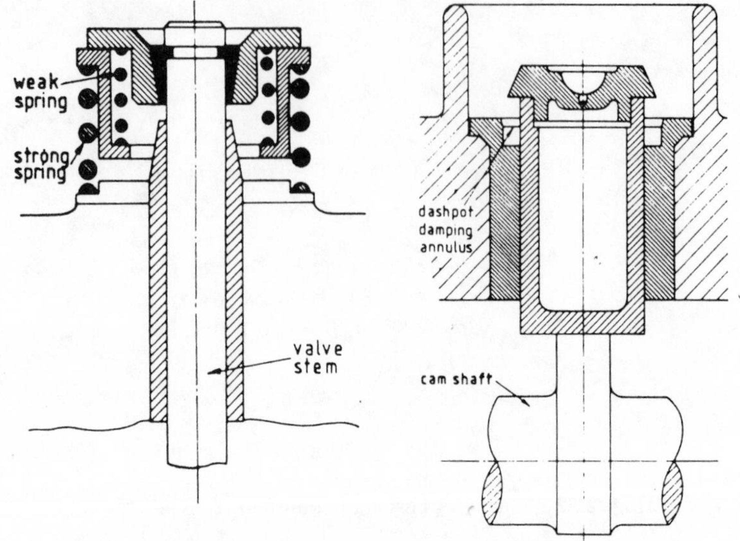

Fig. 2.53 Ballistic system for variable valve timing

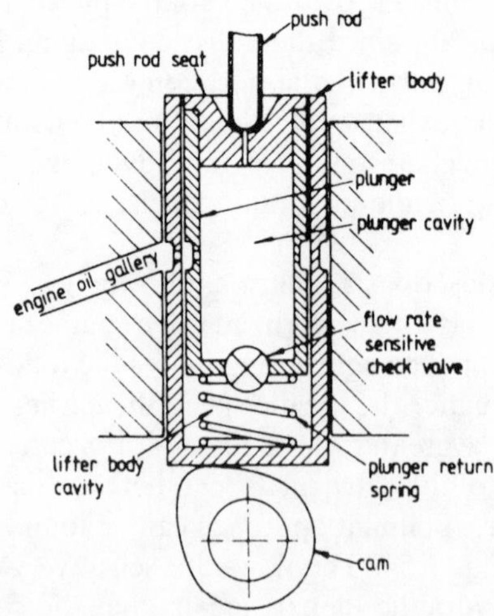

Fig. 2.54 Hydraulic tappet for controlled lost motion

Table 2.10 Performance of a lost-motion variable valve timing system

Engine speed (r/min)	*(deg) opening btdc*	*Inlet valve closing abdc*	*(mm) lift*	*(deg) opening bbdc*	*Exhaust valve closing atdc*	*(mm) lift*
450	5	53	9.76	50	9	9.67
800	9	64	9.97	55	20	9.97
2000	13	74	10.12	58	30	10.12
4000	15	78	10.15	59	34	10.15

This leads to a degree of lost motion that will be inversely proportional to the engine speed. Thus, at low speeds, the valve opening will be retarded and its closing advanced, in addition the valve lift will be reduced. The variations obtained by this system are shown in Table 2.10.

Variable property camshafts

The term variable property camshafts is intended to cover camshafts that incorporate a mechanism that effects variable valve timing, or camshafts that are driven by a mechanism that produces variable valve timing. Systems that just produce variable phasing, but do not change the valve period will be discussed at the end of this section. Valve timing variation can be achieved by driving the cams at a variable non-constant angular velocity. Two systems work on this principle including the AE Developments mechanism, Fig. 2.55. Both systems control the valve timing by the degree of eccentricity between the driving axis, and the cam rotation axis.

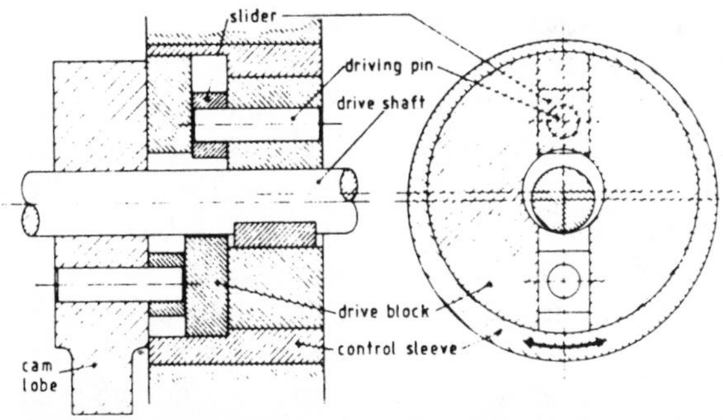

Fig. 2.55 AE Developments variable valve timing mechanism

Each system will also have two points 180 degrees (camshaft angle) apart when there is no angular difference between the input and output. This could be used to provide a fixed valve event; for example, the exhaust valve timing. Neither system can provide independent control over inlet and exhaust valve timing if there is a single camshaft. The AE Developments system contains the mechanism inside the camshaft. This limits the size and possible durability of the components and necessitates precision manufacture. However, the compactness of the system lends itself to existing multi-cylinder engines. In contrast the Mitchell system uses a mechanism that is external to the camshaft, thus enabling the use of larger elements requiring less critical manufacture. However, for engines with more than two cylinders there may be difficulty in finding space for the additional timing mechanisms.

The most common way of varying the phase between the drive and the camshaft is to use a helical spline to connect the drive wheel (gear, cog, pulley) to the camshaft; for example, Alfa Romeo (*Automotive Engineer*, 1984, **9** (2), 5). In the Alfa Romeo system the phasing is either at the advanced position or at the retarded position (so there is no continuous control), and the change over occurs at a predetermined engine load. The mechanism is applied commercially to the Alfa Romeo 2.0 litre twin overhead camshaft engine. Variable valve timing is only applied to the inlet valves, to give a variation in valve overlap from 0 to 32 degrees. This enables Alfa Romeo to maintain the power output at high speed with acceptable performance at low load (low emissions, high torque, good fuel economy).

Development potential:
spark-ignition and diesel engines compared

Professor Dr-Ing H. Förster[14] of Daimler-Benz has discussed the potential for further development of spark-ignition engines, based on a broad knowledge of the diesel engine as well. He argued that the spark-ignition engine has achieved its dominant position worldwide as a drive unit for passenger cars for some very important reasons: high brake mean effective pressure and high maximum speed; therefore, high torque and high power-to-swept volume ratio; low weight and low bulk volume for a given power demand; smooth and low-noise operation; low cost per unit of power output; and ability to operate with different fuels such as gasoline, LPG, alcohol (methanol, ethanol), NG, LNG, hydrogen. No other known power plant can compete in all these areas with the spark-ignition engine, he explains.

Emission control

When the regulations for the more and more stringent emission standards were issued in the USA to comply with the Clean Air Act, there were no known technical solutions to meet the requirements of the law, and for this reason many activities to develop alternatives existed. However, a surprisingly positive solution to the exhaust emission problems of the spark-ignition engine has been developed, which can be applied only to this engine type. This solution utilizes precise electronic mixture control by an oxygen probe combined with exhaust gas treatment in a three-way catalytic reactor, which is able to reduce nitrogen oxide to nitrogen and oxygen and to oxidize the unburned hydrocarbons and the remaining carbon monoxide.

It must be mentioned that the system is complicated since the air/fuel mixture has to be regulated within very narrow limits and the catalytic material demands lead-free gasoline (Fig. 2.56). Due to this fuel's lower octane number, the compression ratio has to be reduced, with the consequence of lower thermal efficiency. Therefore, several versions of

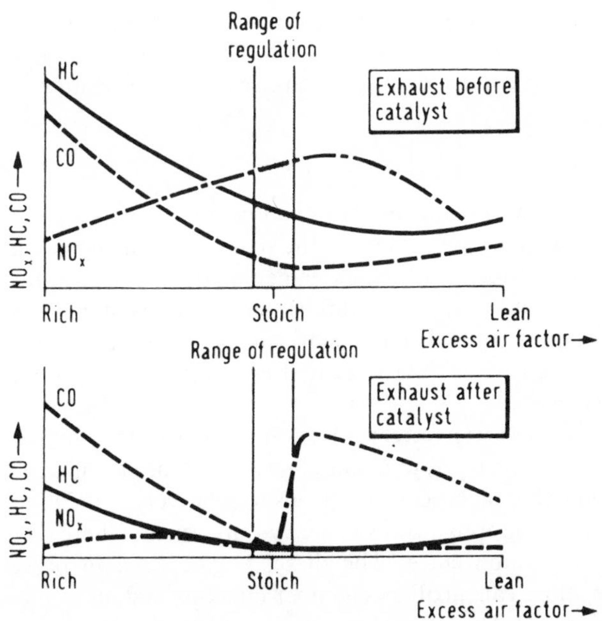

Fig. 2.56 Effect of catalyst on exhaust pollutant reduction

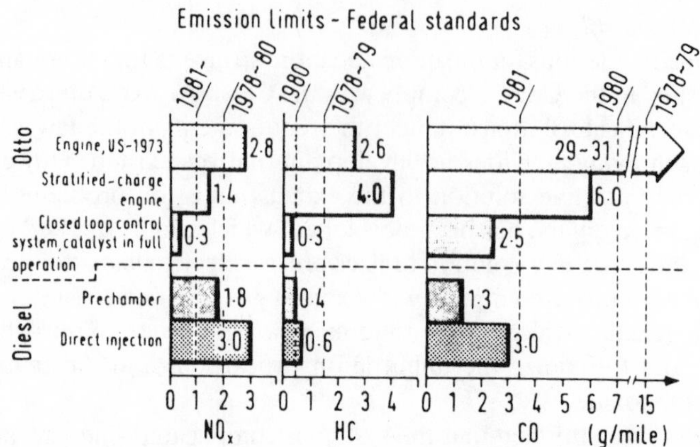

Fig. 2.57 Reduction in exhaust pollutants demanded by regulations over the years

stratified-charge engines were also developed. The results were rather promising regarding NO_x and CO emissions, but the emission of HC had a tendency to increase and, therefore, some form of after-oxidation by thermal or catalytic reactor is necessary. The extremely good results with the closed-loop control mentioned above, however, cannot be attained (Fig. 2.57).

Efficiency: spark-ignition and diesel compared

In view of the many advantages of the spark-ignition engine, increases in diesel engine sales can only be explained by the dominant role played by energy conservation since the difficulties in energy supply started. The only, but today extremely important, advantage of the diesel engine over the gasoline engine is its better fuel economy in everyday traffic and during FE tests.

The diesel process in direct-injection form gives the highest efficiency of all known thermal power plants (Fig. 2.58). Since the high efficiency of this thermodynamic process also leads to relatively high NO_x emissions, this system can be applied only in heavy-duty trucks, where fuel efficiency is of primary importance. The diesel engine used in passenger cars generally utilizes an auxiliary chamber injection system (pre-chamber or swirl chamber). This results in comparatively low emissions without necessitating additional measures, and also reduces combustion noise;

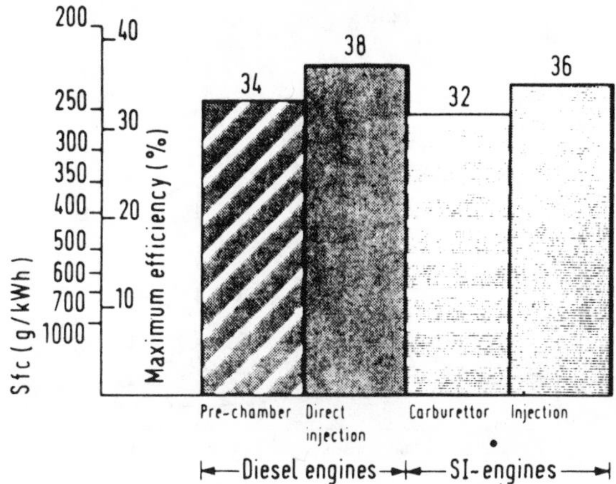

Fig. 2.58 Maximum efficiency of engines for passenger cars

however, with this system, fuel consumption is increased by about 15 per cent.

Consequently, the maximum efficiencies of the spark-ignition engine and the diesel engine with auxiliary chamber are comparable, whereas the efficiency of the spark-ignition engine at part load drops much more rapidly than in the case of the diesel. The spark-ignition engine's fuel consumption in the unsteady state form of driving is also worse than that of the diesel engine. The reason for this is inherent in the thermodynamic cycle of the spark-ignition engine, where the load is controlled by variation of the quantity (mass) of the air/fuel mixture. This means that at part load and idle the engine works at a very low manifold pressure with the consequence of low efficiency and high losses associated with the gas cycle.

This overview alone indicates the main problem to be addressed in the development programme for the gasoline engine of the future: improving its efficiency particularly at part load. In principle, there are four different methods of solving the problem: avoid operation in the areas of poor efficiency; improve the efficiency, particularly at part load; avoid operation at low manifold pressure, and decrease the power demands of the car. Whichever method is used, it must not have a negative effect on exhaust emissions. This applies universally and should be understood to apply to all proposals improving the part load efficiency of spark-ignition engines.

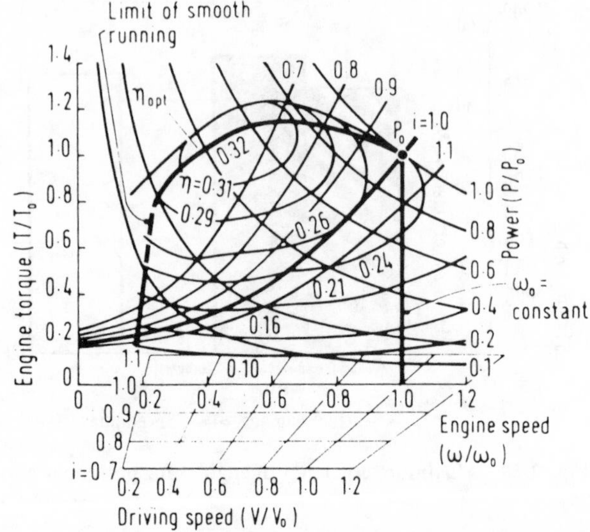

Fig. 2.59 Influence of operating conditions on performance

Influence of operating range

The poor efficiency at part load results from the fact that operation at road load, particularly at low and medium speeds, involves operating ranges with low efficiencies (Fig. 2.59). By choosing a more direct ratio between engine and wheel speeds, the operating area can be brought nearer to that with highest efficiency. This long-known possibility forms the basis for the use of overdrives and all proposals for CVR transmissions. In practice, there are two main problems. If no deterioration in hill climbing ability can be accepted, the overall range of ratios of the transmission has to be nearly doubled for engine operation near its maximum efficiency. This makes the transmission more expensive and, in the case of a CVR transmission, further reduces the transmission efficiency.

Operation of the engine near its maximum efficiency means operation near the full-load line. Consequently, climbing and acceleration capability upon depressing the accelerator pedal is poor (Fig. 2.60). In the case of a manual transmission, the driver would, therefore, often avoid top gear and, in the case of a CVR transmission, each demand for acceleration would cause a change in engine speed, which is undesirable. With modern automatic transmissions, a working compromise can be found. The

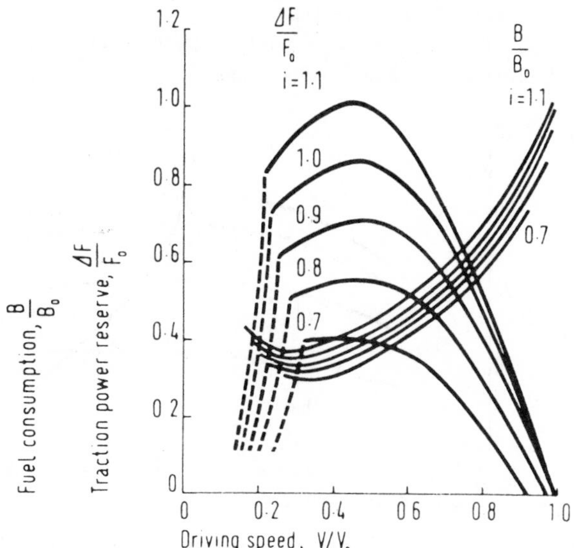

Fig. 2.60 Fuel consumption and tractive power reserve for different transmission ratios

transmission ratio has to be enlarged by one additional gear and – this is important – an automatic shift programme which combines low fuel consumption and high performance has to be selected.

Improving part-load efficiency

There are three approaches for improving the efficiency at part load: increasing the compression ratio, which can be done by several methods, turbocharging, and improvements at idling speed and during low-speed operation.

Compression ratios. Today the compression ratio is chosen so as to avoid engine knocking at full load. Since tolerances of the engine and of the octane number of the fuel have to be taken into account, the compression ratio must always be lower than optimum (Fig. 2.61). By using a knock sensor to control the ignition timing, one can increase the compression ratio to the knocking limit at full load. However, since the optimum compression ratio at part load can be substantially higher without knocking, the engine could still be better optimized for this working area by

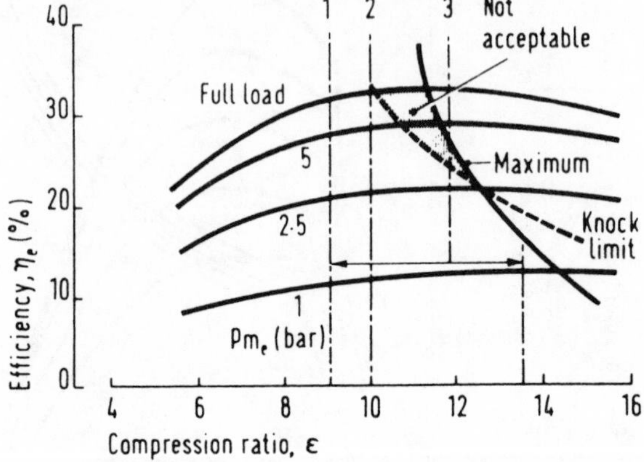

Fig. 2.61 Fuel economy with knock sensor and variable compression ratio

sacrificing efficiency at full load, provided the fuel economy in the tests and in practical driving can be improved as a result. Another possibility is to vary the compression ratio between full load and part load, for instance by using Biceri VCR pistons, where the piston length is varied as a function of the maximum working pressure (Fig. 2.62).

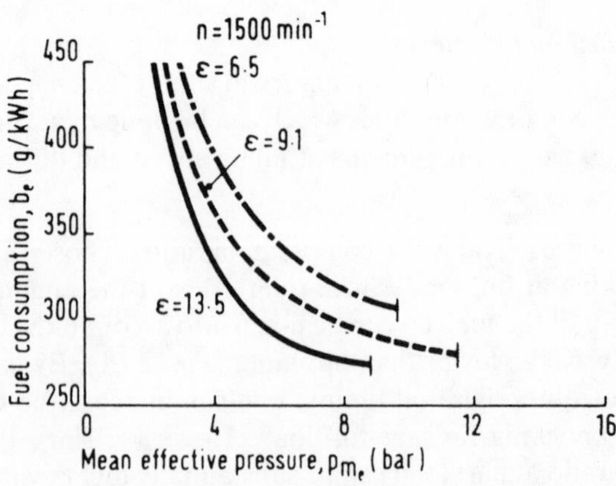

Fig. 2.62 Fuel consumption for variable compression ratio engine

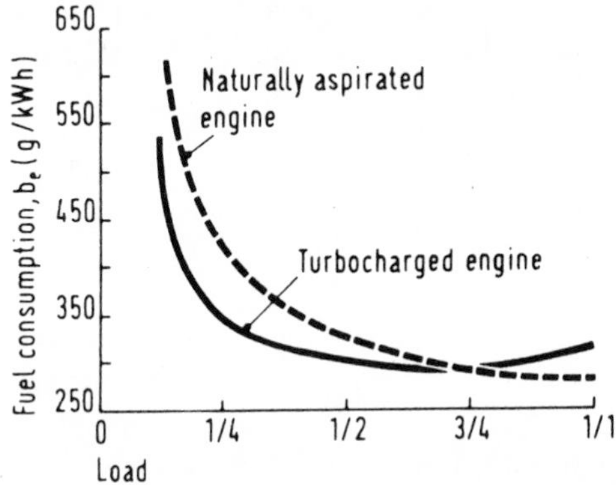

Fig. 2.63 Fuel consumption comparison: turbo and NA spark-ignition engines

Turbocharging. The turbocharging principle, as used in the diesel engine, can of course also be applied to the spark-ignition engine. However, the improvement in part-load efficiency is smaller than with the diesel engine because the compression ratio of a turbocharged spark-ignition engine has to be reduced to avoid knocking at full load. Nevertheless, a reduction in the swept volume of a turbocharged engine for the same power output leads to a decrease in the friction losses. The gain resulting from less friction at part load is greater than the drop in thermal efficiency (Fig. 2.63). Although considerable improvements have been made to turbocharged engines as regards their torque characteristics, their torque under transient conditions will never be quite as good as that of a large-displacement naturally aspirated engine (Fig. 2.64).

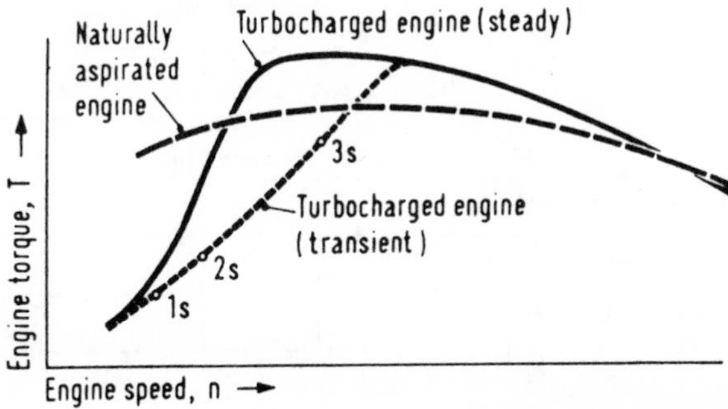

Fig. 2.64 Engine torque comparison: naturally aspirated and turbocharged

Idling and low-speed operation. Since a substantial part of actual city driving, and also of the city driving cycle test, involves idling and low-speed driving, fuel economy can be improved by optimizing the engine to these working conditions. Since a low and stable idling speed can only be obtained using a small overlap of the intake and exhaust valves, some sacrifices at higher speeds have to be accepted.

Cylinder cut-off

An engine-related means of avoiding operating points where the intake air is throttled down extensively is to vary the working displacement by cutting off some of the cylinders. If only the fuel supply is switched off the improvement of fuel economy remains small (Fig. 2.65, column 2). When in addition the gas cycle is interrupted by disengaging the valve control mechanism, higher fuel savings can be accomplished (column 3). The difference between a V8 engine with four working pistons and a four-cylinder engine is rather small (column 4). Of course, cylinder cut-off can be applied more satisfactorily to engines with six and eight cylinders than to four-cylinder versions. If, for example, four cylinders of an eight-cylinder engine are cut off, the four working cylinders operate under more than double the amount of load. This is because they not only have to produce the output to drive, but also have to provide the driving power for the auxiliary units.

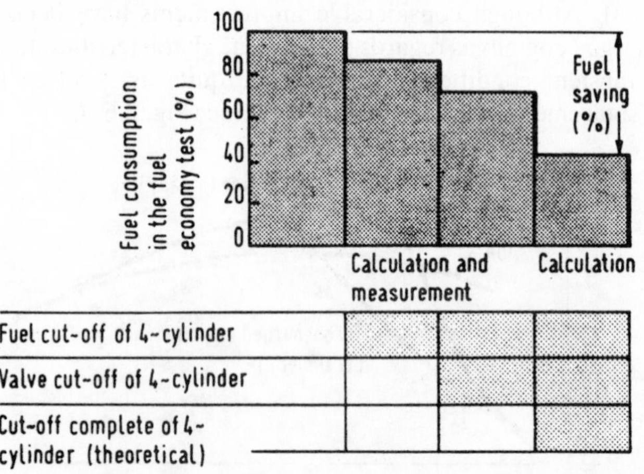

Fig. 2.65 Fuel consumption improvement by fuel, valve, and cylinder cut-off

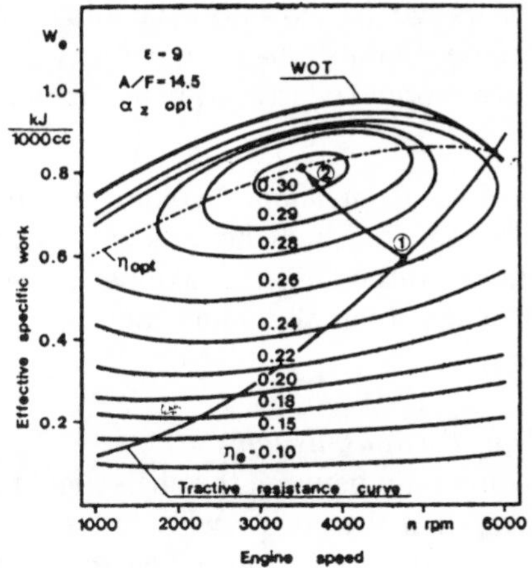

Fig. 2.66 Regions of operation on engine efficiency map

A gasoline engine to challenge the diesel

Dr Ing Karlheinz Radermacher[15] explains that specially designed gasoline engines can challenge the diesel in terms of operating efficiency. BMW, he points out, has been concentrating on two objectives: to reduce the tractive resistance of its cars and, at the same time, to optimize the efficiency of the drive systems. It would appear particularly advantageous to shift the normal running conditions (1) to an area of higher initial efficiency (2), he asserts (see Fig. 2.66).

Modifications under specifically defined conditions

Thermodynamic improvements. To ensure an efficient combustion process, we require a compression ratio, ε, which, depending on the fuel used, should be between 10:1 and 13:1. Wherever conditions allow (for instance exhaust pollution limits), partial load air/fuel ratios of approximately 17.5 are already state-of-the-art with BMW engines. Increasing use of electronic fuel injection systems and high-quality ignition concepts is already providing these benefits.

When running an engine on such a lean air/fuel mixture, it is always advantageous to create a suitable turbulence of the mixture in the combustion chamber. Hitherto, the combustion of heterogeneous mixtures in stratified-charge engines, which was still considered preferable by many people a few years ago, has now turned out to be a less acceptable solution. This is because it emits a larger quantity of unburned hydrocarbons and it is much more expensive to manufacture than a conventional engine.

Changing the engine's level of operation. A completely different approach is to improve fuel economy by changing the engine's operating points. Our primary objective in this case is to make better use of the existing efficiency potential of the internal combustion engine by introducing suitable modifications.

Modifications involving the transmission

Optimizing transmission ratio between the engine and the drive wheels, is the way of ensuring that the engine will always provide the power required with optimum efficiency. Figure 2.66 shows the optimum efficiency curve (η opt) connecting all of the optimum transmission points.

We can apply the following solutions, among others, to reach these objectives: overdrive gears with an extra 'long' transmission ratio, both for mutual gearboxes and automatic transmission; driving the car in the highest possible gear – a feature that can be built into automatic-transmission models by determining the gear shift points accordingly or by the use of continuously variable transmissions (CVT), which allow the car to be run almost the whole time along the optimum efficiency curve.

Once again, electronics will play an important role in this context: a so-called drive computer will not only supervise the functions of the engine but will also select the optimum transmission ratio for minimum fuel consumption. As a result we have a system that is far superior to a simple gear-shift. Theoretically, this concept can improve fuel economy by up to 20 per cent in a combined cycle (city traffic, motorways, and country roads).

The ETA engine

Car makers, so far, make the engines of their cars as small as possible for the performance to be achieved by a specific model. Among other things, this trend towards small engines has been supported by the calculation of road tax as a function of engine displacement in certain countries. Thus, petrol engines have a high specific output at engine speeds of approximately 6000 r/min and a mean piston speed of roughly 14–16 m/s. The overall transmission ratio was then chosen accordingly in order to convert this power into the highest possible top speed. With the car running under

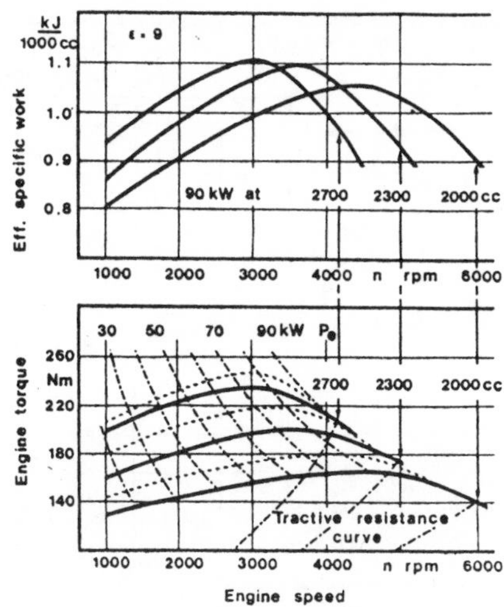

Fig. 2.67 WOT curves for the ETA concept engine

partial load, this inevitably caused a reduction in efficiency at low torque and high engine speeds ((1) in Fig. 2.66). By contrast, the concept of the ETA (the Greek letter η which is used in physics for efficiency) is to provide the power required under all running conditions at far lower engine speeds and accordingly, with a corresponding higher torque. The objective is, therefore, to constantly run the engine with a maximum efficiency, that is, in the most efficient range (Curve (2) in Fig. 2.67). The maximum engine traction available as surplus power over and above the tractive resistance must be maintained at all speeds and in all gears; otherwise improvement in fuel economy would be eliminated by drivers often shifting down to a lower gear.

As shown by Fig. 2.68 (lower half), this produces a selection of alternative torque curves starting with the bottom curve representing a current-production 2000 cm^3 fuel injection engine. The overall transmission ratio assumed in this diagram is such that the tractive resistance curve on a flat surface intersects the full load curve of the engine at maximum power. Figure 2.68 (top half) shows some examples of how an increase in engine displacement, and a modified charge creates full load engine characteristics that comply with the target curves shown below. Basically, this modification of engine characteristics is achieved by the following changes: the intake valves close earlier, the intake manifold is

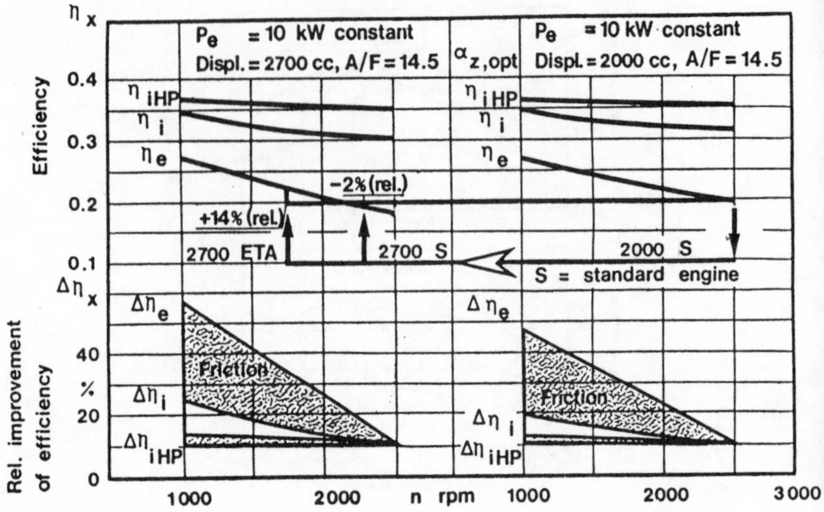

Fig. 2.68 Analyis of engine efficiency at different operating points

longer, the diameter of the intake manifold and intake valves is smaller, and the exhaust system is modified accordingly. Next must be considered which combination of engine displacement and modified torque fits together and gives optimum fuel economy. Here there are two primary factors relevant to fuel consumption: first, the positive influence of the desired change in normal running conditions from relatively high engine speeds and a low torque to relatively low engine speeds and a high torque, second, the negative influence of a larger engine displacement.

Running efficiency. The optimum combination of these factors with minimum fuel consumption can only be found experimentally. It is best to start by determining the consumption ranges of various engines on a dynamometer, as this provides the basis for subsequently simulating various running conditions with the help of a computer. To provide a better understanding of the integral result to be established later, one should at this point carefully analyse the efficiency of the engine by way of an indicator-diagram, followed by thermodynamic evaluation.

Referring to a current-production 2000 cm^3 engine, Fig. 2.69 (right-hand) shows the efficiency curves of interest in this context as a function of engine speed and at a constant effective output of 10 kW. As the engine speed decreases and the torque goes up accordingly, we have an almost linear increase of the engine's effective efficiency which, as the graph shows, results largely from a reduction of the losses caused by friction and the power for driving the auxiliaries. The remaining increase in efficiency is attributable to a reduced loss in the charge and a slight improvement in running efficiency under high pressure. Correspondingly (Fig. 2.69 left-hand), the analysis of a 2700 cm^3 engine with the characteristics already shown in Fig. 2.67: here, the modification of the wide-open throttle characteristics has practically no influence on the indicated and effective efficiency of the engine. The efficiency curves themselves are very similar to those of the 2000 cm^3 engine. With approximately the same degree of efficiency when the engine is running under high pressure, the higher throttle losses and the somewhat greater frictional loss caused in particular by the increase in piston friction at the same engine speed, result in a less efficient engine.

This loss of effective efficiency is already minimized to about 2 per cent, when the speed of 2700 cm^3 standard engine is reduced by combining it with a longer overall transmission ratio corresponding to its higher maximum power. By reducing engine speed according to the ETA

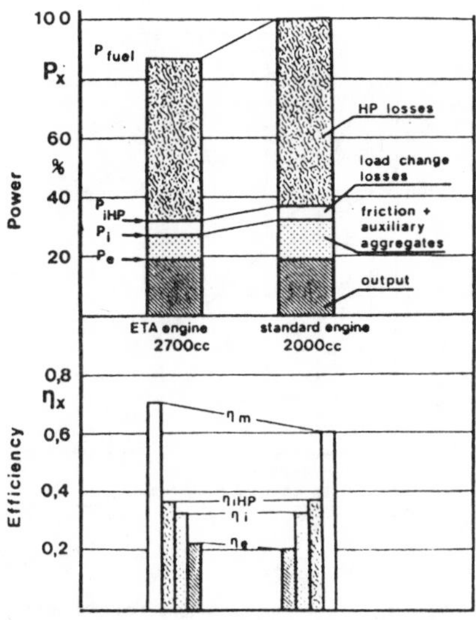

Fig. 2.69 Comparison of engine power and efficiency at different operating points

concept, that is, by 30 per cent in the present case, we obtain a considerable improvement in engine efficiency: 14 per cent at the 10 kW level. Figure 2.69 again shows that this improvement results almost exclusively from the higher mechanical efficiency of the engine. With the indicated efficiency remaining almost unchanged, the substantial reduction in engine speed provided by the ETA concept has a very positive effect on frictional losses which not only offsets, but by far exceeds, the negative effect of the increase in engine displacement.

Improvement in fuel economy. Both fuel consumption simulations and practical measurements in the same car with various engines show that the largest engine – that is, the 2700 cm^3 unit in the present example – provides the highest standard of fuel economy when running under typical conditions in town, on motorways, and on country roads. As is clearly shown by Fig. 2.70, this engine is 12 per cent more economical in terms of fuel consumption than the basic 2000 cm^3 version.

Figure 2.71 compares the different full load characteristics of an ETA engine on the one hand and a smaller, conventional engine on the other, which provides the same road performance and maximum output. A comparison of the efficiency curves shows, furthermore, that the advan-

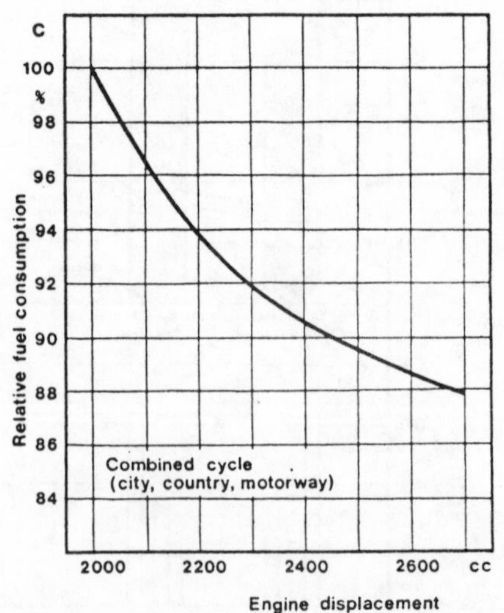

Fig. 2.70 Improvement in fuel consumption by the ETA concept

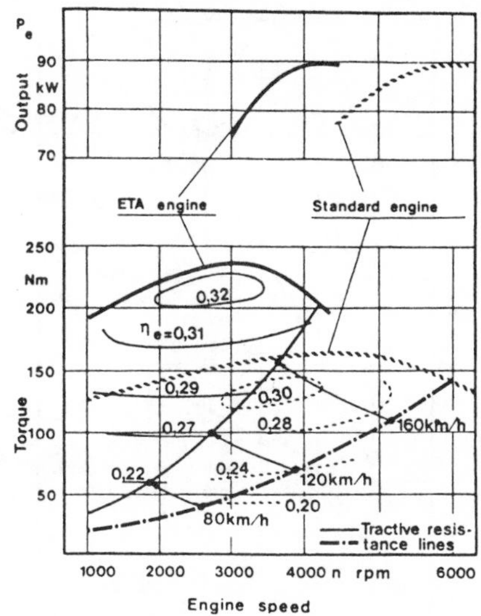

Fig. 2.71 ETA and standard engines compared

tage offered by the ETA engine is largely independent of running conditions and engine operation.

Subsequent modifications. Conventional car engines are designed and built to provide the features already mentioned: a high output at high engine speeds. So while the greater efficiency of the ETA engine results almost exclusively from the reduction of frictional losses, the considerable decrease in engine speeds allows a complete revision of the engine's mechanical features and, as a result, a further reduction of losses caused by friction. As an example, Fig. 2.72 shows how frictional losses in the cylinder head can be reduced by a lower valve spring pre-tension and the omission of three of the seven original camshaft bearings. Further possibilities are to reduce the pre-tension of the piston rings, the amount of oil required for the smaller number of camshaft bearings, and so on. By making consistent use of these possiblities, we can again improve fuel economy by approximately 3 per cent. In all, therefore, the ETA concept reduces fuel consumption by about 15 per cent.

Noise levels and exhaust emissions. As one may expect, the considerable reduction of engine speed of the ETA unit also means a considerable

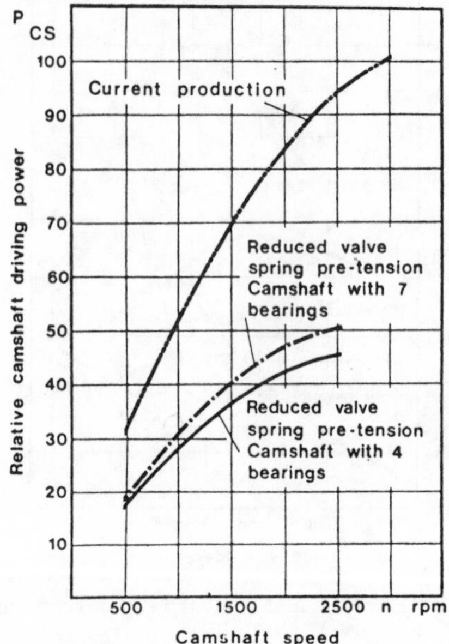

Fig. 2.72 Reducing losses by lowering valve spring rate

reduction of the noise level. ISO noise measurements with the same car but with the different engines show that the ETA engine reduces the noise pressure level by 3.5 dB(A), as one can see in Fig. 2.73.

The emissions of the three exhaust gas components subject to official limits are hardly influenced by the concept. The concentration of CO, which depends almost entirely on the air/fuel ratio, is not affected at all by the changed operating level of the engine. The concentration of HC at the end of the exhaust system, in turn, depends on both the air/fuel ratio and the exhaust gas temperature. Due to the somewhat higher efficiency of the engine when running under high pressure and the reduction of the indicated engine output, the exhaust gas emitted is at a slightly lower temperature than exhaust gas emitted by a conventional engine, which results in a slight increase in the concentration of HC. The NO_x concentration remains unchanged.

The ETA engine vs *the diesel*

Besides the often-quoted advantages of the diesel, the higher standard of fuel economy offered is, nevertheless, also attributable to another factor

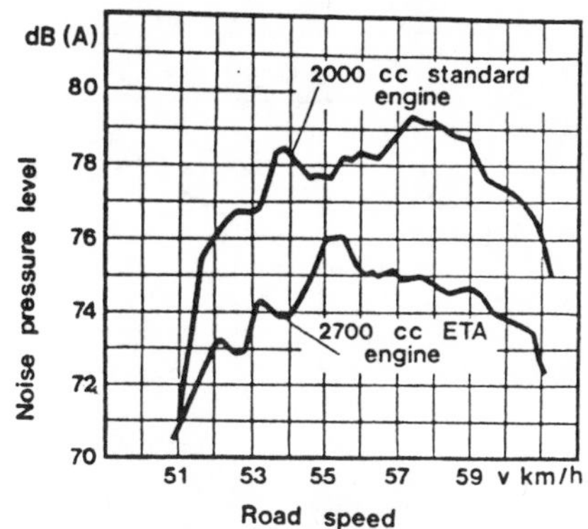

Fig. 2.73 Car exterior noise measurement according to ISO R362

which has often been overlooked: to ensure smooth and efficient combustion, the diesel engines currently used in cars run at engine speeds about 15 per cent lower than their petrol counterparts and have a correspondingly higher torque. This alone may improve fuel economy by about 5 to 7 per cent. In this respect the diesel engine is similar to the ETA concept.

The fuel economy offered by a diesel engine depends to a great extent on its running conditions. Figure 2.74 (top) provides a relative comparison of fuel consumption as a function of road speed, showing a conventional petrol engine, an equally powerful ETA engine, and an equally powerful IDI and DI diesel engine. All of these engines were run in the same car, a BMW 5 Series model. In comparison with the petrol engine, the advantage offered by the diesel engine declines from about 25 per cent at very low speeds and even becomes a disadvantage at top speed. In other words, therefore, the diesel engine consumes more fuel at high road speeds. The ETA engine, by comparison, retains its higher standard of fuel economy throughout, almost irrespective of road speed. Figure 2.74 (bottom) presents a comparison of the fuel energy required to overcome the same tractive resistance. As the graph shows, about 9 per cent of the greater fuel economy offered by the diesel engine is attributable to the higher energy content of diesel fuel.

From the perspective of fuel economy, the diesel engine would, therefore, appear to be worthwhile in relatively small cars used mainly for short-distance transport. Large saloons used mainly for longer journeys, on the other hand, can possibly be run more economically with optimized petrol engine concepts such as the ETA engine.

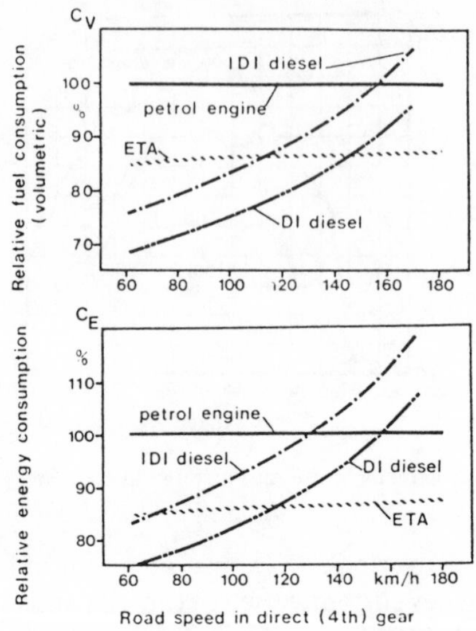

Fig. 2.74 Comparison of fuel and energy consumption: ETA and diesel

Heat flow analysis

Work reported recently[16] on predictive techniques referred to a paper by Munro *et al*. First stage is a cycle simulation which provides the gas-side thermal boundary conditions required for temperature and thermal stress analyses, and the gas pressures throughout the cycle required for piston movement and oil film thickness calculations. The technique can also be used for evaluating the effect of design changes on engine performance; for example, the effect of increasing combustion chamber wall temperatures on thermal efficiency and heat rejection.

The calculation model treats the cylinder contents as being gaseous and homogeneous (single-zone model) and the exhaust manifold and turbine by a filling and emptying approach. The turbine and compressor work can be automatically balanced by adjusting either the compressor delivery or the turbine size. Heat transfer in the cylinder is described by an empirical formula but the calculation includes step by step computer solution of the First Law of Thermodynamics, the Perfect Gas Law, and the flow equation for adiabatic expansion through nozzles. The input data requirements are derived from either engine geometry and measured performance data on existing engines or from design and performance targets on new engine projects.

The output from the computer program includes at each crank angle such items as gas pressure and temperature in cylinder; mass and volume in cylinder; cumulative heat loss from and mass flow into cylinder and also cumulative mass flow out of cylinder. In addition, single values are given for such quantities as indicated mean effective pressure and specific fuel consumption.

Single zone models involve gross simplifications of the processes and conditions within the cylinder. Successful use of such models relies heavily on empirical heat release data and local heat transfer factors. A more fundamental approach to the processes is favoured in which fuel penetration, air entrainment, mixing, heat release, and pollutant information theories are interlinked, and local gas temperatures are used to determine heat transfer. Considerable progress in the development of 'multi-zone' combustion models has been made, the authors explain.

In FE analysis, the loading conditions describe the mechanical and thermal loading cases. Mechanical loading is caused by gas pressure acting on the crown and the top land, and is specified by the maximum cylinder pressure. The thermal loading is much more difficult to obtain and requires the specification of the boundary conditions appropriate to the combustion gases on the crown (gas-side), the cooling oil on the underside and surfaces of the gallery (oil-side) and the sliding lubricated surfaces in contact with the cylinder bore and the gudgeon pin. The gas-side boundary conditions can be calculated by the cycle simulation program, while the remainder are based on data derived from heat transfer measurements and previous experience. Alternatively, when measured piston temperatures are available for about 20 locations, the cycle simulation stage can be by-passed, and the thermal loading can be determined by calculating the temperature distribution throughout the piston using 'trial and error' boundary conditions until the measured temperatures are closely matched. Constraints are applied to a FE model when that model represents part of the structure only. In 3-D analysis of the piston and gudgeon pin assembly, there is generally sufficient symmetry to use a model of a 90 degree slice. The nodes on the planes of symmetry are constrained to have zero movement normal to the plane. For the thermal stressing case, it is assumed that the piston is not constrained by the gudgeon pin, and the pin is effectively cancelled by specifying a zero value of Young's modulus. In 2-D axisymmetric analysis, a model of the piston radial section above the pin is used. The model is constrained to have zero radial movement on the axis of symmetry. Rotational freedom is permitted at the pin bearing and only one node is

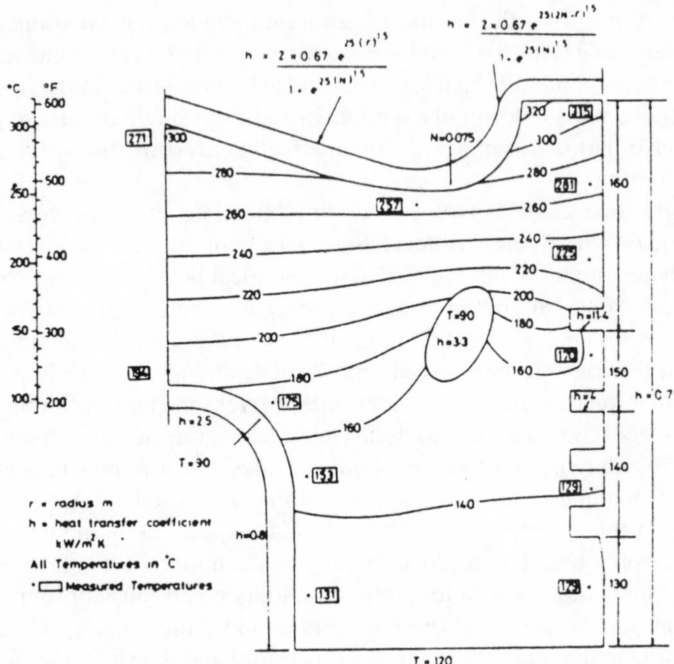

Fig. 2.75 Comparison of computed and measured piston temperatures

constrained. Material properties required are the thermal conductivity, density, specific heat, Young's modulus, and the coefficient of thermal expansion. Calculated piston temperatures are shown in Fig. 2.75, together with the boundary conditions used. Measured piston temperatures are also shown for comparison. The calculated heat balance of this piston is given in Table 2.11. The heat flow to the cooling gallery is 48 per cent of the heat input to the crown and is used when assessing the cooling oil flow requirements through the gallery.

Table 2.11 Piston heat balance for 240 mm soluble core unit at 20.7 bar

Results	*Heat* (kW)	*Heat* (%)
Heat input to crown	18.336	100
Heat to gallery	8.807	48
Heat to undercrown	5.230	28.6
Heat to liner	4.299	23 4

A new approach to combustion analysis

Anthony Dye[17] of Epicam Ltd points out that meeting the challenge of forthcoming emissions legislation in Europe must focus the attention of combustion systems engineers on the need to improve the consistency of the combustion process on a cycle to cycle basis. Dilution of the air/fuel charge mainly to reduce NO_x emissions either by using exhaust gas (EGR route) or with excess air (lean-burn route) exacerbates the problem of intercyclic variation, especially at road-load engine operating conditions. A new hardware/software package (EPICAS) has been developed as an analytical tool to meet the needs of combustion systems engineers in evaluating the effects on combustion of their own design improvements.

This study has revealed the existence of combustion system performance differentials of a high order including half to almost four-fifths in the level of intercyclic variation as measured by 10 key combustion parameters. These include an early combustion phase burn-time variation differential of 48 per cent; intercyclic variation of IMEP differential of 79 per cent; mean burn-time for the early combustion phase differential of 38 per cent, and for the main combustion phase a differential of 29 per cent. An 11 per cent differential in optimum SFC is due to differentials in combustion burn-time, consistency, and lean-burn capability. Informal reference has often been made to i.c. engine combustion as a 'black art'. It is apparent that the hardware/software package referred to in this study can be used to shed new light to dispel the blackness and raise the art to an engineering science with considerable development potential.

Main features of the test method

The content of the system is based on the need to incorporate three main elements of capability.

Continuous high speed data acquisition. The combustion process can be effectively monitored by interpretation of the continuous pressure signal from a piezo-electric transducer mounted in the combustion chamber. In view of the high level of intercyclic variation in combustion and the fluctuations in the pressure diagram which arise in consequence, it is essential that a large sample of consecutive cycle diagrams is accumulated before attempting to characterize the performance of a combustion system. Accordingly, the database consists of cylinder pressure records digitized at intervals of one degree of crankshaft rotation over a minimum of 600 revolutions, thus containing the pressure histories of at least 300 consecutive firing cycles. The data frequency can be increased to 0.1 degree CA intervals, but for steady-speed, road-load evaluation, this is

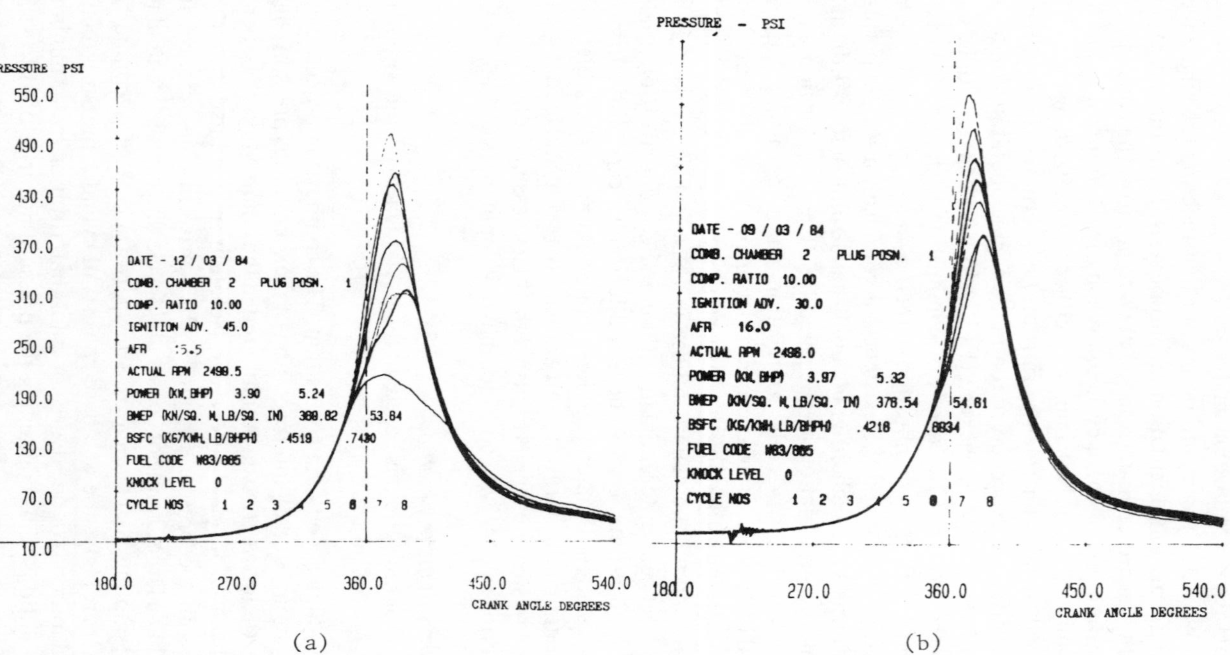

Fig. 2.76 Pressure comparisons for eight cycles (a) standard engine and (b) with turbulence induced by TCI system

not generally necessary. The system has the capacity to transfer pressure records to magnetic disc at a maxiumum frequency of 100 KHz continuously per module of data transfer function. Additional capacity can be gained by adding expansion modules as required according to the number of cylinders to be monitored simultaneously.

Data reduction. The system software extracts key elements from two sets of data for monitoring selected parameters on each combustion cycle. Firstly the original combustion pressure diagrams, as illustrated by the diagrams shown in Fig. 2.76. This yields values for parameters such as the maximum rate of pressure rise, the peak cycle pressure, together with the respective crank angles at which those events occurred and the IMEP and BMEP values for each cycle. Secondly, the normalized pressure diagrams which form the basis for generating burn-time diagrams as illustrated for the eight cycles shown in Fig. 2.77. These are constructed from incremental elements of pressure rise calculated at constant volume and due to combustion only, independent of piston motion and associated with time increments throughout the combustion process. The method used follows that established by Rassweiler and Withrow equating proportional elements of combustion pressure rise with proportional elements of charge mass burned on a time scale monitored via crank angle. Data extraction from the normalized pressure diagrams provides additional monitoring of combustion and its phasing relative to crank angle position: for example, maximum rate of normalized pressure rise, peak normalized cycle pressure, time taken to burn the first 10 per cent of charge mass, time for 0–50 of charge mass burned, time for 0–90 of charge mass burned, time for 10–90 percentages of charge mass burned.

Statistical analysis and display. The performance of specified combustion parameters over the whole database sample of 300 or more cycles is subject to statistical analysis for measurement of intercyclic variation. Maximum, minimum, mean, and standard deviation values are recorded for each parameter. Elementary statistical tests can then be applied to determine the significance of mean differences between database samples in which the within-sample variation may substantially exceed that occurring between the samples being compared. Selected elements of the database or of the calculated values can be displayed via printer, multicolour plotter, or colour graphics unit. In this way, the package can be used to identify and display large or small differences in combustion system performance between samples recorded under controlled conditions.

Computer print-outs

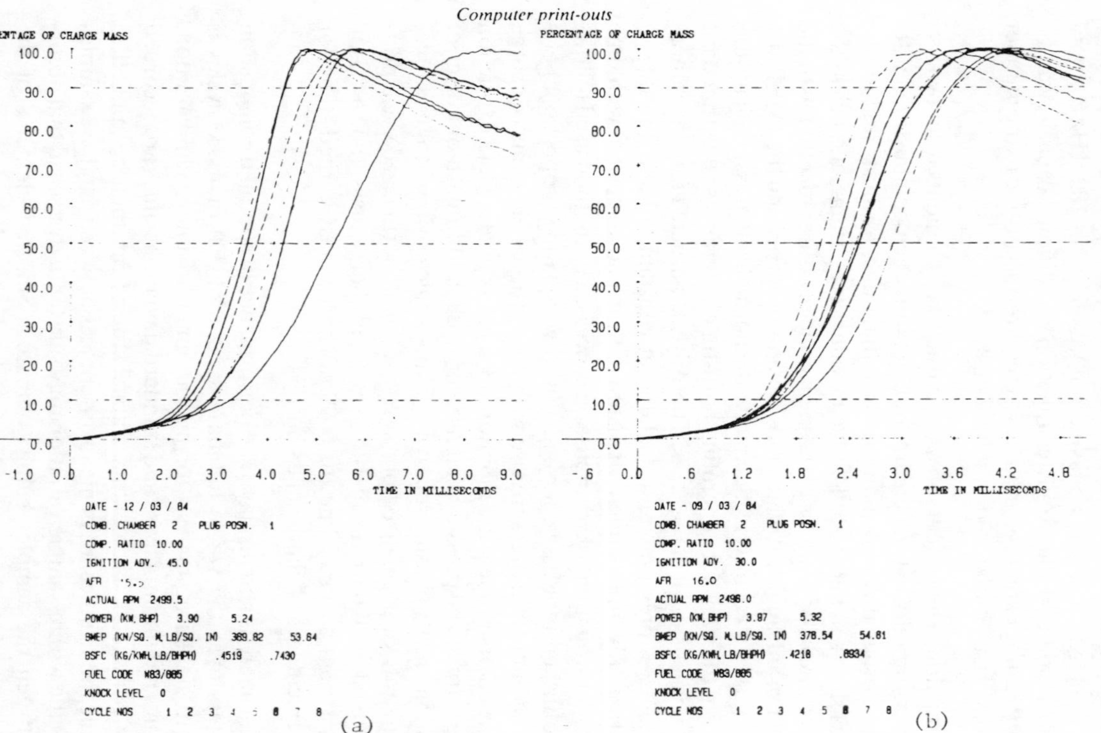

Fig. 2.77 Mass charge burn rate (a) standard engine and (b) with induced turbulence

Evaluation of grid-generated chamber turbulence

A test programme using a Ricardo E6 single cylinder research engine was designed to illustrate the analytical capability of the package by evaluating the performance of the TCI experimental combustion system when compared with the standard engine combustion system at a single, mid-range speed/load condition. In this example, the combustion performance differences between the standard and experimental systems are highly significant and apparent even from visual comparison of small sub-samples of consecutive cycles. The example therefore serves well the purpose of illustrating the capability of the analysis to explore in detail, the areas of potential improvement which are available to the combustion systems engineer/designer.

Test conditions. The engine operating condition was selected to represent a road-load condition commonly occurring with automotive engines. Commercial grade 97 octane gasoline fuel was used and the compression ratio was fixed at 10.0 : 1. Fuel metering was by standard carburettor with variable main jet which provided means for continuously varying the air/fuel (A/F) ratio. SFC was measured at a series of points spanning the A/F range from approximately 12 : 1 to the lean limit for both combustion systems while maintaining the constant power output required by the selected operating condition. MBT spark timing was obtained at each test point before measuring SFC and loading cylinder pressure data. In this example, the turbulence-generating grid remained outside the cylinder during combustion. This enabled comparisons of combustion system performance to be made independently of fuel type, fuel metering system, spark location or energy, combustion chamber shape, volume or compression ratio, all of which remained constant throughout.

Best fuel economy with the standard combustion system was achieved at 15.5 : 1 A/F – point (A) in Fig. 2.78. Cylinder pressure diagrams and burn-time diagrams for the first eight cycles of the sample taken at point (A) are shown in Figs 2.76 and 2.77, respectively. An interesting comparison is thus possible between the point and point (B) where the experimental system is operating with a similar A/F ratio (16.0 : 1). Comparable cylinder pressure diagrams and burn-time diagrams for the first eight cycles of the sample recorded at point (B) are shown in Fig. 2.76. Engine variable monitoring showed that MBT spark advance was 15 degrees CA later with the experimental system, exhaust temperature was 50°C lower and SFC was 8.5 per cent lower in this comparison with the standard

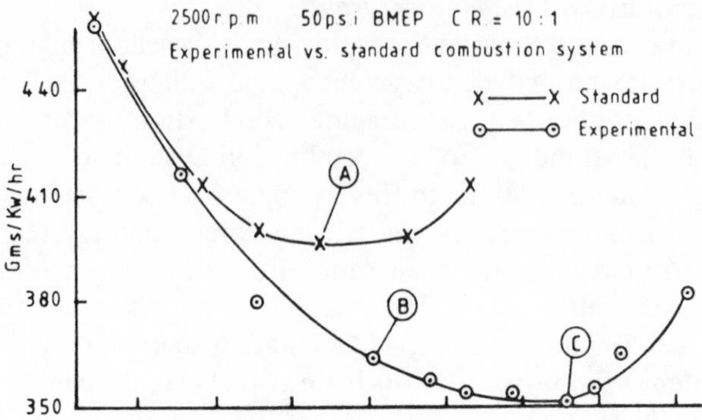

Fig. 2.78 Specific fuel consumption comparison

Table 2.12 Combustion performance percentage differences between experimental and standard combustion systems at comparable A/F ratio

Mean maximum cylinder pressure	+20
Coefft of variation of max cyl pressure	−50
Mean crank angle at which max press occurs	1.1 (deg CA)
Std deviation of CA of max pressure	−53
Mean max rate of pressure rise	+46
Coefft of variation of max press rise rate	−36
Mean IMEP value	−1.2
Coefft of variation of IMEP	−79
Mean normalized max pressure	−3
Coefft of variation of max normalized press	−67
Mean crank angle of max normalized press	−10 (deg CA)
Std Deviation of CA of max normalized press	−66
Mean burn-time for 0–10% of charge burned	−38
Coefft of variation of 0–10% burn-time	−48
Mean burn-time for 0–50% of charge burned	−35
Coefft of variation of 0–50% burn-time	−61
Mean burn-time for 10–90% of charge burned	−29
Coefft of variation of 10–90% burn-time	−56
Mean burn-time for 0–90% of charge burned	−34
Coefft of variation of 0–90% burn-time	−61

system. Results of the combustion analysis based on 300 cycle samples recorded at these points are summarized in percentage terms of 20 combustion parameters in Table 2.12. The 20 per cent increase in mean maximum cylinder pressure occurring with the experimental system compared with the standard is consistent with the 29 per cent reduction in main combustion phase burn-time, where it is evident that a high proportion of the charge mass is burning within a given close interval of TDC. This also affects the maximum rate of rise of cylinder pressure in a similar way, as indicated in Table 2.12.

CHAPTER THREE

Mechanical performance and noise reduction

Following detail design of crankshaft, cylinder block, piston assembly, connecting rods, and bearings, described in the opening section, consideration is given to computer prediction of engine imbalance, followed by designing engine mountings. A section on design for noise reduction commences with exhaust system analysis and introduces gas dynamics theory. This is followed by sections on predicting both air flow in inlet/exhaust ducts and engine structural vibration behaviour. Engine structural noise is next considered followed by a review of the current evolutionary approach to exhaust silencer design. Finally a section is devoted to using gas dynamics theory for exhaust system analysis, including those for multi-cylinder engines.

AN APPROACH TO DETAIL DESIGN

John Hartley[18] provided a valuable series of articles on the likely steps in design forming a useful procedure for a 'clean sheet' design. He argues that engine displacement is normally the first parameter to be chosen by the automotive designer as it has the main influence on engine weight and, hence, vehicle size and weight. While specific output is proportional to speed × thermal efficiency × volumetric efficiency × mechanical efficiency, it is more convenient to use empirical data in deciding on displacement. Typical brake mean effective pressures, the basic measurement of engine performance, are as in Table 3.1. Precise values depend on the level of engine development; however, the levels given reflect compression ratios of 8.5–9.2 : 1 and European emission levels in the mid 1970s.

The need for high engine speed in the power output equation calls for multi-cylinder layout with the high friction associated with too many cylinders becoming the limiting factor, particularly in respect of fuel economy at part load. Questions of primary and secondary balance can affect the cylinder configuration; see Table 3.2. Overall size of engine in relation to configuration may be estimated from those of particular types in Table 3.3. High speeds normally involve small stroke/bore ratios, down

Table 3.1

	With one carburettor to three or four cylinders	*With one carburettor to two cylinders*	*Twin ohc and four valve units*
Maximum bmep	130–140 lbf/in^2 (9.2–9.9 bar)	140–147 lbf/in^2 (9.9–10.1 bar)	150–160 lbf/in^2 (10.35–12.0 bar)
bmep at maximum power	105–110 lbf/in^2 (7.45–7.8 bar)	120–130 lbf/in^2 8.5–9.2 bar)	140–150 lbf/in^2 (9.65–10.35 bar)

to 0.6 : 1 in road vehicles. Mean piston speed (= stroke × engine speed) often used in maximum safe speed calculation is corrected by (stroke/bore)$^{1/2}$ to account for inertia forces. Inertia loadings on reciprocating parts must be calculated for very high speed engines. Current stroke/bore ratios vary from 0.85 to 1.0, being moderated to meet emission standards and provide a wider torque spread. Families of engines using a common bore size can experience wide variation in S/B ratio for small and large capacity versions.

Crankshaft design

For in-line four cylinder engines, all bearings are likely to be in one plane for reasons of balance; with vee engines, notably the 90 degree V8, a cruciform crank is the likely choice to obtain freedom from vibration. In general there is little point in using a material of higher strength than 830 MN/m^2 as no worthwhile improvement in fatigue strength is gained above

Table 3.2

Engine type	
90 degree V twin	Complete primary balance
In-line twin	Primary forces balanced with 180 degree crank, but primary couples and secondary forces unbalanced
Horizontally-opposed twin	Primary and secondary forces balanced, but horizontal couple proportional to the magnitude of the offset between the two cylinders
Horizontally-opposed four	With flat crank and journals spaced as is conventional on in-line four, complete primary and secondary balance obtained

Table 3.3 Engine configurations – overall dimensions

Engine / *Overall dimensions*	*VW 1.3 litre flat-four*	*Ford 2 litre in-line four*	*Ford 2 litre V4*	*Porsche 2 litre flat-six*	*Ford 3 litre V6*	*Austin 3 litre 6 in-line*	*Rover 3.5 litre V8*	*Aston Martin 5.3 litre V8*	*Jaguar 5.3 litre V12*
Length, in (mm)	18 (457)	25 (635)	19 (482)	28 (710)	25 (635)	33 (837)	29 (736)	34 (862)	38 (965)
Width, in (mm)	34 (862)	25 (635)	25 (635)	35 (890)	27 (685)	23 (584)	26 (660)	42 (1070)	39 (990)
Height, in (mm)	16 bare (406)	28 (710)	25 (635)	27 (685)	27 (685)	32 (812)	29 (736)	29 (736)	27 (685)

this point. A 4 : 1 safety ratio is recommended to allow plenty of material for regrinding.

In diesel engines where reliability criteria outweigh those of first cost, forgings in En 16 and En 19 are common, with generous crankpin/main journal overlap. On a 16.5 kg shaft some 3.5 kg of metal would have to be machined off in the forged case, against round 2 kg when cast. Draft angle on castings need only be ½ to 1 degree, whereas on a forging 3 or 5 degrees is required. With SG iron cast units, however, lower levels of bending and torsional stress reversal must be designed for. Fatigue failure must be designed against and careful attention paid to crank geometry, with the avoidance of stress raisers and notches a priority.

Stress reversals in the normal running range arise as a result of combustion pressure and are experienced by the shaft as a change in bending moment. Below 6000–7000 r/min inertia forces are less than combustion forces. At a given speed inertia forces will provide a given level of bending moment relief. Torsional stress reversals are generally lower than bending stress reversals on engines of the short stroke 4- and 6-cylinder type; on 8-cylinder engines torsion predominates. The latter effects can be alleviated by main journal/crankpin overlap. Fatigue stress calculations should be based on maximum combustion pressure figures as follows:

Private car (petrol)	*Car with knock*	*NA diesel*	*Turbo diesel*
40–60 bar	52–80 bar	80–100 bar	110–140 bar

With SG iron the total equivalent stress range, due to combined bending and torsional loadings, must be below the fatigue limit of the material (notched), varying between 130 and 170 MN/m^2. Critical areas are the web cheek and the crankpin-to-web fillet. Experience has shown that a 42 per cent increase in web cheek width increased fatigue strength by 20 per cent and a mere 13 per cent cheek thickness increase gave a 45 per cent increase in indurance limit. Fillet radius as an absolute minimum of 2.5 mm has been suggested for SG iron. For forged steel a value of 5 per cent of crankpin diameter is recommended, with which the safe nominal fatigue stress for forged steel is about 12 per cent UTS. A computer program available from Ricardo and Company calculates nominal bending stress at the crankpin fillet and across the crank web, from the

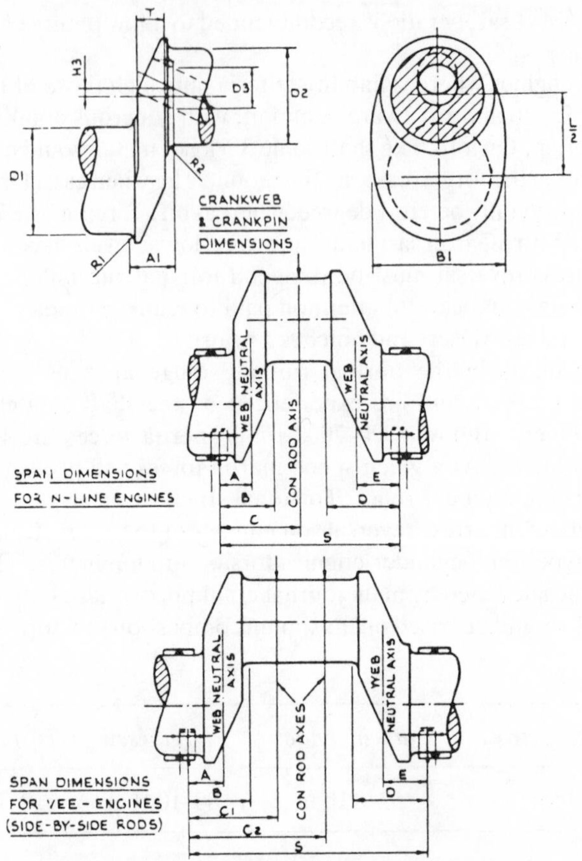

Fig. 3.1 Computer prediction of crankshaft bending stress can be carried out given the basic dimensions shown here

dimensions shown in Fig. 3.1 and offers valuable savings in design time. Values are figures of merit rather than stresses, but plotting these on a Goodman diagram enables a close estimation based on past experience to be made.

Figure 3.2 shows typical engine crankshaft proportions (bmep's bracketed): (a) in-line naturally aspirated (100 bar), (b) in-line turbocharged (140 bar), (c) V8 naturally aspirated (100 bar), and (d) V8 turbocharged

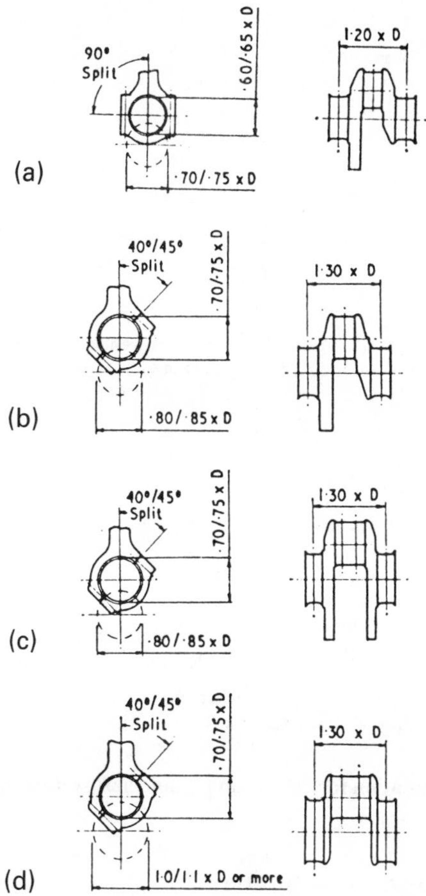

Fig. 3.2 Typical proportions for engine crankshafts under different operating conditions

(140 bar). *D* denotes cylinder bore. The following method can be used for determining section modulus in bending for the web of the crank with pin/journal overlap. It is based on a formula due to W. Ker Wilson published by Ricardo and Company. For the notation shown in Fig. 3.3, inclination of section

$$\theta = \tan^{-1}(S + rp + rj)/A$$

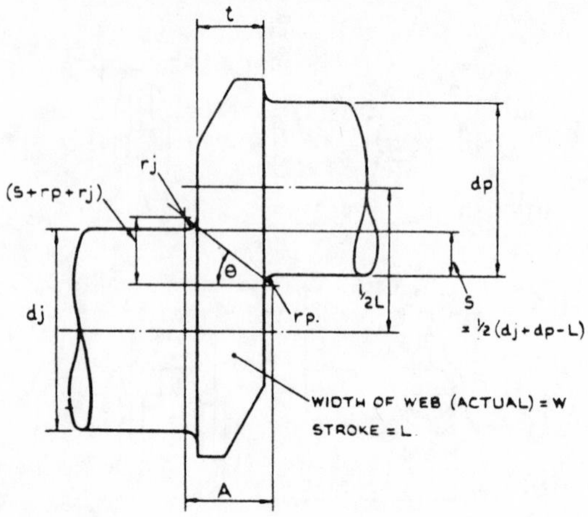

INCLINATION OF SECTION $\theta = \mathrm{TAN}^{-1} \dfrac{S+rp+rj}{A}$

EFFECTIVE THICKNESS OF OVERLAP PORTION, $t_e = A-(rp+rj)\cos\theta$

EFFECTIVE WIDTH OF OVERLAP PORTION

$$We = \tfrac{1}{2}\left\{dp+dj+2(rp+rj)(1-\sin\theta)\right\}\cdot\sqrt{1-\left\{\frac{L}{dp+dj+2(rp+rj)(1-\sin\theta)}\right\}^2}$$

EFFECTIVE MODULUS OF WEB IN BENDING

$$Ze = 1/6\left\{We.t_e^2\left(\frac{2\sec\theta+1}{3}\right)^2+(W-We).t^2\right\}$$

Fig. 3.3 Notation for calculating crankshaft section modulus

Effective thickness of overlap portion

$$t_e = A - (rp + rj)\cos\theta$$

Effective width of overlap portion

$$W_e = \tfrac{1}{2}\{dp + dj + 2(rp + rj)(1 - \sin\theta)\}$$
$$\times \sqrt{[1 - L^2/\{dp + dj + 2(rp + rj)(1 - \sin\theta\}^2]}$$

Effective modulus of web in bending

$$Z_e = (1/6)\left\{W_e \cdot t_e^2\left(\frac{2\sec\theta + 1}{3}\right)^2 + (W - W_e)\cdot t^2\right\}$$

Cast crankshafts are tending to be made by the green sand moulding method often employing the 'INMOULD' process for converting the grey iron into SG iron as the iron is poured into the mould. The molten iron is typically enriched by the addition of magnesium to pockets in the mould. In accordance with good casting design practice, section thicknesses should be kept as near uniform as possible with smooth changes of section to avoid stress concentration. Very thin sections should also be avoided because they draw metal from the thick sections to give a spongy structure and shrink-holes. Various after-treatments can enhance fatigue life. Fillet rolling of big ends and mains is carried out in high performance engines. The introduction of compressive stresses into the area of the fillet provides up to 100 per cent improvement in endurance limit. On a higher production scale, nitriding is commonly employed. Such a process is Tuftriding by ICI employing a number of low temperature salt baths. Up to tenfold increases in service and wear life have been reported for the method.

Cylinder block design

Over 30 per cent of the total cost of an engine can be accounted for by the cylinder block so that manufacturing techniques must be carefully examined. Costly cores must be kept to a minimum and material must be allowed for future increases in cylinder capacity, when inevitable increases of power output are demanded, by arranging generous bore spacing. Minimum recommended spacing between bores is 13.5 mm made up of 4.5 mm metal, 4.5 mm water and 4.5 mm further metal. Extra bore allowance of 5 mm only adds 15 mm to engine overall length and allows for future development. With wall thickness of 4.5 mm it will be possible to reclaim blocks with porous cylinders by sleeving.

Provision of water all round the cylinders reduces coring problems and thus reduces scrap while reducing thermal expansion problems. Cylinder block height depends on crank throw, con-rod length, and piston height, and 3.5 mm should be added to allow for future development. Crankcase skirts are often extended below the crankshaft centre line by 65–90 mm to improve crankshaft bearing location. On diesels this practice is often eliminated to prevent noise amplification. Wall thicknesses should be kept to 4.3–5 mm, with the top deck up to 6–7 mm on iron blocks for petrol engines; on diesels, a much thicker top deck – up to 25 mm – is normal. With aluminium diecast blocks wall thickness can be held to 4 mm. With

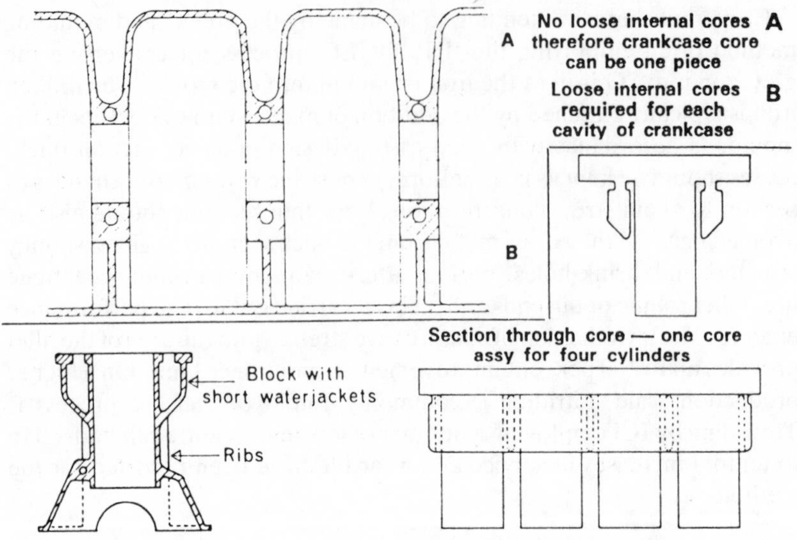

Fig. 3.4 Ribbing configurations for cylinder blocks

careful attention to foundry techniques it is possible to reduce the number of cores for a pair of blocks cast in the same moulding box to only five. If short water jackets are used their core can be formed integrally with a top slab core and with the cores for the cylinders. Large holes will be necessary in the top deck so that sand joining the jacket and the slab cores is strong enough. Additional benefit of this layout is that core holes are not needed in the sides of the water jackets. Another saving follows from avoiding ribs on the main bearing bulkheads, using instead double walls. This is illustrated in Fig. 3.4, and avoids the use of extra cores, appearing solid from the inside but being relieved at the back.

Elimination of end cores, which core out engine front and rear to form timing chests and clutch housings, is also desirable. The result is that all external cores are eliminated. Savings also follow from removing the need for unduly complex cores. This is a question of attending to detail around the fixings and mounting faces. Grouping of the ancillaries together in a separate diecasting is recommended. Oil drainage should also be carefully considered so that adequate passages needing only simple coring are required. In the case of aluminium diecast cylinder blocks, though, uniform wall thicknesses should be the aim; sharp changes in shape or section should be avoided, otherwise turbulence will occur as the metal flows into the die. To obtain thin walls, and uniform sections, it will be necessary to core out in all directions. However, undercuts can be incorporated without resort to sand cores, impossible with high pressure casting. Additional recommendations for aluminium blocks include the termination of the block at the crankshaft axis and fitting of a separate lower crankcase incorporating the main bearing caps – giving an outstandingly stiff assembly. If the block is diecast without internal cores, then a new way of dividing the engine must be considered. For example this might be integral block and head and open water jackets, or the cylinders made self supporting. Two ways of arranging self-supporting cylinders are by siamesing the cylinders or by using wet liners supported up to half their height; in the latter case the liners are clamped between their supporting cylinders and the head face. On diesel engines dry liners are recommended because they give a stiffer structure which is less likely to generate high noise levels.

Piston and ring design

Basic dimensions to be assessed are: diameter and thickness of the gudgeon pin, width of the lands, and the number and width of the rings, skirt proportions relative to the axis of the gudgeon pin, and thickness of the crown and skirt. In establishing these basic data, some assumptions must be made. The first one applies to the maximum cylinder pressure, which will be taken as that indicated in the crankshaft design section below, namely 6.2 MN/m^2 for petrol engines and 14 MN/m^2 for diesels. The diameter of the gudgeon pin is determined by the maximum permissible bearing pressure on the aluminium piston bosses, quoted as 40 MN/m^2 for petrol engines, and as 55 MN/m^2 for diesels by workers in this field.

However, the gudgeon pin diameter is also influenced by the need for oval deformation of the pin to be limited. It has been calculated that deformation should be limited to 0.025 mm and this figure is generally accepted. Oval deformation is calculated as follows

$$\text{Deformation} = \frac{0.041 p d_g^3 b^2}{l E t^3}$$

where p is the maximum gas pressure, d_g the gudgeon pin outside diameter, b the cylinder bore, l the pin length, E Young's Modulus (205.6 GN/m^2), and t the gudgeon pin wall thickness

$$t = \frac{d_g}{171}\left\{\frac{b^2 p}{l \times 0.025}\right\}^{1/3} \text{(mm)}$$

With these factors taken into account, as well as any limitations imposed by manufacturing techniques (for high volumes, pins are usually cold extruded) the proportions can be determined.

Crown thickness

Empirical methods are used to determine the thickness of the crown and, as a general rule, it is advisable to err on the thickness side, even though this may appear to give unnecessary reciprocating mass. The formulae most commonly used to determine crown thickness are, for petrol engines

$$t_c = d/\mathrm{k}(p)^{1/2}$$

where d is the piston diameter, p the maximum pressure, and k a constant, and for diesel engines

$$t_c = d\mathrm{B}/8.3\ (\mathrm{bmep})^{1/2} (\mathrm{mm})$$

where d is the piston diameter and B a constant, varying from 1.4 for a flat topped piston to 1.0 for a piston with a deep combustion chamber bowl. In practice, the thickness tends to be 0.07–0.1 times the piston diameter in petrol engines.

Ring lands

Since much of the heat is transferred from the piston to the cylinder through the land above the top ring, this should not be too narrow, but is an important consideration since the depth influences the overall height of the piston and engine. In petrol engines, the land tends to be similar to the thickness of the crown, or 0.08–0.10 times the cylinder bore. On diesels, the land needs to be deeper and is often 0.2 times the piston diameter. If the combustion chamber is in the piston, the top ring needs to be low enough to ensure that the thickness of metal between the groove and the bowl is sufficient to avoid creating a bottleneck to the transmission of heat away from the crown.

The thickness of the land below the top ring is determined by mechanical stresses, since it has to withstand the full compression load. The thickness is obtained as follows

$$\text{Ring land depth, } w = g\left(\frac{3p}{f}\right)^{1/2}$$

where g is the groove depth, p the maximum gas pressure, and f the safe stress permitted in the piston material. Suitable values for f are 52.7 MN/m^2 for cast iron, 103.5 MN/m^2 for steel, and 25.5 MN/m^2 for aluminium. Thus, for aluminium, $w = g/2.9(p)^{1/2}$ (mm).

Skirt length and thickness

In recent years, the trend has been for pistons to become shorter (Fig. 3.5) as designers sought to keep engine height to the minimum. The distance between the piston crown and the axis of the gudgeon pin is determined by the ring lands and the skirt from the axis of the gudgeon pin to the bottom edge was about 0.5 times piston diameter on engines a decade ago, but now this figure is being reduced, without ill effects, to 0.33–0.37. One researcher recommends that the skirt should tend to be 0.5–0.6 times the piston diameter on petrol engines, the higher figure being desirable.

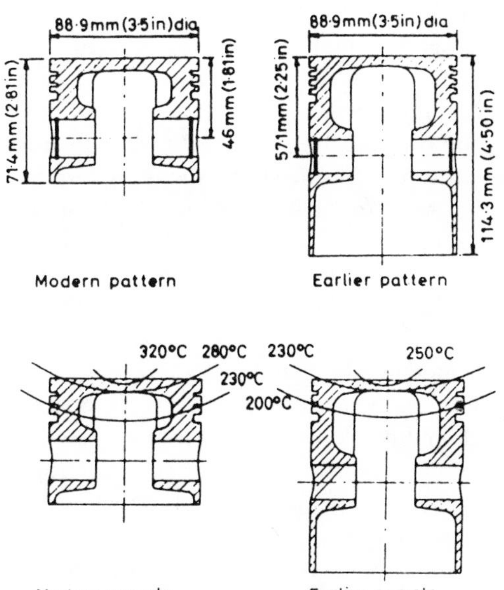

Fig. 3.5 Trend to squat designs for contemporary pistons showing old and new temperature levels

The length of taper from a thickness of *d*/25 at the top of the skirt, just below the gudgeon pin axis, to *d*/35 at the bottom. To some extent the use of diecast pistons ensures this taper, which is needed to give the dies draft.

With diesels, much longer skirts are needed. In any engine, the friction force resulting from the side thrust of the piston creates a tilting motion on the piston, and on a diesel there is always the danger that the resulting high pressure will cause scuffing. This can be counteracted by using pistons with adequate skirt length *above* the gudgeon pin boss. Ideally, the skirt should be equally long above and below the gudgeon pin, ignoring the area occupied by the rings. Whereas the combination of two compression and one oil control rings is standard on petrol engines, three compression rings are still used generally on diesels, although the trend is towards the use of only two rings.

Expansion control

Since the thermal coefficient of aluminium is about 1.7 times that of cast iron, the maintenance of small skirt clearances has always presented problems. The need to minimize the noise of piston slap at low speeds in petrol engines has led to a number of different approaches to the control of expansion, and now similar methods are needed on diesels to reduce overall noise levels.

To prevent seizure in petrol engines, it has become normal practice either to make the skirt flexible, so that it gives way under pressure and does not seize, or to prevent the skirt expanding too much by the use of some form of thermal barrier. The simplest form of flexible skirt is the split skirt or T-slot design, in which the skirt can expand to close the split. The thermal barrier design incorporates a form of circumferential groove or slot in or below the oil ring groove, extending for the width of the thrust and non-thrust axes.

An alternative method of preventing seizure is to limit the way in which the piston expands, by the use of steel inserts. Hoops or struts are cast in and, when the aluminium cools, these are compressed so that they exert an outward force on the skirt. As the piston warms up, the force exerted by the struts progressively reduces, so that overall expansion of the skirt is much less than normal.

Diametral clearance between the piston and bore (Fig. 3.6) is always greatest near the bottom of the skirt when the engine is cold. With a solid skirt, the clearance tapers from 0.085 mm at the bottom to 0.12 mm at the top of the skirt, in a typical case.

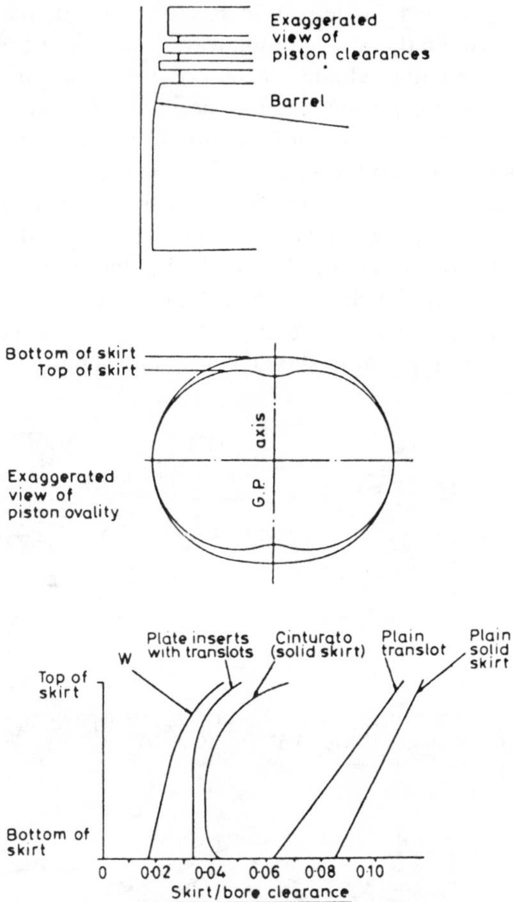

Fig. 3.6 Exaggerated view of piston/cylinder clearances for different types

Cooling

As the temperature of the piston increases, so the permitted stress levels are reduced. An AE Group researcher considers that the undercrown temperature should always be limited to 250°C, at which level he recommends a permissible stress of 52.5 MN/m^2, and points out that at 150°C the permitted stress is 105 MN/m^2.

According to work carried out by Ricardo and Company, there is little to choose between the two common ways of oil-cooling pistons; spraying oil from jets in the cylinder block, or from the gudgeon pin boss. However, the company recommend the 'cocktail shaker' in which a plate is attached to the underside of the piston crown to provide a shallow chamber in which the oil is agitated, if the oil supply is limited. For highly rated engines there is also the Wellworthy piston with solubly cored passages. These passages are cored behind the ring belt, so that local hot spots can be avoided. Oil can be supplied to the galleries in a number of ways, ideally through drillings in the gudgeon pin and then through passages in the piston bosses to the gallery. In this case a pressurized flow is provided, and the pistons should be capable of performing reliably at a

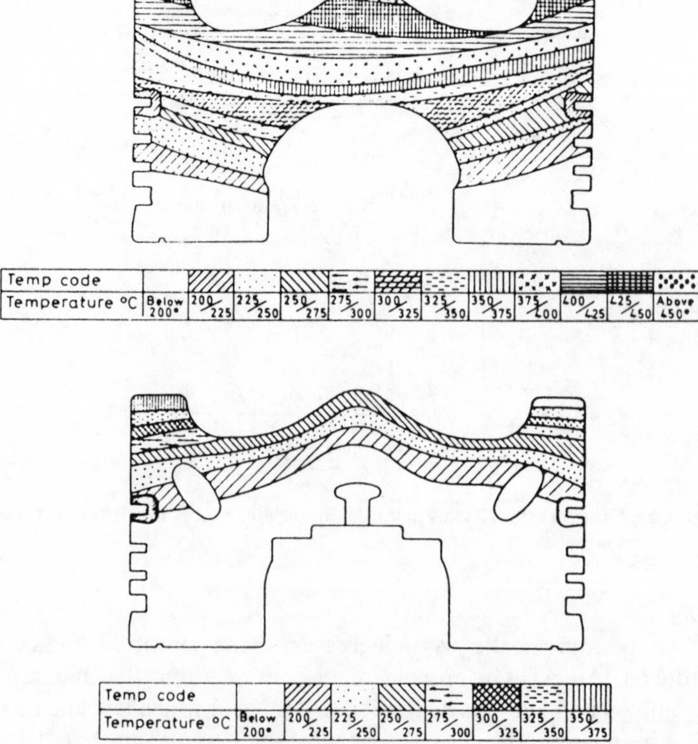

Fig. 3.7 Piston temperature reduction by means of oil galleries

bmep up to 1.4 MN/m^2; the effect of oil cooling galleries is shown in Fig. 3.7.

For higher bmeps, built-up pistons with crowns of a more heat-resistant material are normally recommended. Materials such as En 19 or En 52 are normally used for the crown, En 19 being favoured for its improved thermal fatigue resistance. For automotive use, Associated Engineering are looking into ways of attaching a heat resistant cap to the piston at lower cost. Plasma spraying is one possibility, while eutectic bonding is another.

Piston rings

The basic functions of piston rings are to provide a gas seal; to conduct heat from the piston to the cylinder wall; and to meter the amount of oil that is passed up the bore for lubrication. The rings must, of course, have good resistance to thermal fatigue and wear, and should apply radial pressure uniformly distributed around the cylinder wall. They should also be able to conform to the shape of cylinders that are distorted out of true circular form.

On high speed engines, the thickness of the ring is limited by the need to keep the inertia forces within safe limits. The mass of the ring must be low enough to ensure that the force exerted downwards by gas pressure at the end of the compression stroke exceeds the product of ring mass and acceleration. According to Hepworth and Grandage, the limiting accelerations of the piston that maintain acceptable conditions are:

Ring width (mm)	*Piston acceleration (m/s^2)*
3.0	3 800
2.5	4 800
1.5	7 500
1.0	10 000

Piston acceleration is calculated as

$$\text{Piston acceleration} = \frac{N^2 S}{598\,480}\{1 + (1/r)\}(\text{m/s}^2)$$

where N is r/min, S is the stroke and r is the ratio of connecting rod length to crank *throw*.

Connecting rod design

Traditionally, the connecting rod has been made as long as practicable, so that secondary forces are reduced and cylinder-filling is theoretically improved. In practice, it has proved possible to reduce the length of the rod from the 3.8–4.5 times crank throw formerly preferred to 3.5–3.75 without any adverse effects. Ratios as low as 3.2:1 have been used successfully.

Crankpin diameters also tend to follow trends, averaging 0.58–0.60 times the cylinder bore for petrol engines, and 0.65 for diesels. On highly turbocharged diesels, the trend is towards even higher ratios, in some cases over 0.70. Naturally, bearing sizes are determined by computer programs devised by the specialist suppliers, but these ratios indicate the proportions of modern connecting rods.

Stress calculations

I-section shanks are generally used on connecting rods, except on some lightly stressed two-strokes, which have oval-section shanks. The cross-sectional area of the shank must be sufficient to withstand the compressive load applied by the gas pressure and the tensile load resulting from the inertia force of the piston and rod, while the rod must be stiff enough to resist the bending loads and to prevent excessive deflections around the big and little ends. Designers tend to use comparative stress techniques when designing new rods, since the overall shape is dictated by basic dimensions and the production method. Some formulae used are

$$\text{Compressive stress} = \frac{\text{gas pressure} \times \text{piston area}}{\text{cross-sectional area of shank}}$$

$$\text{Tensile stress} = \frac{Y \times \omega^2 \times R \times W}{12g}$$

where Y, the inertia constant, equals $1 + R/L$, and R is the crank throw, L the connecting rod length, W the weight of the piston, gudgeon pin, and small end, and ω the angular velocity of the crankshaft.

On racing engines, of course, the inertia load is the limiting factor. In diesels, the gas pressures can exceed nominal values, so it has been recommended that a maximum cylinder pressure 50 per cent above the normal value should be used when the compressive stress is calculated.

For the best results, the big end should be split perpendicularly to the axis of the cylinder. Obliquely split rods have been used where the bore is too small for the big ends of the rod to be withdrawn through the cylinder, but since this type of eye is more prone to distortion it should be used only where there is no alternative. Because it is difficult to ascertain the stiffness of the big end, Vandervell Products have installed a special rig on which the deflections can be measured and the stiffness evaluated. If the eye distorts too much, fretting of the back of the bearing takes place; this can lead to fatigue failure of the rod, and it is possible for the bearing to spin in the eye. The use of eccentric shells can help here.

Cap retention

Setscrews, bolts and nuts, or studs and nuts can be used to retain the cap. One advantage of the bolt and nut arrangement is that the shank of the bolt can be used to locate the cap to the rod. If part of the shank is also serrated – as on the Jaguar V12 – positive location in the rod during assembly is also obtained. If setscrews are used, location is normally effected by serrations, tenons, or dowels. Vandervell have found that with serrations it is impossible to obtain contact over the complete area, so that there is a spring effect opposing the bolts when they are tightened. As a result, it is difficult to obtain the correct clamping force and the correct load on the setscrews. Typical cap location is shown in Fig. 3.8.

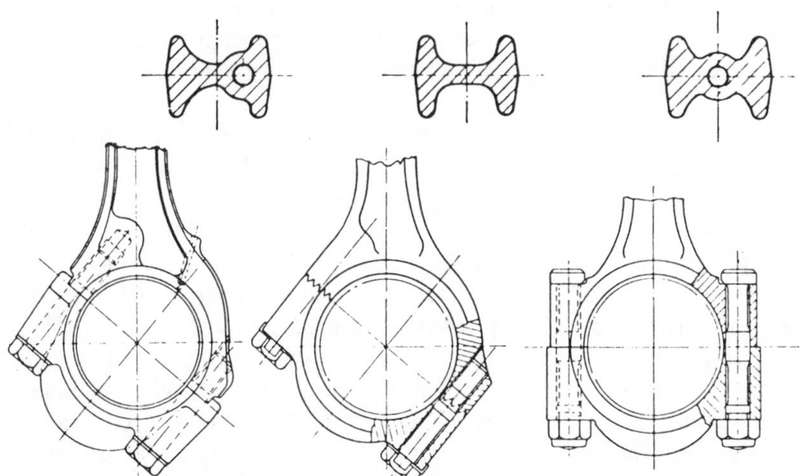

Fig. 3.8 Alternate methods of big-end cap location

High tensile steel, such as En 16T, is used for the bolts. These must be tightened so that the load is sufficient to 'crush' the bearings and to overcome the inertia force of the piston and rod. To obtain consistent torque values, the face on which the nut or head of the setscrew bears must be machined to a consistently fine surface finish. In addition, the bosses must be robust and hard enough to withstand the load without distortion or local compression, while the threads of the fasteners must always be in the same condition – lubricated or dry.

Manufacture

Traditionally, connecting rods are steel forgings, but the trend in petrol engines at least is to the use of malleable or spheroidal-graphite (SG) iron castings or sinter forgings. Materials such as En 15S or En 16T steel tend to be used generally for these forgings, although En 24 is used on many racing engines and titanium alloys are used in some cases.

In volume production there are many advantages in the use of one-piece forgings, both in part costs and in control of supplies. Of course, the big end eye has to be formed to allow for the cap to be sawn off, and the cap must be formed with one reinforcing rib between the bosses instead of two. Although the minimum draft angle on forgings is usually quoted as 7 degrees, it is only on the outer periphery that the draft angle needs to be held to the minimum. The junctions between the web and flanges can have greater draft and generous radii should be used. Any theoretical losses will be more than made up by the more consistent forgings that will result.

Allowance for some variation in the length of the forging should be made, in that the little end boss should be formed as an oval, with radii struck from points about 1.5–2 mm apart. The shank should blend smoothly into the big and little end bosses, in the interests of both manufacture and strength. Finally, there is the question of balancing lugs. These lugs are usually incorporated at the top of the little end boss and at the bottom of the big end cap. They are normally machined automatically to give the required matched, overall, big end and little end weights. These lugs are often larger than they need be, so that suppliers' estimates of tolerances and previous experience should be taken into account when sizes are determined. If space is limited, the upper lug can be placed on the side of the little end boss.

Where a bush is installed at the little end, it is normal practice to rely on a hole at the top of the boss to allow oil to reach the bush, and the hole is normally stepped so that it is shaped like a funnel. In this case 'squeeze' lubrication is used, in which some oil flows into the gap each time the inertia force gives maximum clearance above the bush and is then squeezed around the bush. On some highly turbocharged diesels the little end boss is tapered, so that the area of the bush below the gudgeon pin is greater than that above the pin. Wherever the gas load predominates, the design is to be recommended since the useful bearing area can be increased – in the bush and piston bosses – although, of course, the bearing area that opposes inertia forces is reduced.

The alternative to the use of a bush in the little end is for the gudgeon pin to be pressed in. Although this method has obvious cost benefits, it imposes extra loads on the piston bosses where lubrication is usually marginal. Another feature found in many connecting rods is a small drilling from the big end through which oil is sprayed on to the cylinder bore. This drilling, which is by no means universal, must be positioned accurately.

On turbocharged engines it is normal practice to cool the underside of the piston crown with oil, either by spraying from jets in the cylinder block, or by taking lubricant up through the shank of the connecting rod and little end. The use of separate jets is to be preferred for several reasons: first, the drilling of a hole through the shank of the rod is complicated, especially since, if the bearing is not to be starved of oil, the drilling cannot be straight up the middle of the rod. The problem here is that the column of oil in the drilling has such inertia that it tends to continue to pull oil upwards, and if the drilling is on the central axis of the rod it reduces the bearing area when loads are highest – at tdc. At least one company found that, owing to these combined effects, the load capacity of the bearing is redued by 60 per cent by a vertical drilling. To avoid this disadvantage the hole can be drilled at an angle from as far in advance of tdc as is practicable, but even then the lubrication of the bearing is likely to be impaired.

Materials

To reduce costs a significant change in manufacturing methods of connecting rods must be made. The alternatives to conventional forging are malleable iron, SG iron castings, or sinter forging. Most of the early cast connecting rods, such as those used in the Fiat 128 and 130 were of malleable iron, but the trend here is to SG iron. Whether rods can be cast with such precision that no balance lugs are needed remains to be seen, but this is confidently predicted by foundry experts. Certainly GKN consider that sinter forged rods should require no balance weights. Since the sinter forging needs only 14 machining operations compared with 23 for a normal forging, there is a possibility here of reducing costs, reducing weight by about 10–15 per cent and decreasing material usage.

Crankshaft bearing design

Traditionally, bearing loads are obtained from polar load diagrams (Fig. 3.9), but the essential feature is that an adequate oil film thickness be maintained over a sufficient area to withstand the load. As a basic guide for determining bearing sizes for preliminary studies, Vandervell quote the following design capacities for their products (MN/m^2):

			Copper–lead with lead–indium overlap	
Babbit	*Micro-Babbit*	*Aluminium/tin*	*VP1*	*VP2*
13.5	14	28	62	41–45

Precise dimensions are derived from computer programs based on film thickness studies, and basic design criteria are fatigue resistance, corrosion resistance, and embeddability. In determining the bearing sizes,

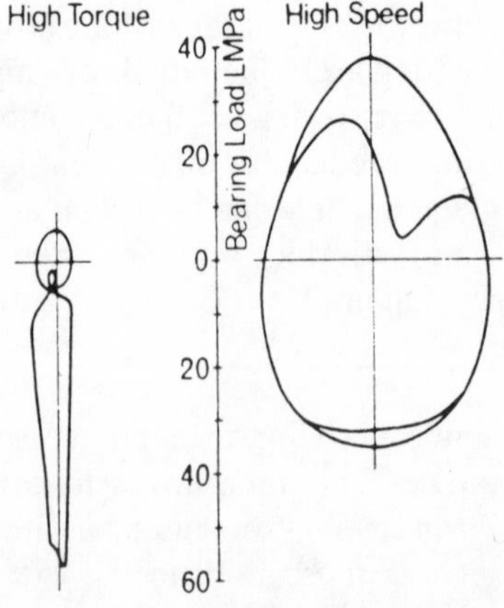

Fig. 3.9 Polar load diagrams for diesel (*left*) and petrol (*right*) engine crankshaft bearings

the maximum pressure in any part of the oil film, the minimum film thickness, and the maximum temperature of the film are taken into account.

Since fine dirt is virtually always present in engines, embeddability is an important characteristic, and this tends to be inversely proportional to hardness. Thus, white metal or Babbit has very good embeddability, while aluminium-tin is better than copper. However, Vandervell claim that their lead–indium overlay is as good as white metal, although of course it can only deal with particles that are smaller than the overlay thickness – about 0.018 mm.

As always, conflicting requirements – of cost and the need for greater load capacities – affect the choice of bearings, and on the basis of cost and availability of materials the trend is towards the use of aluminium–tin, developed originally by Glacier. One problem affecting all materials in petrol engines is high temperature. Temperatures have continued to creep up in the past decade and, whatever oil is used, by the time a temperature of about 170°C is reached the viscosity is very low and bearing lubrication is endangered. It is important to realize that the temperature of the oil film in the big end is usually some 30–40°C higher than that of the oil in the sump. Therefore, if the bulk oil temperature reaches 130°C, bearing failure may take place prematurely.

With the increasing use of turbochargers, bearing loads are becoming much higher, and for this application silicon–aluminium bearings are recommended owing to their superior fatigue and wear characteristics. A typical material contains 11 per cent silicon and 1 per cent copper, with the remainder aluminium, and with a 0.013–0.020 mm thick lead–tin overlay. On the Glacier Sapphire fatigue test rig, the fatigue strength of this material is shown to be 30 per cent higher than that of aluminium–tin alloys.

Big end bore deflection under load is shown in Fig. 3.10 and pressure distributions for aligned and misaligned shafts are shown in Fig. 3.11.

COMPUTER PREDICTION OF ENGINE IMBALANCE

Dr Brian Law[19] of Perkins Engines has shown that the requirement for rapid engine balance assessment, accounting for actual engine geometry, can be met by a computer program. Use of this program can further the understanding of engine excitation and balance. All reciprocating engines are, of course, potential generators of dynamic excitation due to the nature of the piston, connecting rod, and crankshaft motions.

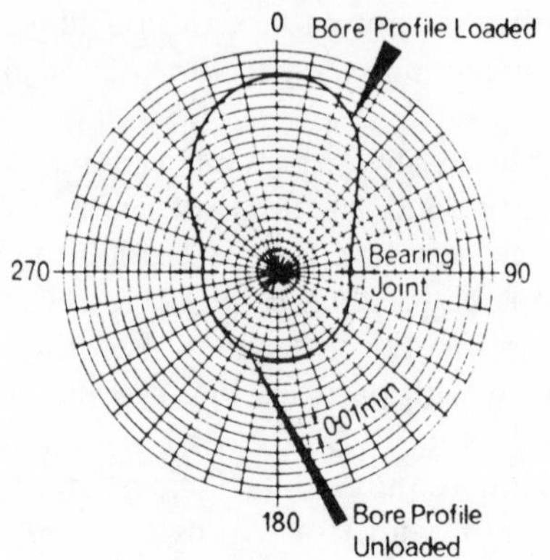

Fig. 3.10 Big end bore deflection under load

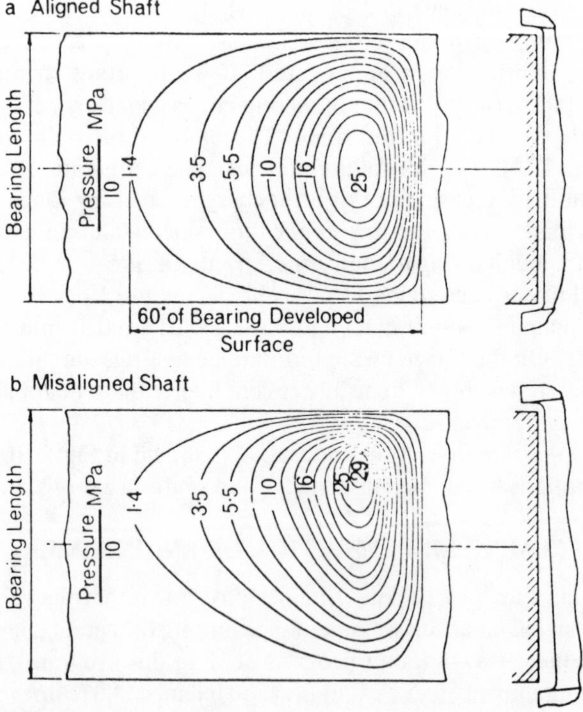

Fig. 3.11 Oil film pressure distribution for aligned and unaligned crankshafts

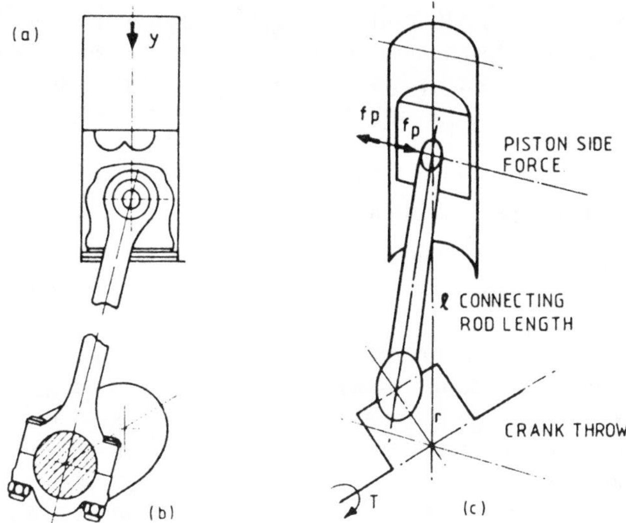

Fig. 3.12 Idealization of a single cylinder for balance analysis

A section through one cylinder of an engine is shown in Fig. 3.12. The dynamic excitation comprises these three components: (a) the piston and connecting rod small end execute a reciprocating motion along the cylinder axis which requires a continually varying force of $m\ddot{y}$; (b) rotation of the large end of the connecting rod and the crankthrow about the crankshaft axis generates centrifugal forces equal to the combined mass–radius product $\times\ \omega^2$; (c) a varying torque acts on the crankshaft and engine which depends on the inertia forces and gas pressure forces acting on the piston and connecting rod. The torque is equal to the moment of the piston-to-cylinder force about the crankshaft axis.

The centrifugal and reciprocating excitation cannot be eliminated (practically) in a single-cylinder engine. In a multi-cylinder engine, however, some components of the excitation produced in one cylinder may cancel excitation produced from another cylinder, but some of the excitation components may be additive. The nature and magnitude of the nett excitation will depend on the engine configuration. In engines which are not completely balanced, engine mounting forces and vibrations will be produced by the unbalanced forces, couples and torques. Engine vibration can cause failure of components in the vehicle as well as causing significant operator discomfort. The engine structure, particularly in the regions influenced by engine mounting forces, must withstand the excitation produced by engine operation. Knowledge of engine balance characteristics are, therefore, important when designing a suitable mounting for the engine.

Calculation of engine excitation

Rotational forces and couples

The connecting rod may be idealized as a massless link connecting lumped masses at the small and large ends. The large end lumped mass (usually 60–70 per cent of the rod mass) is considered to rotate with the crankpin centre. The mass radius products of all portions of the crankshaft (crankwebs, crankpins, counterweights, large end masses) are evaluated first. The out-of-balance forces and moments (about the centre of the crankshaft; are then derived for a unit rotation speed of the crankshaft (Fig. 3.13). Graphical methods, involving vector polygons, are usually favoured for the summation task. Subject to space and weight limitations it is always possible to effect complete rotational balance by adding mass to (or removing mass from) the crankshaft or pulley and flywheel.

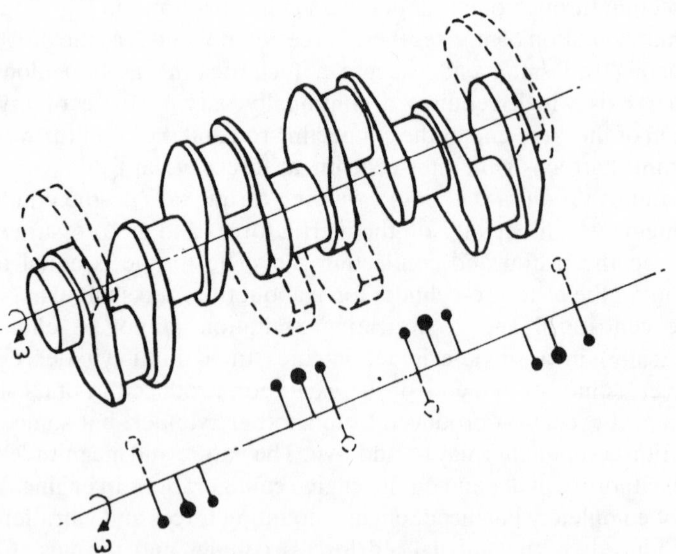

Fig. 3.13 Rotational balance – 4-cylinder crankshaft

Counterweights are added to some crankshafts (usually opposite the crankpins) to reduce the main bearing force component due to centrifugal loading. For example, the flat-four crankshaft illustrated in Fig. 3.13 does not require masses for rotational balance, but counterweights may be added to reduce the loading on the centre bearing due to rotation of the adjacent crankthrows. Additional counterweights must be added elsewhere on the shaft to restore rotational balance – these counterweights are often positioned on the extreme crankwebs.

Reciprocating forces and couples

The instantaneous piston acceleration at crank angle θ is given (approximately) by

$$\ddot{y} = r\omega^2 \cos \theta + r\omega^2 \frac{r}{l} \cos 2\theta$$

The first term results in an alternating force along the cylinder axis of $mr\omega^2 \cos \theta$ at engine frequency which is called the primary force. The second term describes an alternating force on the piston r/l (typically 0.3) times the primary force in magnitude and at twice engine frequency. An equivalent dynamic system comprising forward and reverse rotating eccentric masses is often used to represent the reciprocating force produced by one piston. The four masses are illustrated in Fig. 3.14 and all are assumed to rotate about the crankshaft axis at the point where the cylinder axis intersects the crankshaft axis. The nett excitation from a multi-cylinder engine may be derived by first constructing the system of eccentric masses. The cylinder phasing relationship is represented in the equivalent system by positioning the rotating masses according to the

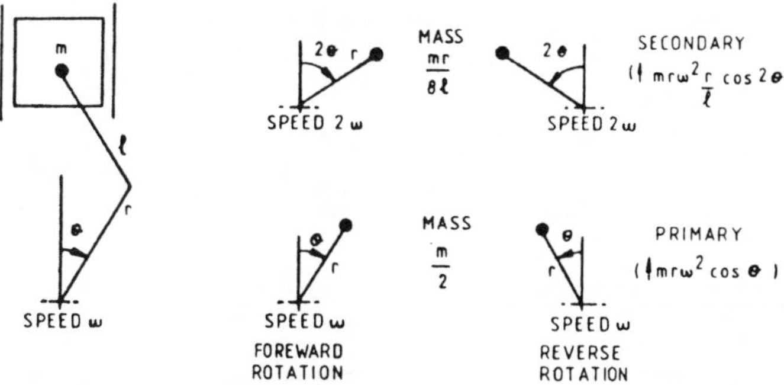

Fig. 3.14 Reciprocating balance – equivalent rotational system

crank mechanism angle for each piston relative to its cylinder centreline. The nett forces and couples due to the primary and secondary equivalent masses may then be derived using a vector polygon to sum the force and moment components.

Crankshaft torque

The torque components due to cylinder pressure (often called the gas torque) and inertia forces (often called the inertia torque) are usually calculated separately. Expressions are derived in the literature for the instantaneous torque due to gas and inertia loadings in a single cylinder. The engine torque *vs* crank angle diagram can be constructed by summing the appropriately phased torque from each cylinder. The engine speed variation caused by the cyclically varying torque may be calculated for a given driven polar inertia (crankshaft, flywheel, and directly driven equipment). Such calculations are used to specify suitable flywheels to meet engine application requirements.

Calculation of engine balance by a computer program

The approach to engine balancing outlined above is suited to the general examination of balance characteristics and provides general results like those in Table 3.4. Analysis of specific engine designs can present a more involved problem. The crankthrows in multi-cylinder engines are seldom identical. Certain crankwebs are deliberately made stronger than others to withstand the greater bending loads imposed on them. The cylinder spacing may not be uniform due to cooling and bearing length require-

Table 3.4 Inherent balance characteristics

	Cylinder Arrangement 4-stroke			
	3-Cylinder in-line	4-Cylinder in-line	6-Cylinder in-line	8-Cylinder 90° Vee
Crank diagram (clockwise rotation looking at no. 1)	 1, 2, 3	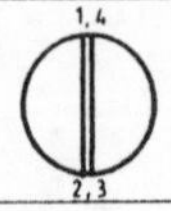 1,4; 2,3	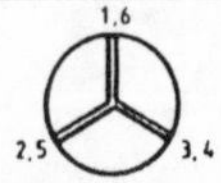 1,6; 3,4; 2,5	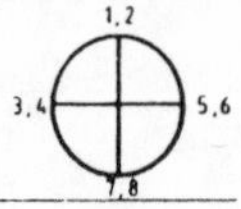 1,2; 5,6; 7,8; 3,4
Firing order	1-2-3	1-3-4-2	1-5-3-6-2-4	1-8-7-5-4-3-6-2
Firing impulses per engine rev	1½	2	3	4
Engine balance				
External forces: Primary	Balanced	Balanced	Balanced	Balanced
Secondary	Balanced	Unbalanced	Balanced	Balanced
External couples: Primary	Unbalanced	Balanced	Balanced	Balanced
Secondary	Unbalanced	Balanced	Balanced	Balanced

ments. The engine designer may wish to examine the effect on engine excitation of the tolerance band on reciprocating mass. Even when the engine is sufficiently close to the 'textbook case' the application of the above procedure can be tedious especially when awkward cylinder arrangements are concerned (60 degree vee). In response to a need for a more rapid and general approach to engine balancing one company has generated computer programs for the examination of engine excitation and balancing. Initially a balance program was written in BASIC for use on a Tektronix 4051 desk-top computer. Subsequently a FORTRAN version has been generated with additional facilities to examine engine balancing by combinations of rotating balancing shafts. The FORTRAN program runs on an IBM 370/168 mainframe computer.

All practical engine configurations which have a single crankshaft may be analysed by these programs. No restrictions are placed on the uniformity of reciprocating masses, crankthrow geometry, or cylinder spacing and inclination. In the first phase of the program the vertical and horizontal components of the reciprocating forces in each cylinder and the vertical and horizontal components of the centrifugal forces are computed at 5 degree intervals of crankshaft rotation angle. The total horizontal and vertical forces and their moments about a chosen datum are evaluated to provide the engine imbalance at these 5 degree intervals of crank angle. The imbalances are displayed to the user in graphic form. In the second phase of the program the user may interactively see the effect of adding balancing masses to the crankshaft and/or balancing shafts to the engine.

The reciprocating balancing of a 3-cylinder in-line engine will be examined to demonstrate the operation of the program. The data supplied to the program is given in Table 3.5. The resulting primary and

Table 3.5 Record of balance program input data

3 cylinder in-line demonstration			
Cylinder centreline to datum (mm)	104.8	0.0	−104.8
Crank angle from No. 1 TDC to give piston TDC in each cylinder	0.0	240.0	120.0
Angle of cylinder to vertical	0.0	0.0	0.0
Reciprocating mass (kg)	1.83	1.83	1.83
Crank throw (mm) = 63.5			
Con. rod length (mm) = 223.9			
Speed (r/min) = 2500.0			

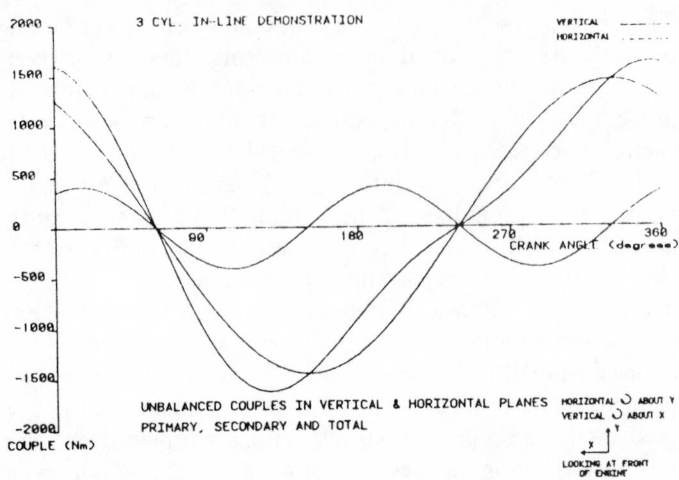

Fig. 3.15 Computer program output for 3-cylinder in-line engine

secondary reciprocating imbalances (Fig. 3.15), are couples in the vertical plane which have peak values at 330 and 15 degrees of crank angle, respectively. These imbalances cannot be eliminated by means of balance weights on the crankshaft. The magnitude of the vertical primary couple can be reduced using crankshaft balance masses, but an unbalanced horizontal couple, generated by the balance masses themselves, will result. The secondary couple cannot be influenced at all by masses attached to the crankshaft. Complete balancing of the primary excitation, however, can be achieved using a reverse rotating balance shaft in conjunction with masses attached to the crankshaft (Fig. 3.16). Balancing of the secondary couple requires two shafts rotating at twice crankshaft speed. The primary couple is about 3.5 times greater than the secondary couple. The primary couple would generate an amplitude of vibration about 14 times that produced by the secondary couple in an unrestrained engine. Balancing of the secondary couple is, therefore, far less important than elimination of the primary couple. The balancing program can calculate the size and position of the masses on the crankshaft and balancing shafts which are required for balancing of the excitation forces and couples.

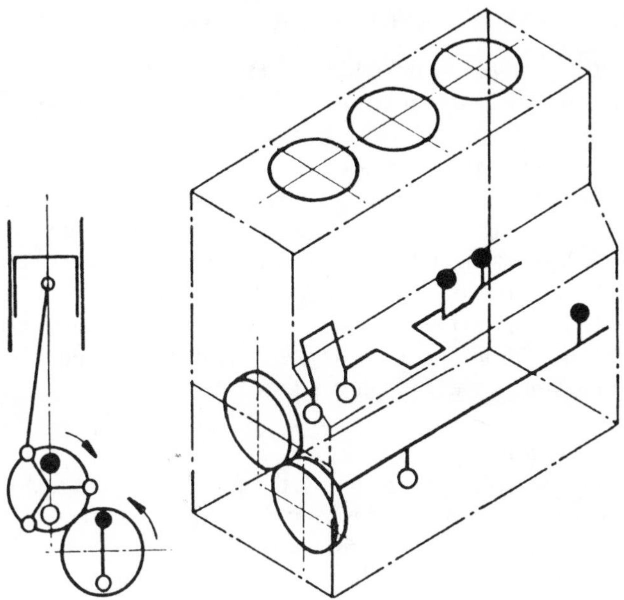

Fig. 3.16 Primary couple balance for 3-cylinder in-line engine

Balancing of engine rolling excitation

The intermittent firing and the crank mechanism motion produce crankshaft speed fluctuations which, as described above, can be reduced by increasing the polar inertia of the flywheel (and/or drive train). Equal changes in angular momentum, but in the opposite sense, are produced by the fluctuating torque loading on the engine block structure. The amplitude of engine rolling depends on the polar inertia of the engine about the crankshaft axis and on the engine mounting arrangement. The rolling motion is the only significant excitation in some engine configurations (in-line 6 cylinder engine). Elimination of the rolling excitation may be effected by a counter rotating shaft of equivalent polar inertia to the crankshaft/flywheel system. The shaft must be supported by the engine and driven by the crankshaft such that it has angular accelerations equal (in magnitude) to those of the crankshaft system. Such a balancing arrangement was the subject of a US patent.

ENGINE MOUNTING

A. J. Reed[20] explains that flexible mountings are now universally accepted for all normal internal combustion engines in road vehicles. The basic design principles involved are quite straightforward, and in the majority of cases only minor adjustments are found to be necessary when the actual installation in the vehicle is tested. However, owing to the ever-improving standards regarding noise and vibration, it is now usually considered necessary with passenger cars to study the vibration system of the vehicle as a whole, and analogue computers are sometimes used to solve the complex equations involved when the flexibilities of the engine and body structures, in addition to the characteristics of the suspension systems, have to be taken into account at the design stage.

Basic design requirements

In most engine types in use today there are, in theory, no primary out-of-balance forces or couples, and the first requirement is to ensure that the mounting system will satisfactorily insulate the torsional impulses produced by the power unit. In practice, some primary out-of-balance can occur due to production tolerances on individual engine components, and it is usually advantageous for a degree of flexibility to be provided in planes at right angles to the crankshaft axis, to give adequate insulation in the normal running range.

From time to time, however, engines with unusual cylinder arrangements – for example, three cylinders in line, or V4 or V6 arrangements – are designed or adapted for vehicle use. These layouts sometimes involve relatively large primary or secondary forces or couples. While it is usually possible to design flexible engine mounting systems that have relatively small vibration amplitudes, it must be remembered that the out-of-balance forces are proportional to the square of the speed. Consequently, the forces transmitted at higher speeds, even with only moderate damping in the mountings, may well be excessive. In contrast, the gas torque harmonic components stay practically constant throughout the speed range and the forces transmitted diminish rapidly above the resonant speed range. With these unusual layouts it is, therefore, sometimes desirable to accept uneven firing intervals in order to reduce the out-of-balance forces.

Having ensured good vibration insulation, the designer still has to check that the engine movement is not excessive under the worst running conditions, and that the mountings have sufficient dynamic load capacity.

Forces of plus and minus three times the static loads are typically obtained during rough track running on proving grounds. A 'centre of percussion' position is sometimes used for the front mountings; this arrangement reduces not only the tendency to pitch on bad roads, but also the shock loads on the rear mounting or mountings. Unfortunately, though, the exceptionally small static load on the rear mounting can cause problems in obtaining satisfactory vibration insulation. Adequate control of fore-and-aft movement of the engine is necessary to avoid troubles on accelerating and braking; it is not unknown for a cooling fan to foul the radiator during test running. Since the majority of cars now have hydraulic operation of the clutch, the engine mountings no longer have to withstand the operating forces involved.

An otherwise satisfactory engine suspension system may cause 'front-end shake' owing to the amplification of the movements of the unsprung mass – that is, the wheels, brake drums, and so on – on the tyres. On the other hand, the engine on its mountings can act as a tuned absorber for the body structure, thus reducing body movements on bad roads and making a real contribution to a smooth ride at higher speeds. In general, the unsprung mass frequency, or 'wheel hop' frequency, is in the region of 8–12 Hz on most passenger cars, and it is very desirable that the vertical frequency of the engine mounting system should not be too close to this value. The vertical bending frequency of the body structure on cars of integral construction is usually at least 17 Hz, and if the engine mounting frequency is close to this, the engine can act as a tuned absorber, as just mentioned. However, this is usually only possible on six- or eight-cylinder engines where a fairly stiff mounting arrangement is acceptable because of the relatively small torsional impulses and inherently good balance.

Calculations

A few simple calculations are often adequate to establish the stiffness requirements of the mountings in relation to their possible positioning. In Fig. 3.17, if the moment of inertia about the axis XX is I, then the resultant natural frequency f in cycles per minute is given by

$$f = \frac{60}{2\pi} \cdot \frac{\text{torsional stiffness}}{I}$$

Since I can be measured experimentally, or estimated from the dimensions and weights, the required torsional stiffness of the mountings about this axis can be readily calculated for a given value of f.

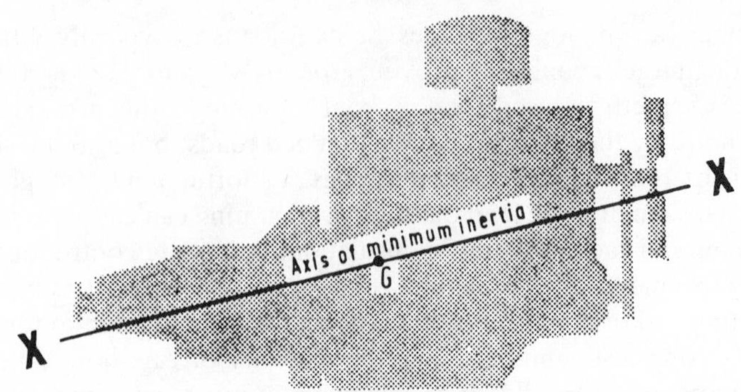

Fig. 3.17 Inclined axis of inertia about which engine should oscillate if mountings are properly designed

This value is generally chosen so that the 'firing order' resonance speed is sufficiently low to avoid excessive amplitudes of movement and transmission of forces at idling. For example, on a four-cylinder in-line four-stroke engine there are two firing impulses per revolution, so the 'firing order' is the second order. At a typical idling speed of 450 r/min, the second order frequency is 900 cycles per minute and the highest value of natural frequency that can normally be accepted is two-thirds of this, or 600 cycles per minute. The second-order resonance will then occur below normal idling, at 300 r/min.

Owing to uneven firing or misfiring, other minor resonances can become important, and it is usually necessary to ensure that the $\frac{1}{2}$, 1, or $1\frac{1}{2}$ order does not occur at normal idling speed. In this example the $\frac{1}{2}$ order is at 1200 r/min, the 1 order at 600 r/min and the $1\frac{1}{2}$ at 400 r/min. Therefore, the first order might cause slight roughness on fast idling, or $1\frac{1}{2}$ order on slower than normal idling; roughness at 1200 r/min, due to the $\frac{1}{2}$ order, might serve as a timely warning to clean the plugs or injectors. It is useful also to obtain a first approximation to the main vertical natural frequency of the engine on its mountings, again in cycles per minute from the formula

$$f_v = \frac{60}{2\pi}\cdot\frac{K_v g}{W}$$

where K_v is the vertical dynamic stiffness and g the gravitational constant.

The theory and practical application of V-mountings have been described in a number of publications. In particular, a detailed analysis of

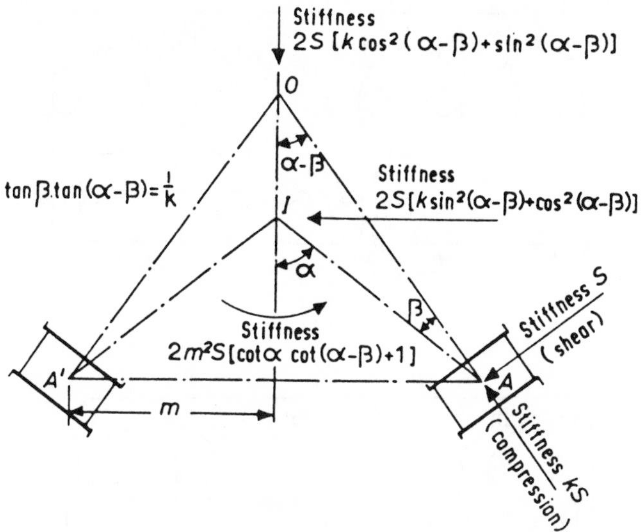

Fig. 3.18 Equations for calculating mounting stiffnesses

the theoretical design basis was given in 1957 to the Institution of Mechanical Engineers by M. Horovitz, in his paper entitled 'The suspension of internal combustion engines in vehicles'. Figure 3.18 illustrates the basic formulae needed by the designer for calculating the various stiffnesses of mountings of this type.

EXHAUST SYSTEM ANALYSIS

In reviewing recent literature, the editor[21] found that, in the introduction to P. H. Smith's *Scientific Design of Exhaust and Intake System* (Foulis), this author explains that no straightforward analytical procedure exists to rationalize the design of the exhaust pipe and silencer assembly. Smith lists the requirements leading to design criteria for exhaust systems as follows: low-cost manufacture, adaptability to a group of engines, quick assembly to engine and chassis, adequate clearance from other components, low engine-compartment temperature, induction 'hot-spotting' facility, durability, acceptable quietness, and adequate scavenging at low speeds commensurate with minimum power loss at high speeds.

The trend in petrol engine design is to arrange for maximum charge induction per stroke and to obtain fullest possible burning from the

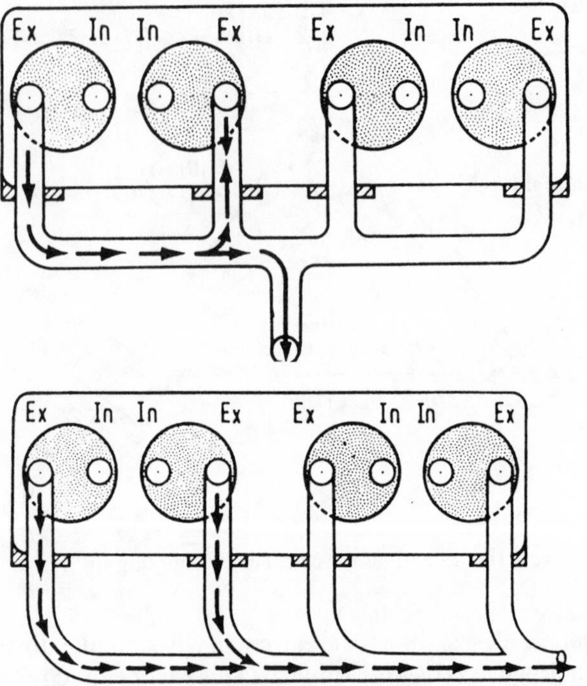

Fig. 3.19 Inter-cylinder interference in a short stub manifold

charge. This usually involves considerable valve overlap and therefore gives the engine a high sensitivity to back pressure and to the exhaust system design in general (for example, the problem of intercylinder interference in a short-stub manifold is an urgent one, Fig. 3.19). Smith warns against the over-simplified approach of designing on the basis of mean back pressure, however.

Cyclic variations in back pressure for a variety of systems measured by co-author Morrison on a single-cylinder research engine are shown in Fig. 3.20(a). (One ideal curve, Fig. 3.20(b) is shown for comparison.) It should be noted that, of the two straight-through pipes compared, although the shorter (full line) had a lower mean back pressure, the particular length of the longer (dotted line) caused a negative pressure at the end of the exhaust stroke, at the engine speed concerned. An interesting point mentioned is that the depth of the velocity gradient in

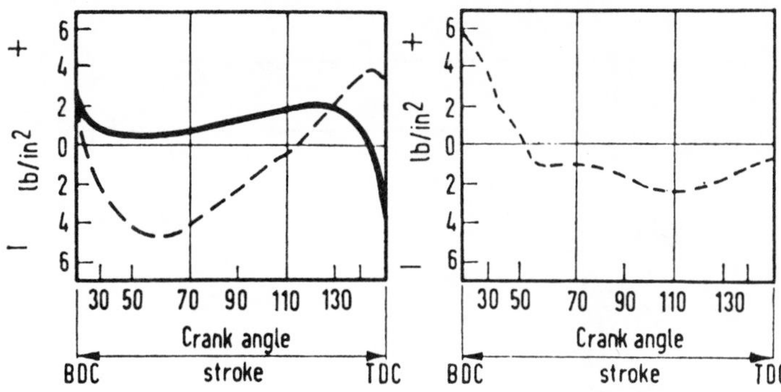

Fig. 3.20 Typical (*left*) and ideal (*right*) cyclic variations in back pressure for a single cylinder engine

exhaust pipes – due to wall friction – is by no means insignificant and is measurable in millimetres.

Smith postulates that, whereas early stationary engines having very long exhaust pipes were the first to exploit pulsating exhaust flow for cylinder scavenging, the modern high-speed multi-cylinder engine lent itself to a similar approach. He shows the basic kinetic energy theory of scavenging – before superimposition of sonic wave effects – as a pressure 'slug' acting as shown in Fig. 3.21, the gas column moving at an average speed of 60–90 m/s. The superimposed waves – moving at about 400–500 m/s – arise from the sudden pressure impulse brought about by the particular combination of valve opening rate and timing, in association with the cylinder pressure. There was also evidence to suggest that the high-velocity head of the gas as it enters the pipe is partially converted into a static head in order to accelerate the residual gas.

Gas dynamics theory

It is probably in the field of gas dynamics – rather than any other theoretical aspect of the internal combustion engine – that a concerted study will lead to a thorough understanding of existing combustion processes and to a logical development of others. The further development of the turbocharged diesel engine is a good example, where the need for a proper understanding of air movement through the cylinders is necessary to achieve turbine/compressor matching for a range of load conditions. A. B. Cambel and B. H. Jennings in their book *Gas Dynamics*

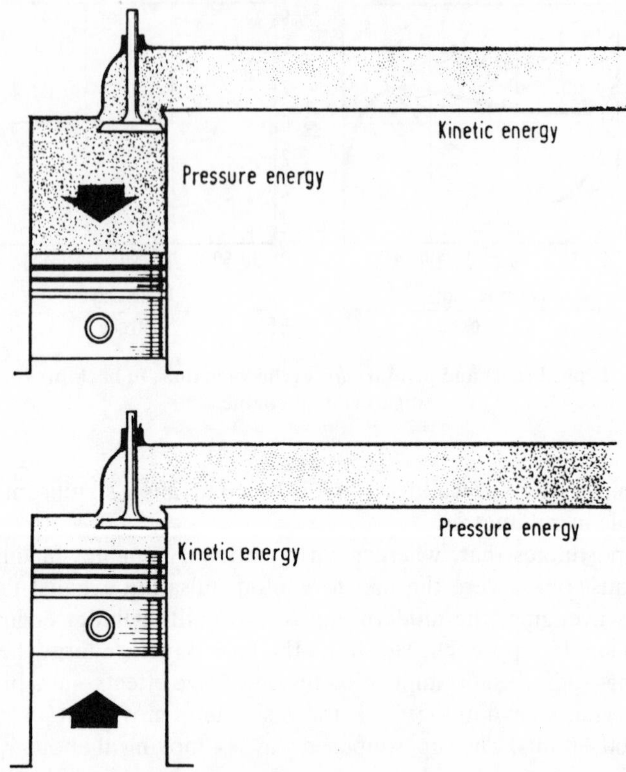

Fig. 3.21 Pressure slug assumed in kinetic-energy scavenging theory

(McGraw-Hill), have pointed out that the science of gas dynamics – which is receiving continuously increasing recognition – is of particular relevance to the analysis of engines, since it deals with compressible fluids, and includes thermodynamic, aerodynamic, and thermochemical considerations, in addition to the classic fluid-dynamic ones. The importance of the fundamental density and pressure relations with temperature is emphasized, namely

$$\left(\frac{P}{P_0}\right)^{(n-1)/n} = \frac{T}{T_0} \tag{3.1}$$

and

$$\varrho = \varrho_0\{1 - \alpha_{\mathrm{m}}(T - T_0)\} \tag{3.2}$$

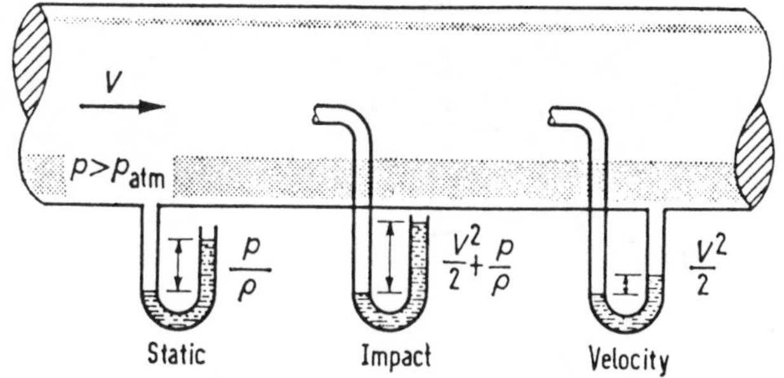

Fig. 3.22 Three different pressure measurement methods

where n = index of compression/expansion and α_m = mean coefficient of thermal expansion. In referring to pressures, the authors point out the necessity to differentiate between the three types of pressure as measured by the methods shown in Fig. 3.22. They also underline the premise that at high gas velocities (120 m/s for air) the pressure change associated with the generation of velocity is high enough to vary the density and, therefore, the onset of compressible flow and the less appropriate application of equation (3.2) above.

One-dimensional steady flow

The analysis is explained to be generally based on satisfying the four equations of mass, momentum, energy, and fluid state. Since, in accordance with the law of conservation of matter, ϱVA = constant, this can be logarithmically differentiated (that is, differentiated and divided by the original expression), and combined with a similar form of the gas law

$$(P/\varrho = RT)$$

to give

$$\frac{dP}{P} + \frac{dV}{V} + \frac{dA}{A} - \frac{dT}{T} = 0 \qquad (3.3)$$

The momentum equation was derived by consideration of a typical channel (Fig. 3.23(a)) whereby

$$P_1A_1 - P_2A_2 + F_T = \varrho_2 A_2 V_2^2 - \varrho_1 A_1 V_1^2$$

relating pressure, P, area, A, mass flow, F, density, ϱ, and volume, V.

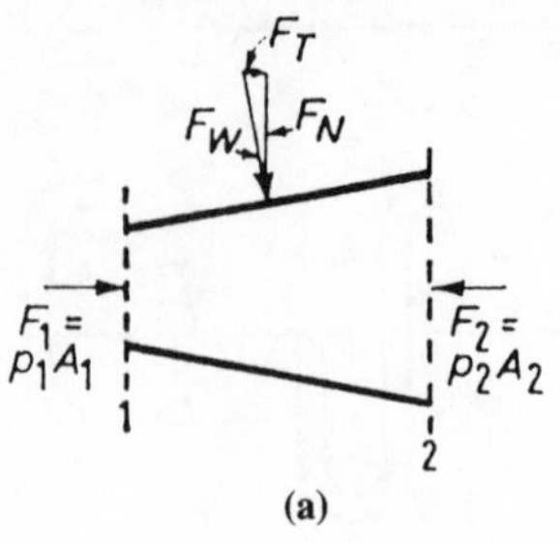

(a)

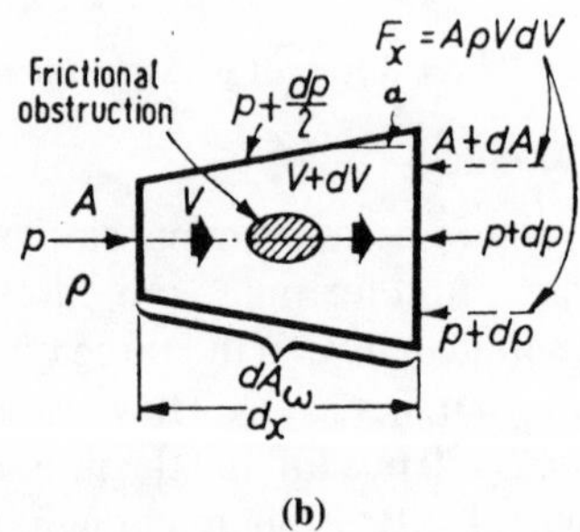

(b)

Fig. 3.23 Nomenclature for deriving momentum equation in a channel, with and without wall friction

Introducing wall friction, F_f, and an obstruction, F_d (Fig. 3.23(b)), mass flow after distance dx

$$F_x = PA - (P + dP)(A + dA) + (P + dP/2)\, dA + dF_1 + dF_d$$
$$= -A\, dp + dF_f + dF_d$$

and

$$= \text{volume} \times \text{density} \times \text{acceleration}$$

$$= (A + dA/2)\, dx \cdot \varrho \cdot \frac{dV}{dx}$$

(time having no influence in steady flow), so $F_x = A\varrho V\, dV$ and therefore

$$dP + \varrho V\, dV - \left(\frac{dF_f}{A} + \frac{dF_d}{A}\right) = 0 \qquad (3.4)$$

Without the bracketed friction terms this is Euler's equation which, on integration between appropriate limits, gives Bernoulli's equation. So by substituting the isentropic pressure–density relation, P/ϱ = constant in differentiated form, then

$$\frac{V_1^2}{2} + \left(\frac{\gamma}{\gamma - 1}\right)\frac{P_1}{\varrho_1} = \frac{V_2^2}{2} + \left(\frac{\gamma}{\gamma - 1}\right)\frac{P_2}{\gamma_2} \tag{3.5}$$

This relates the pressure and kinetic energy for steady flow.

Wave propagation
Since the acoustic velocity can be expressed as

$$a = \sqrt{\frac{\mathrm{d}P}{\mathrm{d}\varrho}}$$

and (3.5) gives $\mathrm{d}P/\mathrm{d}\varrho$ after logarithmic differentiation so $a = \sqrt{(\gamma RT)}$, the speed with which a small disturbance propagates under isentropic conditions. To find whether a pressure wave can move faster than the acoustic velocity, the authors considered a simple system, as in Fig. 3.24.

As the disturbance moving at velocity d'^2 passes any point in the pipe, the fluid there acquires velocity V_2 and pressure P_2, so the wave is a discontinuity over which mass, energy, and momentum equations can be applied, as follows

$$a'\varrho_1 = (a' - V_2)\varrho_2$$
$$P_1 + \varrho_1 a'^2 = P_2 + \varrho_2(a' - V_2)^2$$

and

$$\frac{\gamma}{\gamma - 1}\left(\frac{P_2}{\varrho_2} - \frac{P_1}{\varrho_1}\right) = a'^2 - \frac{(a' - V_2)^2}{2}$$

whereas initially

$$a_1 = \sqrt{\left(\frac{\gamma P_1}{\gamma_1}\right)}$$

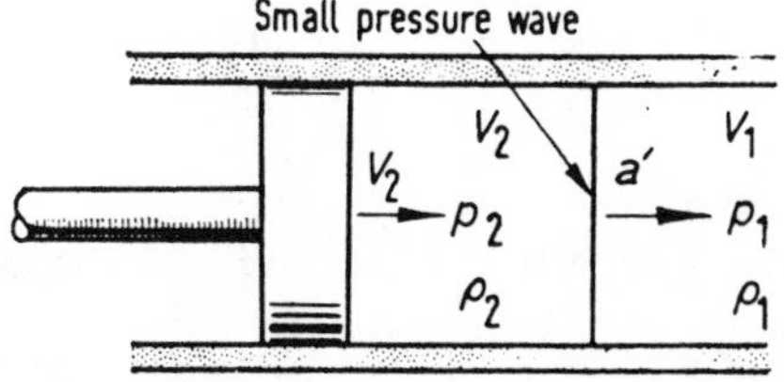

Fig. 3.24 Nomenclature for deriving wave-propagation equation

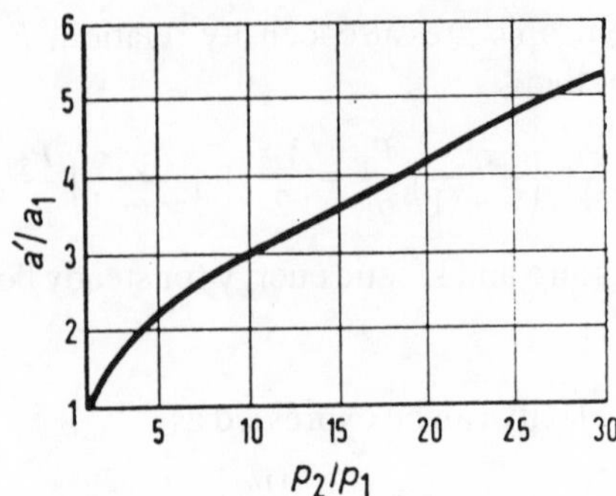

Fig. 3.25 Relationship from combined mass, energy, and momentum equations

Combining these

$$\frac{a'}{a_1} = \sqrt{\left[\frac{1}{2\gamma}\left\{\gamma - 1 + (\gamma + 1)\frac{P_2}{P_1}\right\}\right]} \qquad (3.6)$$

as plotted in Fig. 3.25. So for $P_2/P_1 > 1$ the pressure wave velocity is supersonic (and no longer isentropic, the disturbance becoming a steep-fronted shock wave having a different propagation velocity from the normal non-steep wave), the acoustic velocity being the lowest at which the wave will travel.

Velocity and temperature effect

The Mach number, M – defined as the ratio of local flow velocity to acoustic velocity – is valuable in studying compressibility phenomena in which temperature and velocity are involved

$$M = \frac{u}{a} = \frac{u}{\sqrt{(\gamma RT)}}$$

Squaring and differentiating logarithmically gives

$$\frac{\mathrm{d}M}{M} = \frac{\mathrm{d}u}{u} - \frac{1}{2}\cdot\frac{\mathrm{d}T}{T}$$

for a free stream temperature, T.

In a generalized flow system as depicted in Fig. 3.26 the enthalpy of the gas above a given datum can be written

$$h - h_0 = C_P(T - T_0) = \left(\frac{\gamma}{\gamma - 1}\right)R(T - T_0)$$

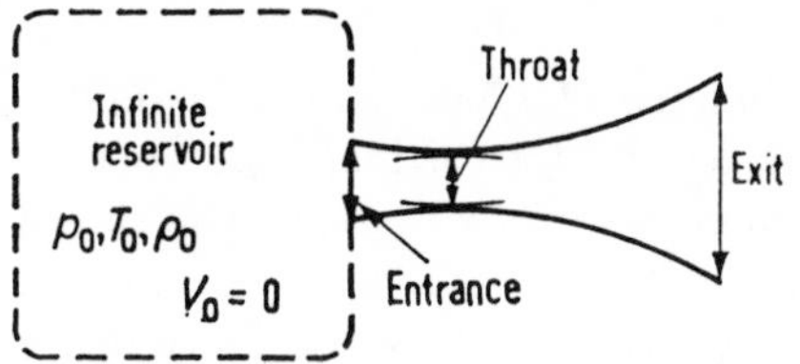

Fig. 3.26 Generalized flow system for deriving enthalpy equation

and differentiated into

$$\mathrm{d}h = \left(\frac{\gamma}{\gamma - 1}\right)R \cdot \mathrm{d}T = -\frac{P\,\mathrm{d}\varrho}{\varrho^2} + \frac{1}{\varrho}\,\mathrm{d}P + \mathrm{d}U \quad (U = \text{internal energy})$$

On combining this with the differential form of the Bernouilli equation

$$\left(\frac{\gamma}{\gamma - 1}\right)R \cdot \mathrm{d}T + V\,\mathrm{d}V = 0 \tag{3.7}$$

Integrating between reservoir conditions and any point in the flow

$$\left(\frac{\gamma}{\gamma - 1}\right)RT + \frac{u^2}{2} = \left(\frac{\gamma}{\gamma - 1}\right)RT_0 \tag{3.8}$$

or

$$\frac{a^2}{\gamma - 1} + \frac{u^2}{2} = \frac{a^2}{\gamma - 1} \tag{3.9}$$

(u being the flow velocity existing in the system). This implies that the acoustic velocity under the reservoir conditions is greater than the local acoustic velocity at any point in the flow. Consider subsonic flow with gradual increase in u until the acoustic value is reached at a certain location. When $u = a$, at the critical condition, then from equation (3.9)

$$a_{\text{crit}}^2 = \frac{2}{\gamma + 1} \cdot a_0^2$$

Introducing the Mach number to equation (3.8) and simplifying

$$\frac{T_0}{T} = 1 + \left(\frac{\gamma - 1}{2}\right)M^2 = \left(\frac{P_0}{P}\right)^{(\gamma-1)/\gamma}$$

thus the critical Mach number at the throat of the flow system is given by

$$M_{\text{crit}} = \sqrt{\left(\frac{2}{\gamma - 1}\right)}\sqrt{\left\{\left(\frac{P_0}{P_{\text{cr}}}\right)^{(\gamma-1)/\gamma} - 1\right\}}$$

Unsteady flows

In his book *Wave Diagrams for Non-steady Flow in Ducts* (Van Nostrand) G. Rudinger points out that the original studies of large-amplitude sound waves was undertaken in 1807 by Poisson, who showed that the velocity of wave propagation is the sum of the acoustic velocity and the disturbance (particle) velocity produced by the wave. Since this implied a change in shape of the wave form as the wave progressed, it complicated the analysis.

The establishment of wave equations for non-steady flow conditions leads back to an early work: *On the mathematical theory of sound* by Earnshaw (*Phil. Trans. Roy. Soc.*, 1860, **150**, 133–148). In considering wave motion accompanied by change in gas temperature, he assumed that the heat developed through a sudden density change in the passage of a pressure wave involved an isentropic state for which

$$\frac{P}{P_0} = \left(\frac{\varrho}{\varrho_0}\right)^{\gamma} \tag{3.10}$$

The basic dynamic equation

$$\left(\frac{\mathrm{d}y}{\mathrm{d}x}\right)^{\gamma+1} \cdot \frac{\mathrm{d}^2 y}{\mathrm{d}t^2} = \frac{\gamma P}{\varrho} \cdot \frac{\mathrm{d}^2 y}{\mathrm{d}x^2}$$

was considered to apply for the system. Integration and rearrangement of the subsequent equation gave the particle velocity

$$u = \frac{\mathrm{d}y}{\mathrm{d}t} = X \pm \frac{2\sqrt{(\gamma P/\varrho)}}{\gamma - 1}\left(\frac{\varrho}{\varrho_0}\right)^{(\gamma-1)/2}$$

where X is evaluated.

$$u = 0 \quad \text{at } \varrho = \varrho_0 \quad \text{gave } X = \pm\frac{2\sqrt{(\gamma P/\varrho)}}{\gamma - 1}$$

therefore

$$\left(\frac{\varrho}{\varrho_0}\right)^{(\gamma-1)/\gamma} = 1 \pm \frac{(\gamma - 1)u}{2\sqrt{(\gamma P/\varrho)}}$$

which, on substituting $a_0 = \sqrt{(\gamma P/\varrho)}$ in equation (3.10), gave

$$u = a_0\{(P/P_0)^{(\gamma-1)/2\gamma} - 1\}2/(\gamma - 1)$$

This fundamental expression can be used in the prediction of exhaust pipe length for optimum performance, as described later in this chapter.

AIRFLOW, NOISE, AND VIBRATION PREDICTION

Work reported recently on predictive techniques[16] has shown how the dynamics of the moving gases and structural displacements on the engine can be estimated.

Inter-cylinder air mass flow rate distribution

I. C. Findlay *et al.* from the National Engineering Laboratory and co-worker J. Orrin from Ford described a flow rate predictive analysis at the 1985 SAE Congress, alongside an experimental propane injection technique used to validate the prediction. The authors used the NEL engine simulation computer program which calculates gas dynamic and thermodynamic events occurring in ducts and cylinders during an engine cycle by solving unsteady gas flow equations.

An appendix gave the equations governing one-dimensional, unsteady, compressible flow of a perfect gas in a duct with varying area, heat transfer, and friction as in Table 3.6.

After rearrangement these equations can be transferred into a system of ordinary differential equations and a solution can then be obtained using the method of characteristics. In this analysis changes in entropy due to friction or heat transfer have been neglected and the flow is therefore treated as being homentropic.

In conjunction with the gas dynamics calculation in each pipe it is necessary to consider flow conditions at the various pipe boundaries in the system. At a pipe entry or exit to atmosphere and at the inlet and exhaust valves, the boundary is modelled as a nozzle between the pipe and a volume is which stagnation conditions prevail. The mass flow rate is calculated using the appropriate nozzle flow equations and assuming quasi-steady conditions.

The carburettor is also represented by a nozzle of suitable effective area at the junction of two pipes and the nozzle flow equations are used to calculate the mass flow. Because the engine simulation program is homentropic it is not possible to allow for entropy changes across the carburettor directly and so correct downstream reference pressure and temperature values must be used.

Flow at the manifold pipe junction is modelled using the following relationships: (a) conservation of mass across the junction, (b) pressures in the pipes in which flow is towards the junction are equal, (c) as flow is deflected when passing through the junction the pressure loss is given by

Table 3.6 Equations for unsteady flow

Momentum

$$\frac{\partial u}{\partial t} + u\frac{\partial u}{\partial x} + \frac{1}{\varrho}\frac{\partial p}{\partial x} + \frac{4f}{D}\frac{u^2}{2}\frac{u}{|u|} = 0$$

Energy

$$q\varrho A\,dx = \frac{\partial}{\partial t}\left\{\varrho A\,dx\left(C_v T + \frac{u^2}{2}\right)\right\} + \frac{\partial}{\partial x}\left\{\varrho A u\left(C_v T + \frac{p}{\varrho} + \frac{u^2}{2}\right)\right\}dx$$

Continuity

$$\frac{\partial \varrho}{\partial t} + u\frac{\partial \varrho}{\partial x} + \varrho\frac{\partial u}{\partial x} + \frac{\varrho u}{A}\frac{\partial A}{\partial x} = 0$$

Notation for mass flow analysis

A	Cross-sectional area
C_l	Loss coefficient
C_v	Specific heat at constant volume
D	Pipe diameter
f	Friction factor
p	Pressure
q	Heat-transfer rate per unit mass
R_{nm}	Air/fuel ratio, n, designates the cylinder number and m the test line point
T	Temperature
t	Time
u	Particle velocity
W_{ai}	Mass flow rate of test gas in the ith line
x	Displacement
ϱ	Density

$\Delta p = C_1(\varrho u^2)$ (downstream) where C_1 is an empirical loss coefficient based on the angle which the flow has turned.

Throughout the cycle conditions in the cylinder are defined by the first law of thermodynamics, a progressive heat release model being used to simulate the combustion process. Input data for the test engine simulation included lengths and diameters of all the engine pipes (Fig. 3.27), junction details, cylinder bore, piston stroke, connecting rod length, valve timing, and, in addition, effective valve area values derived from steady flow test results. Also supplied were data for the combustion model and the intake and exhaust reference conditions. The restrictor plates used in the later experimental tests were modelled as orifices of suitable effective area.

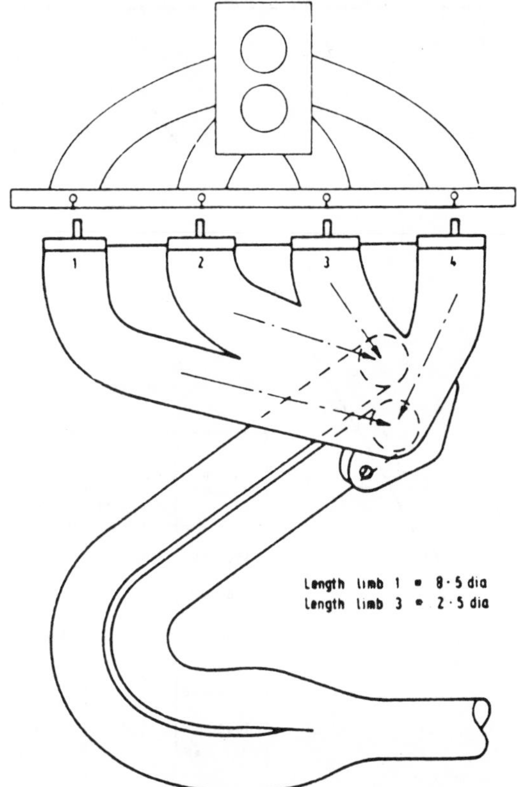

Fig. 3.27 Dimensional data for ducts

The computer prediction of the standard engine air consumption is shown in Fig. 3.28 where the difference between measured and predicted values is on average about 7 per cent. Although this discrepancy is not negligible in absolute terms, it should not impair the program's ability to indicate the changes in the relative air distribution between cylinders, the authors maintain.

Improved branched pipe model

J. F. Bingham from the NEL and G. P. Blair of Queen's University, Belfast have proposed a computer model for a multi-cylinder engine branch pipe, allowing the calculation of unsteady gas flow in engine

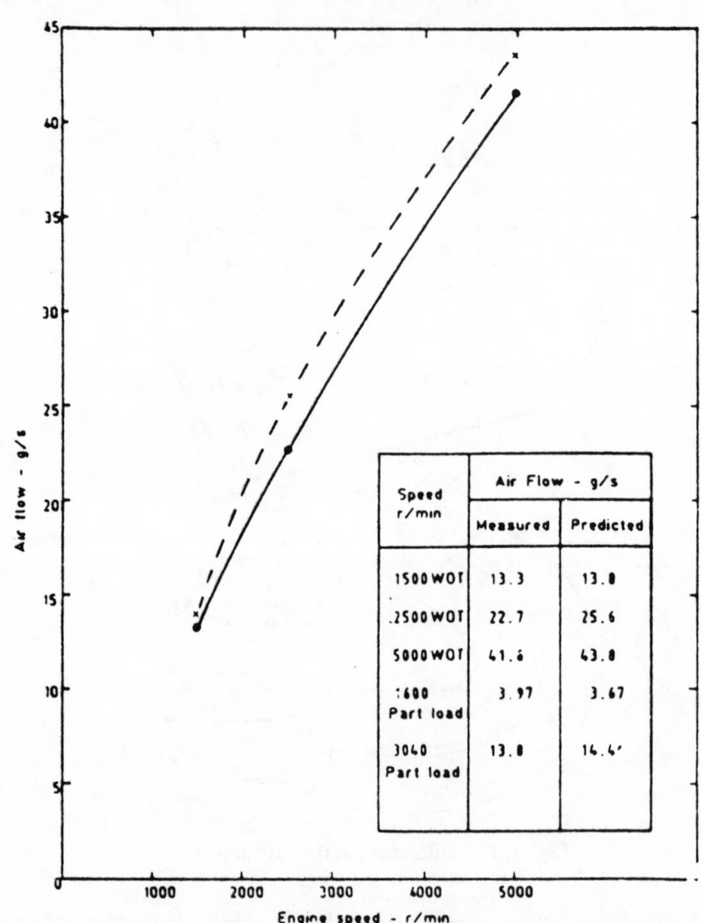

Speed r/min	Air Flow - g/s	
	Measured	Predicted
1500 WOT	13.3	13.8
2500 WOT	22.7	25.6
5000 WOT	41.6	43.8
1600 Part load	3.97	3.67
3040 Part load	13.8	14.4

Fig. 3.28 Measured and predicted air flows compared

manifold junctions. Allowing for pressure losses at the junction, the model is arranged for use with gas-dynamics programs based on the method of characteristics. Empirical coefficients obtained from flow tests are suggested for use with existing momentum theory models, then a generalized approach is suggested based on the model of Fig. 3.29, the solution of the flow continuity equation across the junction and the calculation of an instantaneous pressure-loss term at the junction. Iden-

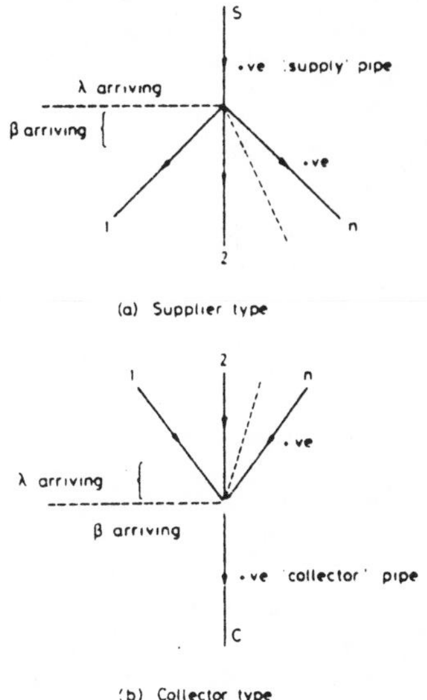

Fig. 3.29 Junction classification for branched pipe model

tification of the flow path is crucial, the authors explain, and junctions in the manifolds are classified as follows.

Supplier type. In this case flow from one 'supply' pipe may flow at different times into any one of a number of downstream branch pipes. In general most inlet manifold junctions are supplier type.

Collector type. In this case flow from a number of upstream branch pipes converges into one 'collector' pipe. Most exhaust manifold junctions are collector type.

The designation of a junction as supplier or collector type is determined by the most likely flow pattern, but the method does not fail if the current flow regime differs from the expected pattern. The sign convention shown in Fig. 3.29 for the two types is required for correct execution of the calculation procedure. A flow diagram of this procedure is shown in Fig. 3.30. Engine simulation computer programs based on the method of characteristics advance in discrete time steps and so the generalized junction model has been arranged to ensure compatibility with this scheme.

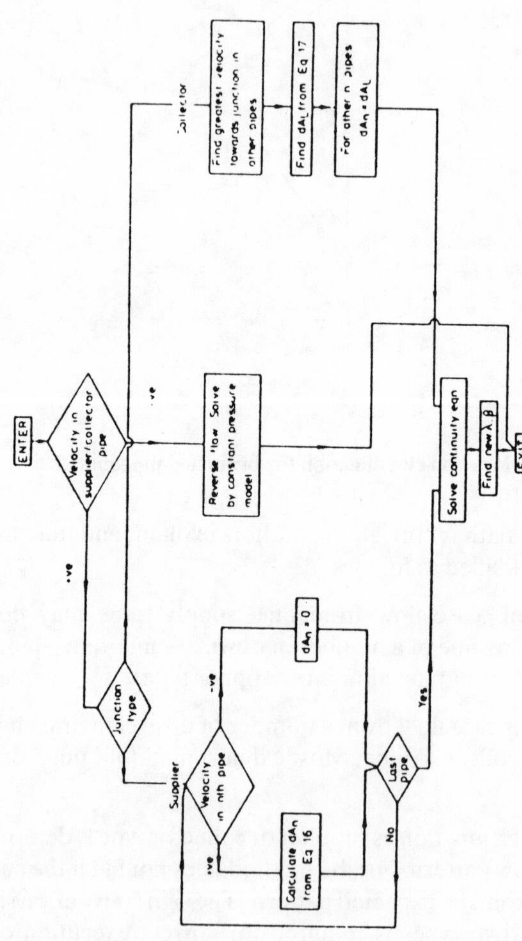

Fig. 3.30 Flow diagram for generalized junction model

The theory is fully developed in the paper and engine test work described which verified the predictions. The generalized method is concluded to be as accurate as existing momentum methods but without the limitations of the latter.

Engine structural dynamics

T. Priede *et al.* of Southampton University ISVR described experimental techniques used to verify predictive methods, in the dynamics area, at the recent Institution of Mechanical Engineers Automobile Division Conference on Noise and Vibration. They point out that the success of methods such as FEM modelling depend upon an understanding of the noise-generating mechanisms in a running engine. Their work relates to simulation of gas-loadings, piston-slap, bearing impact, and inpulsive loading of the engine structure; the importance of defining vibration energy paths through the structure and the transmission of energy at the bearing surfaces is emphasized.

Both static and dynamic combustion simulation is undertaken, the former involving pressurizing the combustion chamber of a test engine to around 8 MN/m^2, hydraulically, and slow relaxation of the structure measured as the pressure is dissipated. The authors say reasonable correlation is obtained with more sophisticated dynamic stiffness tests as well as usefully validating static FEM models. Problems of assessing non-linearities could cause a limitation, and a so-called 'banger' pulse rig can be used for closer dynamic simulation, involving igniting a compressed mixture for one stroke only, giving a valuable method of assessing combustion attenuation characteristics. Eighteen engines of cylinder capacity 0.35–0.6 litres/cylinder, have been tested. By measuring the average surface vibration velocity and then normalizing the results with their respective cylinder pressure spectra, the average mobilities of the various engines were obtained. Figure 3.31 shows the envelope of the results between 500 and 4000 Hz. The range is surprisingly large, and if the levels are corrected for typical running engine combustion spectra, the overall 'A' weighted range would be around 13 dB. With the exception of the Vee type configuration, the general spectral shape was very similar for all the engines tested. Over most of the frequency range considered, the mobility of the structures follow lines of constant effective stiffness – 20 dB/decade of increasing frequency. Between 3000 and 5000 Hz the spectra slopes flatten and in some cases drop. The range in effective stiffness from the weakest to the stiffest engine is a factor of four.

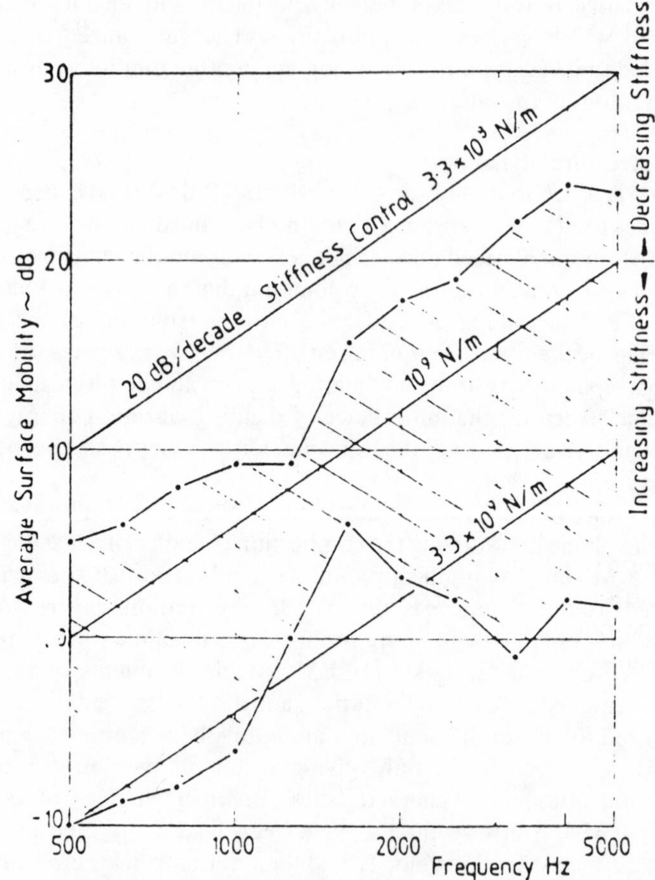

Fig. 3.31 Cross mobility envelope for 18 engines

Vee type engines show a steeper spectral slope, implying decreasing stiffness with increasing frequency. As the banger pulse can be considered as a single event excitation, the rig lends itself to the study of structural wave propagation. By studying the vibration time histories at points over the engine surface, with reference to the common time datum of combustion initiation, a measure of wave propagation velocity can be estimated.

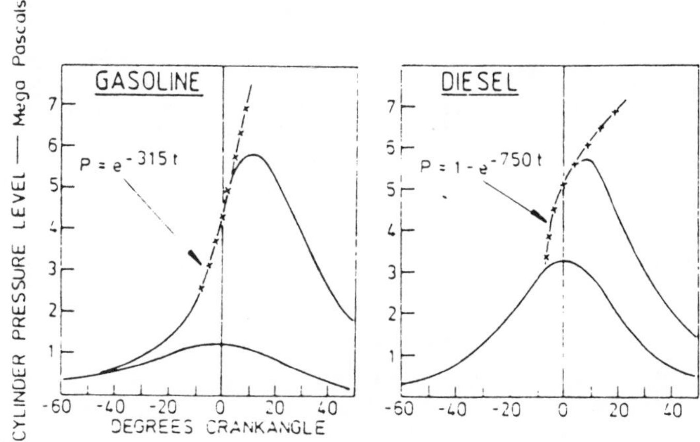

Fig. 3.32 Combustion pressure rise comparison

As previous research has shown a rate of combustion pressure rise to be of the form shown in Fig. 3.32, an hydraulic rig has been built to simulate these and is compared with results from a running engine in Fig. 3.33. An hydraulic ram, described by the authors, was adapted and used for investigation of propagation of vibration. It can be applied between any two points of the engine structure. The design features and hydraulic control systems of the impulsive ram are similar to those for the hydraulic combustion simulation rig. In tests during which the force is applied in the axial direction between the two bearing caps of an engine of 2.1 litres/cylinder, a propagation velocity in both directions of around 0.5×10^3 m/s was about 1/15 of the velocity measured on the banger rig experiments. These two conflicting results imply that the speed of propagation of noise producing energy through a given engine structure is dependent upon how and where that energy is put into the structure.

Although further investigation has as yet not been carried out in any great depth, the results have clearly demonstrated two basic mechanisms of wave propagation present in an engine structure. For future low-noise engine design, it is important that these two mechanisms of propagation be better understood, for although there is interaction between these two forms of propagation, their general control may well be best achieved by separate methods.

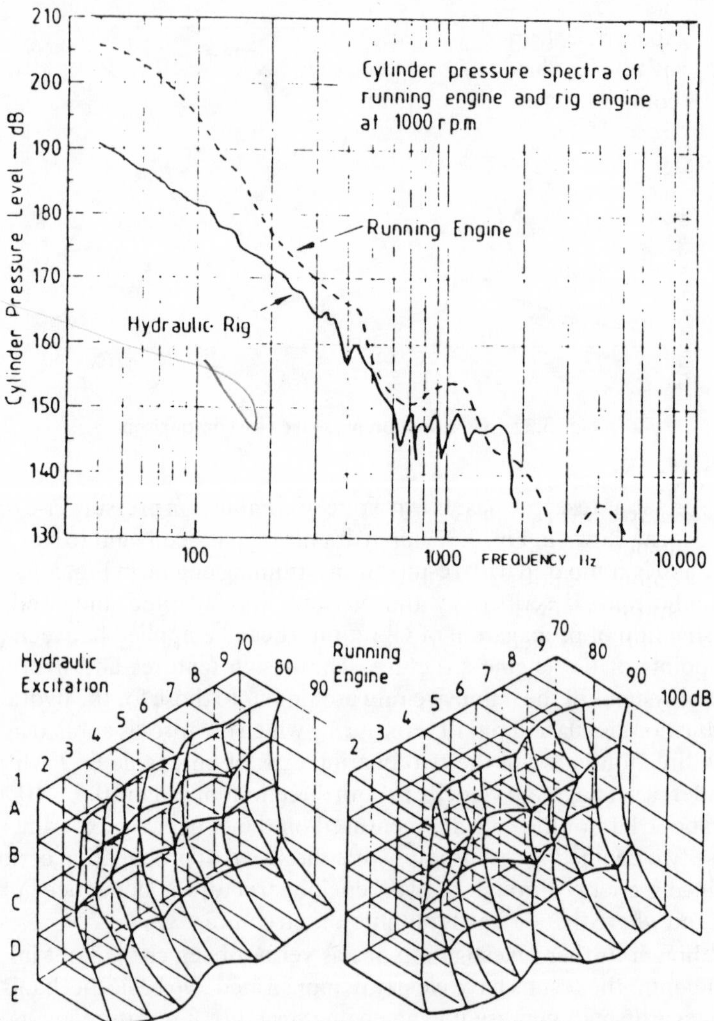

Fig. 3.33 Cylinder pressure spectra and block-shape comparisons

The authors conclude that although previous researchers have demonstrated that engine noise may be successfully predicted by following a known input force through a series of measured component mobilities, to a relevant noise radiator, the connecting characteristics between these various components and the actual routes taken by the vibration energy are far from predictable. Empirical information needed to define the model is still lacking and the authors' work has shown that the inability to simulate the oil film characteristics is a major weakness of many of the rig tests and its likely modification of results must always be considered.

One of the major force paths into the engine structure is via the crankshaft and main bearings. The way in which the main bearings are excited is very dependent upon the stiffness of the crankshaft, and the way in which the structure is driven by the main bearings varies greatly.

Over the important noise-producing frequency region of 500–3000 Hz, the dynamic response of an engine structure may be approximated to a stiffness-controlled one-degree-of-freedom system. Although this is a gross over-simplification, the average mobility measured during the described dynamic rig tests generally follow constant stiffness lines over this frequency range.

The combustion stiffness is obviously very similar to that of the vertical bearing stiffness; however, the horizontal bearing stiffness can be as little as 1/10th of that of the vertical stiffness. Piston-slap stiffness is far lower than combustion stiffness, again as much as a factor of 10 less. Therefore structural modifications possibly hold greater potential for the reduction of piston-slap noise than they do for the reduction of combustion noise.

ENGINE STRUCTURAL NOISE

While engine vibration usually originates in the crankshaft assembly, noise generally comes from piston and valve gear, plus the resonance of engine panels excited by the crankshaft assembly. Besides the means described earlier to achieve best balance of the assembly, other considerations are the overhang of the torsional vibration damper (fitted to smooth out irregular gas loadings) and that of the even heavier flywheel and clutch assembly. Crankcases are usually stiff enough in the plane of the cylinders but may be five times less stiff in a perpendicular plane. Deep skirt designs and sidewall ribbing can alleviate the problem.

The main noise source of the valve gear is most probably the camshaft. It is hit a series of irregularly spaced blows on every lift and drop, yet slender camshafts are commonly provided with bearings only on each side of every two cylinders, or every four cams. Experience has shown that overhead camshafts with bearings straddling every cam, and light valve gear, do not get noisy at high speeds. In a pushrod engine having an unfavourable rocker-arm ratio, the equivalent mass at the cam is equal to the tappet and pushrod plus two valves. With pushrods, too, insufficient ramp on the closing side of the cam to allow for the collapse of the valve gear under the closing load, permits the valve to hit the seat at much more than the designed velocity.

A rocker shaft and its brackets are subject to the sum of the pushrod load and that required to accelerate the valve. For high-speed operation they should be designed on a comparative deflection basis. As a criterion of valve-gear rigidity, the deflections should be small compared with the ramps of quietening curves. They can be measured by applying double the maximum high-speed inertia load to the valve head, with the valve gear in compression. Valve springs with many coils and a low rate, surge and clash at high speeds, and do not remain as quiet as those with only three or four working coils. Rocker-gear covers should be attached directly to the cylinder head rather than to the valve gear or rocker-shaft brackets, to avoid noise amplification. For minimal noise these covers should be made of non-metallic and non-resonating material. Also, integral panels enclosing valve gear should be ribbed or have a compound curvature rather than flat surfaces.

This advice is based on the experience of the contributor to Chapter One.

Lessons from truck noise-reduction programmes

The editor[22] considers that there is much for designers to learn from the UK-government sponsored Quiet HGV programme, since the systems approach to the whole vehicle led to an understanding of which are the best routes to noise reduction.

The 238 kW Rolls-Royce engined 38-ton Foden S83 tractor was demonstrated at the Transport and Road Research Laboratory to have a drive-by noise level of 81 dB(A). In a paper presented to the Institution of Mechanical Engineers on 'The TRRL quiet heavy vehicle project', J. W. Tyler's Table 3 shows the allocation of work to the contracting research organizations and the target sound pressure levels (Table 3.7). The general design philosophy was to reduce sound emission from the engine by means of structural changes and to provide a structural enclosure around the engine and transmission to isolate airborne noise

Table 3.7 Upper limits of sound levels to be emitted by 'source location' and the research organizations involved in work on each one

Research organization	*Source*	*Maximum level dB(A)* at 1 m	at 7.5 m
ISVR	Engine including gearbox	92	77
	Air intake, exhaust system (computer modelling)	84	69
MIRA*	Cab noise	75 dB(A)	
	Cooling system	84	69
	Exhaust system (development of practical systems)	84	69
NEL	Final cooling fan design		(Additional target of 90 dB(C))
TRRL	Tyre road surface noise	—	75–77

*MIRA were to design the vehicle cab to provide 3 dB(A) attenuation of the sound level emitted to the roadside by the power unit.

transmitted from the engine structure and its ancillaries. Thirdly, the cooling system is totally ducted and employs a mixed-flow fan and, finally, the exhaust system was redesigned.

Figure 3.34 shows the reduction in sound pressure level of the ISVR

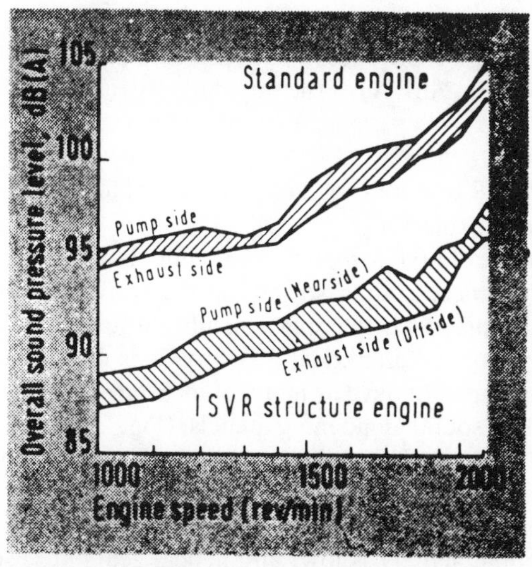

Fig. 3.34 Noise output of ISVR and standard QHV engines compared

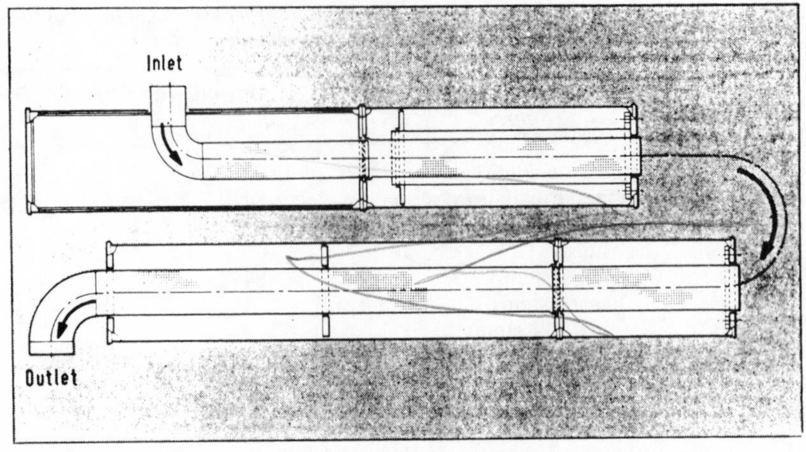

Fig. 3.35 Silencer configuration for the QHV

research engine, which was forerunner to the existing unit used in the QHV, compared with the levels applying to the vehicle prototype.

Figure 3.35 is a drawing of the MIRA ISVR designed silencer assembly built by T.I. Cheswick Silencers Ltd. Final design of the silencer was entirely reactive and consisted of two cylindrical boxes with a total volume of 186 litres. It produces a back pressure of 39 mm of mercury and a noise level within 2 dB(A) of the target of 69 dB(A) at 7.5 m.

The QHV engine

The original ISVR research engine was modified to use gear drives for all accessories and auxiliaries.

Essentially the 13.05 litre unit (135 mm bore × 152 mm stroke) embodies a two-piece crankcase split on the crankshaft axis, the lower 'bed plate' structure of which imparts significant stiffness to the assembly. The upper part of the crankcase is of dry-liner design, and the whole assembly is arranged to have as many of its external surfaces as possible covered by flat sound deadening panels (Fig. 3.36) attached to the crankcase in regions of low vibration levels. These comprise neoprene and perforated steel sandwich panels.

The gear train is situated at the rear of the engine, where the nodal point of torsional activity is low, and helical gears are used. A pair of

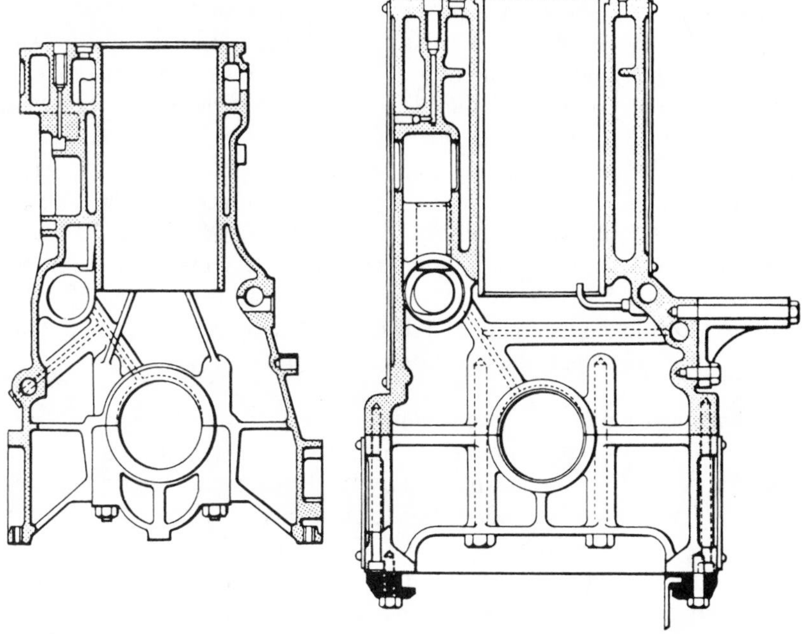

Fig. 3.36 Engine structures before and after ISVR modification

cylinder heads is fitted with four valves per cylinder and integral rocker-box pedestals. The rocker-box lid is a two-piece design having a rubber mounted centre portion aimed at reducing the amount of noise radiated from the valve gear. The sump casing is isolated from the crankcase skirt by suspending it from an extruded rubber moulding trapped against the crankcase by a retaining plate.

In modifying the engine the philosophy was to retain as much as possible of the original running gear, in a strengthened structure. Besides the bedplate design mentioned earlier, the bottom deck of the cylinder block was moved down to the lower end of the cylinder, forming a rigid section to reduce wall flexing caused by crankcase distortion. Other crankcase structure changes were aimed at attaching close-fitting damping panels, including a single flat panel to cover the offside of the structure and arrangement of the ancillaries on the nearside. Panels were also fitted along the sides of the bedplate and oil galleries grouped on the nearside.

Computer predictions suggested that 2 mm offset of the gudgeon pin to the thrust side would reduce impact severity at full load and this was incorporated. Removal of the timing gear train to the rear of the engine involved the use of a compound drive. Pressure angle of the gears was reduced from 25 degrees to 20 degrees, and the teeth were also helically cut to reduce noise. Laminated, damped panels fitted to either side of the cylinder block and bedplate consisted of a central sheet of steel perforate and outer plain mild steel sheets, sandwiched together with neoprene.

Contributions to overall noise reduction from the constituent elements amounted to 2.5 dB(A) for moving timing gear from front to rear, 3 dB(A) for the damped panels, 4 dB(A) for the block structural modification, and 2 dB(A) for the isolation of the covers.

General approach to engine noise control

In a paper (C14/77) presented to an Institution of Mechanical Engineers conference in 1977 it was explained by ISVR researchers that the spontaneous ignition of the fuel–air mixture in a diesel results in almost instantaneous cylinder pressure rise; this excites the higher modes of engine structure vibrations, resulting in the predominant high-frequency impulsive noise. Mechanically-induced noise generated by the crank mechanism is also impulsive and is subjectively indistinguishable from combustion-induced noise. Mechanical noise arises mainly from inertia forces which change direction and cause impact of pistons and bearings across their running clearances.

Extensive investigation of over 50 commercial engines at ISVR has established basic relationships between combustion-induced noise and the engine parameters of speed, size, and load. These are well documented in SAE paper 750795 by Priede and lead to the conclusion that the important noise-related parameters are speed and bore size. Recorded research into ignition process control and quieter combustion systems has shown that only limited noise reduction is possible. Although there are significant reductions in combustion-induced noise in turbocharging, for example, the high peak pressures involved increase mechanically-induced noise.

While there is considerable latitude in selecting design parameters for low noise, economic reasons reduce this latitude in practice, and the really significant area of noise control is, in the main, structural design. In-line engines consist of a box divided into compartments, the top of the box being closed by the comparatively stiff cylinder head and the bottom closed by a relatively flexible sump. Also at the top and about halfway down, horizontal decks support the cylinders, but their stiffness is lessened by the cylinder bores, and the whole structure is flexible in torsion about an axis parallel to the crankcase, as the sump does not effectively close the torque box. Overall bending stiffness is also much stiffer in the vertical than the horizontal plane. When the structure is excited over a range of frequencies, the vibration amplitude at a point will show resonant peaks at each of which the structure will take up a different mode shape. The first vibration mode, occurring at a few hundred cycles per second, is the torsional one, and at about 100 Hz lengthwise beaming takes place. Above this frequency the structure behaves as a solid body and panels forming the sides of the bays vibrate independently.

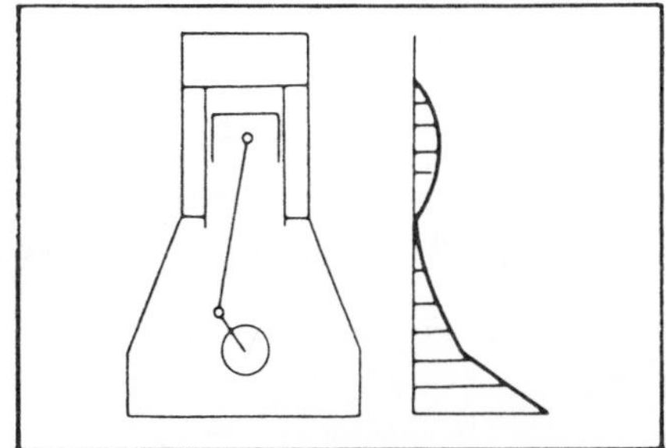

Fig. 3.37 Mode shape resulting from 'flapping' of crankcase walls

Investigation of a variety of structures has been carried out and recorded in a paper entitled 'A review of low noise diesel engine vibration at ISVR' (*J. Sound Vibration,* **28** (3), 403–431). The fundamental 'flapping' mode of a conventional engine crankcase skirt is shown in Fig. 3.37 and the main stages in structural design conceived at ISVR aimed at stiffening the crankcase are shown in Fig. 3.38. It is important to appreciate that structural vibration has referred effects on other parts of the engine, such as the oil pan, timing case, and valve cover, and a useful indication of the sources of noise radiation for an in-line turbocharged diesel of 10 litres capacity is shown in Fig. 3.39.

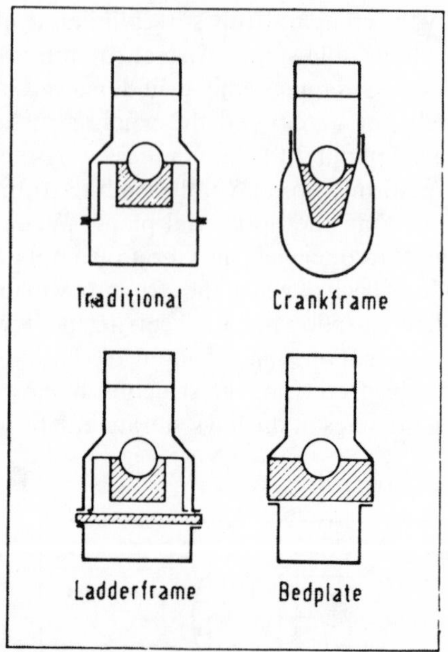

Fig. 3.38 Stages in crankcase design for noise reduction

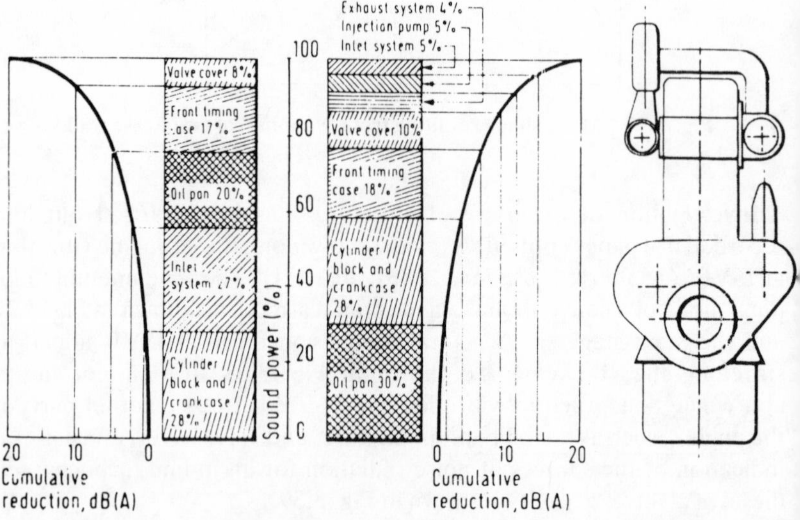

Fig. 3.39 Contributions to noise output from a typical 10-litre truck engine

Silencer development

In the exhaust silencer produced for the QHV shown in Fig. 3.35, the overall length was 4.27 m at 254 mm diameter. This compares with a standard production silencer for the equivalent vehicle of 1.08 m with a 254 × 330 mm oval section, says a recent report.[23]

According to T.I. Cheswick Ltd, the silencer maker normally receives a brief from a vehicle manufacturer governing the physical size, noise level required below the legal maximum, and back pressure limit. The last factor is also governed, in the case of trucks, by engines certified to BS Au 141a, which specifies maximum back pressure and inlet depression values beyond which unacceptable smoke emission could take place; this is at the 4.4 kW/t minimum installed power requirement for the engine in the vehicle.

Principal silencer materials are mild steel, aluminized steel, and stainless steel. Aluminized steel is generally used for rear silencers on cars and for internal components; on trucks and buses both cases and end plates are produced in aluminized steel.

The approach is one of allowing the exhaust gases to expand into chambers of large cross-sectional area. Recent research suggests that the energized column of gas, following an exhaust valve opening, with its noise (or shock wave) front creates the sound emitted at the tail pipe. Associated with this are 'gas noise characteristics' which need to be damped out or eliminated acoustically. Two different methods of approach are required for these different frequencies and types of noise.

Gases leaving the combustion chamber have a velocity of around 100 m/s, generating a shock wave velocity of up to 400 m/s. At these speeds there is an appreciable distance differential between the noise created by the valve opening and the slug of gas with which it was originally associated. Expansion of the relatively low frequency wave fronts, successively, into volumes adding up to one-third engine capacity is general practice.

'Roughness' following in the wake of the wave front is reduced by the use of louvred tubes within closed portions of the silencer, acting as Helmholtz resonators to attenuate the high frequencies.

Configuration

Shape of the expansion chambers is critical. A typical example is a present-day large car having a 14.75 litre single cavity chamber of large cross-section and overall length of 400 mm giving good low-frequency

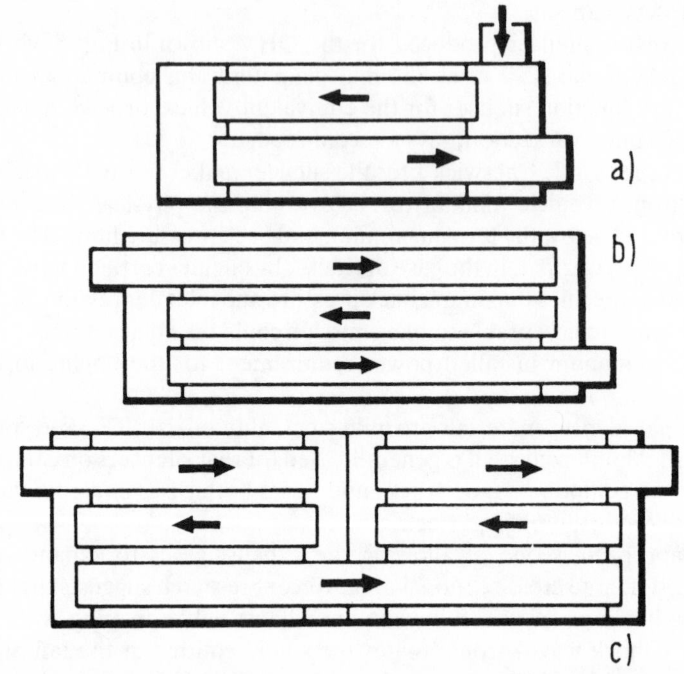

Fig. 3.40 Two-pass (a), three-pass (b), and double-three-pass (c) silencer systems

attenuation with offset pipe connections. Should the case diameter be reduced to 100 mm, the resulting 1830 mm length to achieve the same volume would result in negligible low frequency attenuation. In cars it is not usually possible to obtain all the volume in one place; hence the development of multi-box systems. The first is usually placed a quarter-wave length of the predominant frequency from the manifold. More often a large silencer, particularly for commercial vehicles, has several expansion chambers arranged so that there is a flow reversal between one and the next. The reversals cause out-of-phase pulsing at the same frequency with an accompanying attenuating effect. Figure 3.40 shows two-, three- and double-three-pass systems, indicating how a number of reversals can be achieved.

One peak frequency in the combustion exhaust spectrum is loosely termed 'boom' and is often dealt with by a separate Helmholtz resonator. Here a resonating chamber with a narrow neck or tube inserted can be calibrated so that the resonating frequency is out of phase with the predominant exhaust frequency. It has been shown that the noise response before and after the insertion of a short tuning tube, 50 mm long and 40 mm in diameter is a 10.5 dB reduction in boom peak around 145 Hz, as well as a general reduction in noise level. Given adequate resonator volume, the boom frequency can be completely removed.

The second problem of gas-induced noise is largely one exhibiting itself as 'pipe noise', case rattle, plate vibration, or 'overrun rasp'. The first three are usually overcome by such measures as laminated pipes, damped end-plates, increased number of bulkheads, and insulated case structures. The exact cause of rasp is uncertain, save that it is certainly affected by manifold design in respect of manifold outlet positioning in relation to exhaust ports. Without resource to manifold redesign, the normal approach is to provide a resonator within 1070 mm of the exhaust ports. An alternative approach is to pass the gas through an 'absorption chamber' prior to expansion for eliminating higher frequencies.

Use of absorption materials should be treated with reserve, say T.I. Cheswick, owing to their tendency to deteriorate in service. The process of attenuation involves absorbing the sound waves in mineral wool packing and converting them into heat energy. The problem is to ensure proper packing of the material, type approval regulations requiring a durability cycle of 10 000 km. Long-strand glass silk is the most popular materials, with basalt fibre as an alternative for high temperature applications.

GAS DYNAMICS APPLIED IN EXHAUST SYSTEM ANALYSIS

Exhaust pulses: single-cylinder engine case

Dr Pullman[24] has explained that approximately one-third of the energy supplied to an engine is dissipated in the exhaust gases. By utilizing the waves in the exhaust manifold, it is possible to improve the engine performance considerably. Analytical methods available enable the degree of improvement to be determined and, equally important, help the designer to avoid conditions of impaired performance. The object of this section is to outline the fundamentals of the exhaust gas discharge process and to indicate a method of calculating the exhaust pulse for a single cylinder/pipe system. Information is given on the reflection of this pulse and the effect of the wave reflections upon the subsequent engine cycles in terms of air throughput. Because of the complexity of the process, it is difficult to apply rigorous methods to the multi-cylinder engine case, even with the aid of a digital computer. Consequently, the reader is referred to a qualitative technique for the multi-cylinder case, and to experimental data confirming the influence of wave motion on engine performance. The methods described can be extended to cover evaluation of the exhaust gas energy available to a turbocharger, and the corresponding improvement in engine performance.

When an exhaust valve opens, the products of combustion discharge into the exhaust pipe and generate a positive wave which travels down the pipe. This problem has also been considered by earlier workers, who suggested that the valve/exhaust pipe configuration is equivalent to an orifice and sudden enlargement. It is further suggested that the flow process from the cylinder to the plane of minimum area (vena contracta) is isentropic – frictionless adiabatic. The flow on the downstream side of the valve is, however, non-steady and account of this must be taken by inserting the added parameter that the gas pressures and velocities are related according to the fundamental wave equations for finite-amplitude pressure waves

Particle velocity

$$u = \left(\frac{2}{\gamma - 1}\right) a_0 \left\{ \left(\frac{P}{P_0}\right)^{(\gamma-1)/2\gamma} - 1 \right\} \tag{3.11}$$

Propagation velocity

$$c = u + a$$

and since

$$a = \left(\frac{P}{P_0}\right)^{(\gamma-1)/2\gamma} a_0$$

$$c = a_0 \left[\left\{\frac{\gamma + 1}{\gamma - 1}\right\} \left\{ \left(\frac{P}{P_0}\right)^{(\gamma-1)/2\gamma} \right\} - \frac{2}{(\gamma - 1)} \right] \tag{3.12}$$

With these assumptions and the use of the basic thermodynamic, energy, momentum and continuity laws, it is possible to obtain the plot shown in Figs 3.41 and 3.42, giving the variation of pulse pressure with cylinder pressure for various area ratios. It will be appreciated that the flow in the critical region (vena contracta) can be either sonic or subsonic, that is, the particle velocity in the concentrated gas stream can be equal to the local acoustic velocity. To complete the solution to the problem, it is necessary to calculate the variation of cylinder pressure with time. The

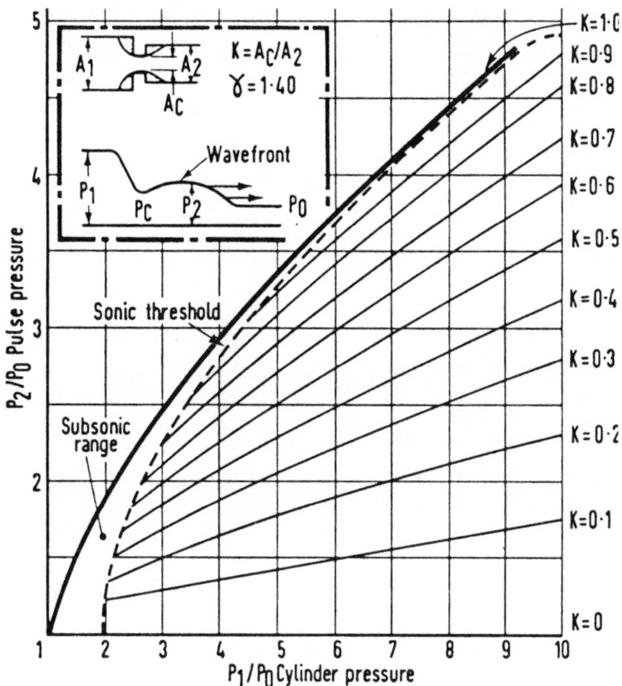

Fig. 3.41 Pressure wave chart: sonic discharge

rate of change of mass in the cylinder must be equal to the rate of mass flow in the vena contracta.

Sonic range

This occurs when the gas velocity at the vena contracta is equal to the local acoustic velocity and the critical pressure is known to be

$$\frac{P_1}{P_c} = \left(\frac{\gamma + 1}{2}\right)^{\gamma/(\gamma-1)} = 1.892; \qquad \text{for } \gamma = 1.40$$

If the pressure and density of the gas in the cylinder at the moment of release are P_R and ϱ_R, and at an instant later they are P_1 and ϱ_1, then the mass M of gas in the cylinder is given by

$$M = V_1\varrho_1 = V_1\varrho_R\left(\frac{P_1}{P_R}\right)^{1/\gamma} = V_1\varrho_R\left(\frac{P_1}{P'}\right)^{1/\gamma}\left(\frac{P'}{P_R}\right)^{1/\gamma} \qquad (3.13)$$

where P' is an arbitrary reference pressure, for example, 1 atmosphere. Writing pressure ratios with respect to reference pressure P' as $\overline{P}$ then

$$\frac{dM}{dt} = -\varrho_R \overline{P}_R^{-1/\gamma} \cdot \frac{d}{dt}(V_1 \overline{P}_1^{-1/\gamma}) \tag{3.14}$$

For the vena contracta of effective area KA_e, the rate of mass flow is

$$\frac{dM}{dt} = -\varrho_c (KA_e) u_c \tag{3.15}$$

In the case when the velocity u_c is sonic, this equation becomes

$$\frac{dM}{dt} = -KA_e a_1 \varrho_1 \left(\frac{2}{\gamma + 1}\right)^{(\gamma+1)/2(\gamma-1)} \tag{3.16}$$

It is mathematically convenient to introduce the term a_E, the velocity of sound at conditions corresponding to isentropic expansion from release conditions a_r and P_R to the reference pressure P'.

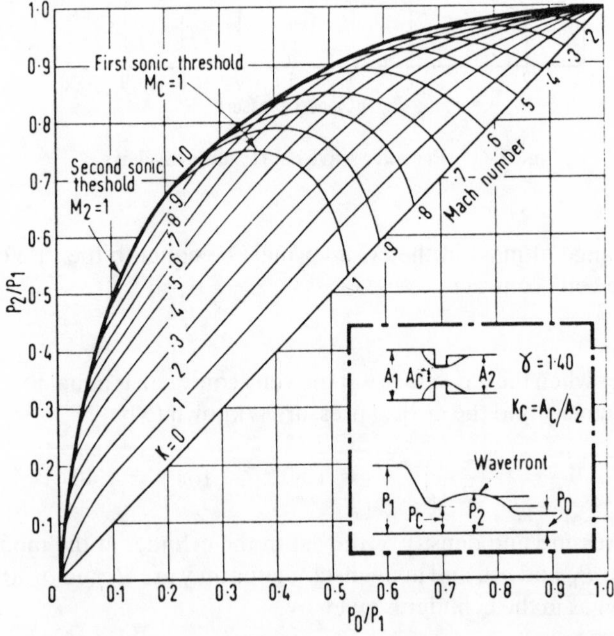

Fig. 3.42 Pressure wave chart: sonic and subsonic discharge

Equation (3.16) can be rewritten

$$\frac{\mathrm{d}M}{\mathrm{d}t} = -KA_e \varrho_R \overline{P}_R^{-1/\gamma} a_E \left(\frac{2}{\gamma + 1}\right)^{(\gamma+1)/2(\gamma-1)} \cdot \overline{P}_1^{(\gamma+1)/2\gamma} \qquad (3.17)$$

Combining equations (3.14) and (3.17), we get

$$\frac{\mathrm{d}}{\mathrm{d}t}(V_1 \overline{P}_1^{1/\gamma}) = -KA_e a_E \left(\frac{2}{\gamma + 1}\right)^{(\gamma+1)/2(\gamma-1)} \cdot \overline{P}_1^{(\gamma+1)/2\gamma} \qquad (3.18)$$

This equation can be simplified by conversion to finite difference form and by introducing the displacement volume, V_D, and the crank-angle, α, time, t, relationship $\mathrm{d}t = (1/6N)\,\mathrm{d}a$ where N is the speed in revolutions per minute. For the case where $\gamma = 1.4$, equation (3.18) becomes

$$\Delta\left(\frac{V_1}{V_D}\overline{P}_1^{5/7}\right) = \frac{-KA_e a_E}{NV_D} \cdot \frac{1}{1.728} \cdot \overline{P}_1^{6/7} \cdot \Delta\alpha \qquad (3.19)$$

Equation (3.19) can be solved by step-by-step methods to give P_1 as a function of crank angle. When this value is substituted for the appropriate value of K from Fig. 3.41, the corresponding pulse pressure, P_2, is obtained. The value of K is found by steady-flow calibration. Examples of such calculations have been published elsewhere, and Fig. 3.43 shows typical results. It will be appreciated that the sonic range is relatively easy to analyse, since downstream pressure variations do not affect the gas discharge. For the subsonic range, however, the downstream pressure is an essential parameter. The gas motion, and, hence, pressure in the exhaust pipe, consists of a train of waves and reflected waves which might arise in the following way. An initial pulse, calculated according to the method described above, travels along the exhaust pipe to the open end. Let us assume that this pulse is the first to be generated on starting the engine, and that the exhaust pipe contains air at rest and atmospheric conditions. The pulse shape changes somewhat during its passage, since the velocity of any wave point is dependent upon its pressure amplitude – equation (3.11). High-pressure wave points therefore tend to overtake preceding low-pressure wave points, thus altering the wave profile.

If an outgoing wave of pressure P_2 and associated particle velocity u_2 is superimposed upon an incoming wave of pressure P_r and velocity u_r, then the net particle velocity u_n is given by

$$u_n = u_2 + u_r \qquad (3.20)$$

due attention being paid to sign.

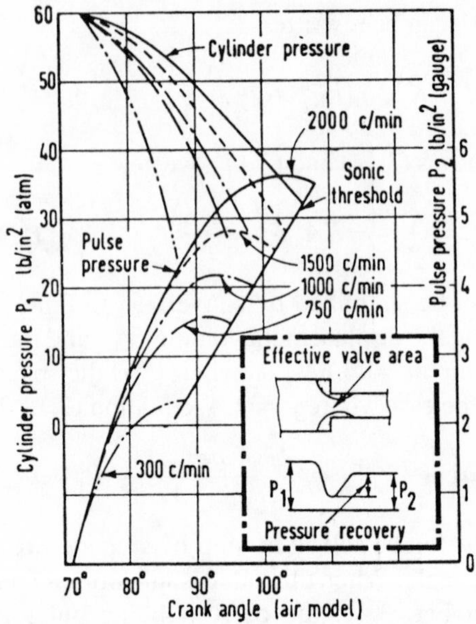

Fig. 3.43 Effect of speed on pulse formation

By deduction, from equations (3.11) and (3.20)

$$\left(\frac{P_n}{P_0}\right)^{(\gamma-1)/2\gamma} = \left(\frac{P_2}{P_0}\right)^{(\gamma-1)/2\gamma} + \left(\frac{P_r}{P_0}\right)^{(\gamma-1)/2\gamma} - 1 \tag{3.21}$$

and

$$u_n = \left(\frac{2}{\gamma - 1}\right)a_0\left\{\left(\frac{P_n}{P_0}\right)^{(\gamma-1)/2\gamma} - 2\left(\frac{P_r}{P_0}\right)^{(\gamma-1)/2\gamma} + 1\right\} \tag{3.22}$$

For a wave reflection at a closed end, the superimposition gives a net particle velocity $u_n = 0$. Hence

$$P_r = P_2 \quad \text{and} \quad \left(\frac{P_n}{P_0}\right)^{(\gamma-1)/2\gamma} = 2\left(\frac{P_2}{P_0}\right)^{(\gamma-1)/2\gamma} - 1 \tag{3.23}$$

At the open end the wave is reflected in the subsonic range according to the expression

$$\left(\frac{P_r}{P_0}\right)^{(\gamma-1)/2\gamma} = 2 - \left(\frac{P_2}{P_0}\right)^{(\gamma-1)/2\gamma} \tag{3.24}$$

and then travels back toward the cylinder. If the exhaust valve is still open, the reflected wave will be superimposed upon the pulse being generated, giving some net pressure, P_n, which determines the rate of discharge. If at bottom dead centre the net back-pressure, P_n, is negative, then good scavenging of the cylinder can be achieved and this leads to higher combustion efficiency. The wave trains so generated will continue their movement being reflected and reduced by friction and charging profile until wave motions in successive cycles are identical. It is clear that such a calculation procedure involves a lengthy step-by-step process and is best carried out by computer.

Subsonic range

In view of the difficulties just mentioned of ascribing an accurate value to P_n without carrying out a step-by-step calculation, it is assumed that residual wave motion is absent and that the back-pressure is atmospheric. It can then be shown that the throat Mach number, M_c, is given by

$$M_c = \sqrt{\left(\frac{2}{\gamma - 1}\right)}\sqrt{\left\{\left(\frac{P_1}{P_c}\right)^{(\gamma-1)/\gamma} - 1\right\}} \qquad (3.25)$$

and equation (3.19) becomes

$$\Delta\left(\frac{V_1}{V_D}\,\overline{P}_1^{1/\gamma}\right) = \frac{-KA_e a_E}{6\;NV_D}\sqrt{\left(\frac{2}{\gamma - 1}\right)}\overline{P}_c^{-1/\gamma}\sqrt{(\overline{P}_1^{(\gamma-1)/\gamma} - P_c^{(\gamma-1)/\gamma})}\cdot\Delta\alpha \qquad (3.26)$$

The extension of the previous calculation into the subsonic range is achieved by the simultaneous use of equations (3.25) and (3.26) and Fig. 3.35. The trial-and-error method of solution is as follows: values of P_c and hence, by equation (3.25), of M_c, are assumed, and the results obtained from Fig. 3.42 are compared with those from equation (3.26) until agreement is reached.

Calculation of the exhaust wave form can thus be carried into the subsonic range. If due care is taken over the magnitude and sign of the actual back-pressure obtaining at any instant, then the method is applicable to the whole of the cycle. It is possible to extend this theoretical method to cover the case of a turbocharged diesel engine; the energy of the exhaust pulse can be found and used to predict the degree of turbocharging possible. In this way, the relative merits of various exhaust systems can be evaluated.

Effect on engine mass flow

Every young motorcyclist who aspires to tune his machine for racing knows that, for a given engine configuration and speed, there are exhaust pipe lengths that enhance the engine performance and others that reduce it. It can be shown by dimensional analysis that the airflow per stroke depends upon a number of factors of which, for a constant scavenge pressure, NL/a is the most important, where N is engine speed, L is pipe length and a is acoustic velocity in the pipe.

A model simulating two-stroke engines has also been described in the literature. Experiments have been carried out using this model to determine, for fixed values of N and a, the variation in airflow with L and the corresponding cylinder and manifold diagrams (Figs 3.44 and 3.45). For the point A of maximum loss, the wave diagram shows a large position wave F at the exhaust port at around bottom dead centre when the port is wide open. The pressure drop across the cylinder is thus reduced, leading to a fall in air throughput. For maximum flow (point B) the indicator diagram shows a rarefaction wave, D, reaching the ports at bottom dead centre. The consequent increased pressure differential across the cylinder

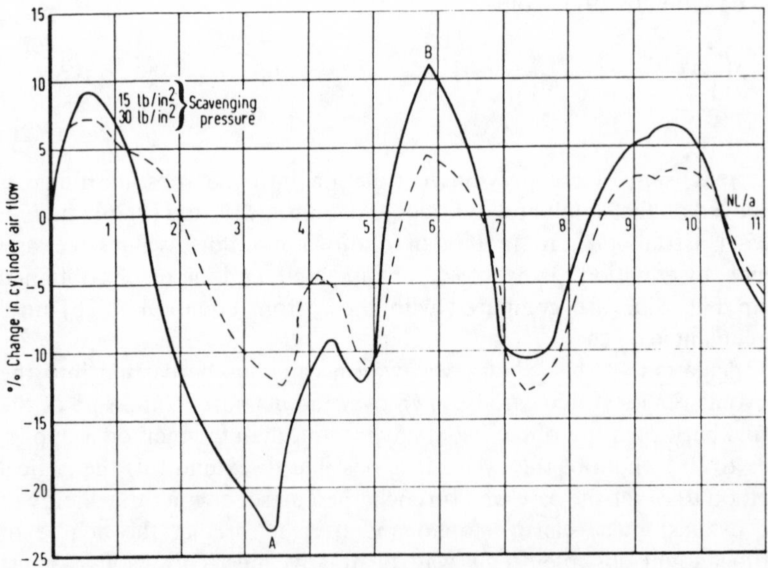

Fig. 3.44 Variation of scavenging air flow

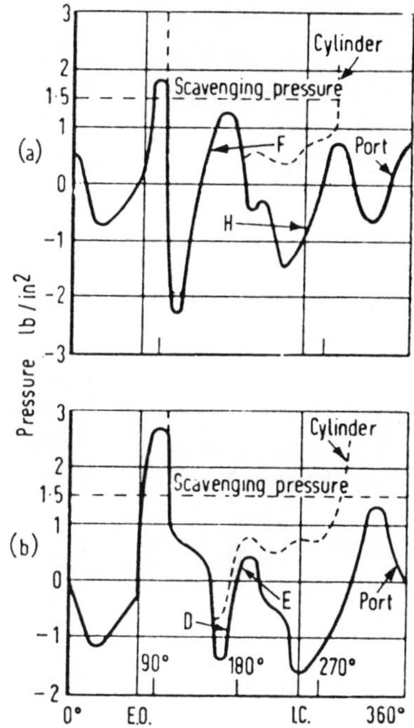

Fig. 3.45 Indicator diagrams for points A and B on Fig. 3.44

causes an increase in mass flow. This increased air throughput improves the purity of the cylinder contents for combustion purposes and can raise the cylinder pressure at port closure; both factors lead to greater power output.

Notation

a Acoustic velocity

$a_E \quad = a_1\left(\frac{P'}{P_1}\right)^{(\gamma-1)/2\gamma} \quad = a_R\left(\frac{P'}{P_R}\right)^{(\gamma-1)/2\gamma}$

A Area of duct

c Propagation velocity of wave point

K Ratio, area of port to area of pipe

L Length of manifold

M	Mass
M	Mach number
N	Revolutions or cycles per minute
P	Pressure (absolute)
P'	Arbitrary reference pressure
$\overline{P}$	P/P'
t	Time
u	Particle velocity
V	Volume of cylinder
γ	Ratio of specific heats
ϱ	Density

Subscripts

0	Undisturbed conditions in pipe
1	Cylinder
2	Pipe
c	Vena contracta
D	Displacement
e	Exhaust
E	Conditions corresponding to isentropic expansion from release conditions to reference pressure P'
n	Net pressure due to superimposition of two wave points
r	Reflected wave point
R	Release conditions

Multi-cylinder engine exhausts

Dr Morrison[25] considers that, apart from its silencing function, an exhaust system should be designed to aid the scavenging of burnt gases and the induction of the fresh charge. Valve design and, in particular, valve overlap must affect the picture, but, nonetheless, a good exhaust system can augment good valve design by providing a low positive pressure or ideally a negative pressure (below atmospheric) at the critical top dead centre piston position.

Analysis has generally been restricted to single-cylinder engines having a single pipe exhaust, and although this has led to an understanding of this case, it has not added significantly to the real case of, say, a four-cylinder engine with a cast-iron manifold. Here the author discusses characteristics of the single-pipe case, but goes on to discuss the interaction effect of multi-pipe manifolds and sheds a little design on what to date has been mainly a development science.

Primary positive wave

With the opening of the exhaust valve and the pressurizing effect of the rising piston, the spent gases expand into the exhaust system at sonic speed and a high pressure wave is generated. This will be referred to as the primary positive wave. The intensity of the pulse varies with engine speed as shown by Fig. 3.46 (the peak pressure at 2000 r/min is 1.34 greater than that for 1000 r/min), the ratio reducing as engine speed increases and until a limiting pressure is reached. The base scale chosen in Fig. 3.46 is time rather than crank-angle, to demonstrate that the rate of rise of pressure increases with speed. It is this effect, magnified, which accounts for the 'crackle' of high-speed engines with quick- lift camshafts.

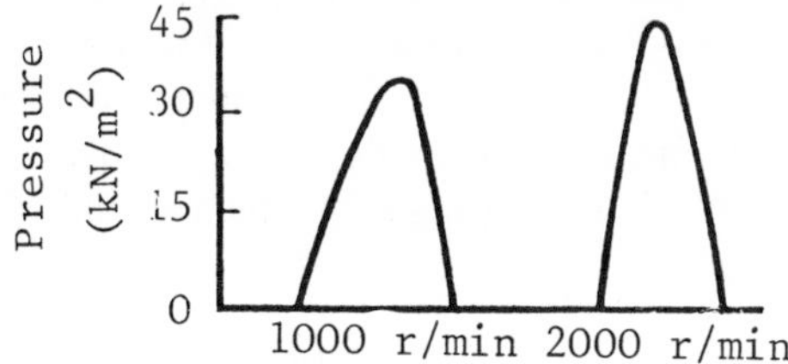

Fig. 3.46 Pulse intensity variation with engine speed

Critical length of pipe

All pipes exhibit a natural peak pressure at a given speed and generally an increase in length will result in an increase in pressure. However, a length is reached beyond which pressure does not rise and this is termed its critical length. This is, of course, speed-dependent, the critical length decreasing as engine speed increases; see Fig. 3.47. Peak pressure also

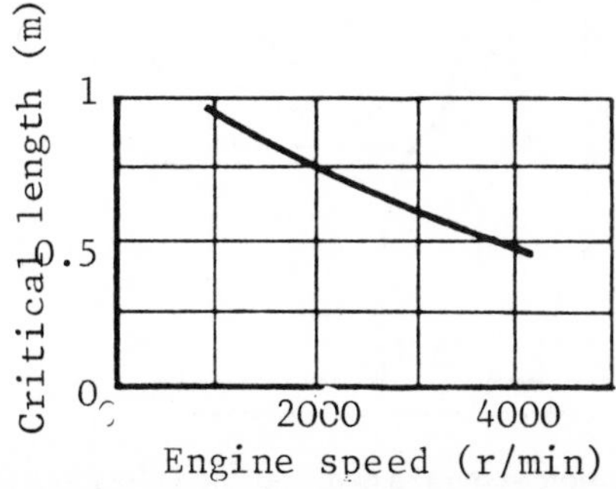

Fig. 3.47 Critical length variation with engine speed

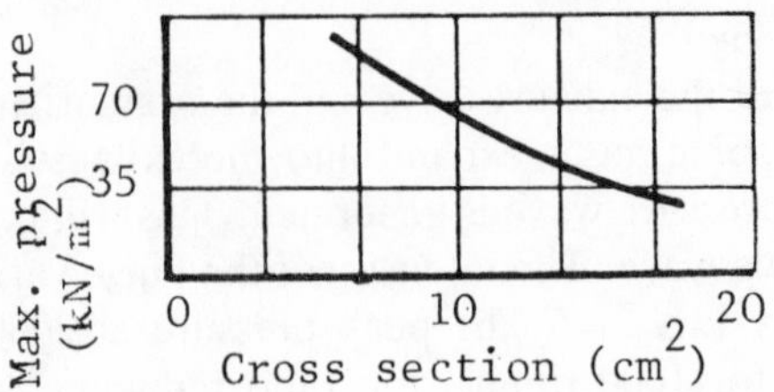

Fig. 3.48 Peak pressure variation with pipe diameter

varies with pipe diameter – pressures are proportional to pipe bore area, except when the bore is large; see Fig. 3.48. Peak pressure should be kept down to a figure around 27.5 kN/m^2, and a pipe size should be chosen to do this. If a smaller size is used, a primary wave of very high pressure will be produced which is difficult to destroy. Below the critical length the wave front becomes steeper as it travels along a pipe and may develop into a shock wave which sounds much louder than the original wave.

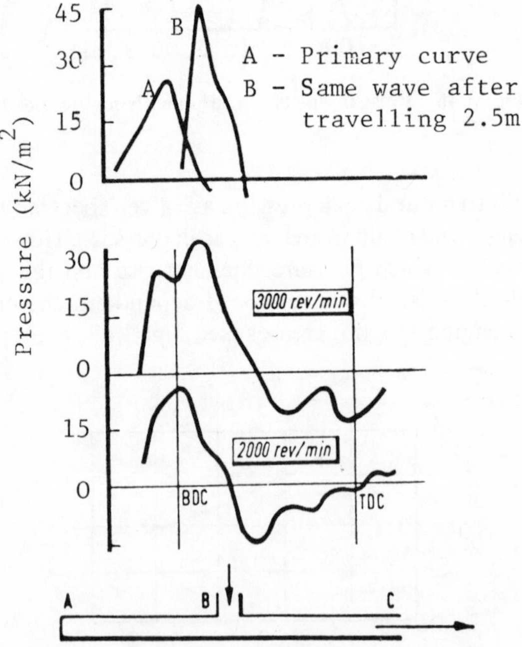

Fig. 3.49 Simple system to demonstrate pulse interference

Interference

Consider an exhaust pipe having an extension with a blind end A (Fig. 3.49). The pulse entering at B spreads in both directions, BA and BC. The pulse is reflected from A and follows the pulse that is already some way along BC, thus extending the period of the positive pulse and afterwards adding to the negative period and also helping to damp the vibrations – see diagrams of 3000 and 2000 r/min in Fig. 3.49. It was found that the best length of the interference pipe was 0.5 m, but this would vary with different engines. Ideally, the pipe should be attached near the engine.

Separate manifolds

It is commonly supposed that downpipes from the two manifolds could be joined together at any convenient distance, say 0.5–1 m from the engine, and that the pulse from one pipe will not greatly affect the other. But while the author has shown that a sufficiently large exhaust chamber (10–15 times the stroke volume) will isolate the vibration of adjacent primary pipes, this is far from the case in the flowed type of pipe connection.

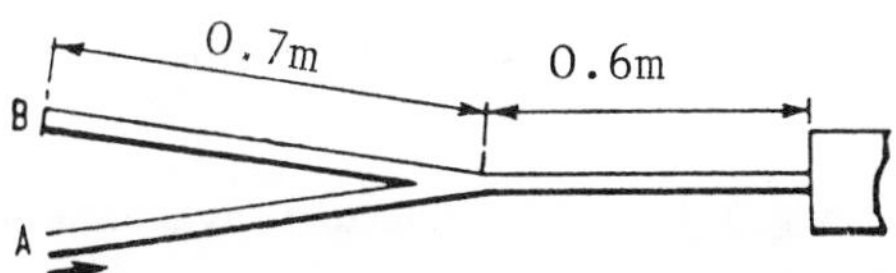

Fig. 3.50 Flowed fork pipe test set-up

This configuration (Fig. 3.50) was tested on a single-cylinder engine, the curve in Fig. 3.51 being taken at the engine stub, A, and Fig. 3.52 at the blind end B. At A the period of the first wave giving the second positive peak corresponds with the period of an open pipe 700 mm long instead of the whole length of 1300 mm. This gives a good diagram at this speed, but the peak will move to the right as the speed increases and will be over top dead centre at high speeds, which is to be avoided. Also, an extremely violent wave is set up in the other pipe which, at this speed, will have a bad effect. What precisely would happen if this arrangement were applied to a four-cylinder engine, it would be impossible to estimate, but the effects would certainly be rather like those found in the above test. Making the pipe, after the join, of larger diameter would reduce the intensity of the wave at B in Fig. 3.49.

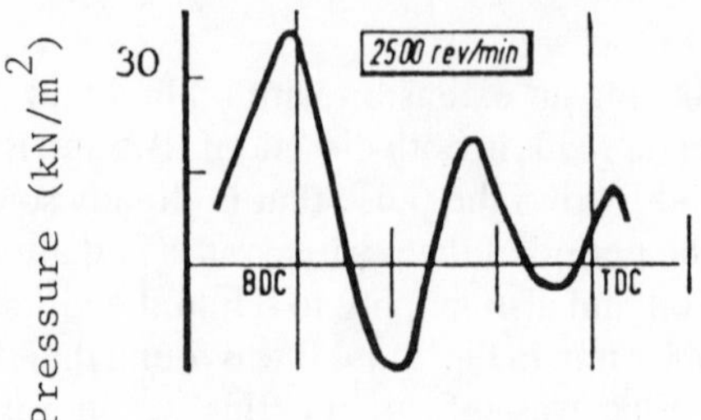

Fig. 3.51 Pressure fluctuation at engine stub A in Fig. 3.50

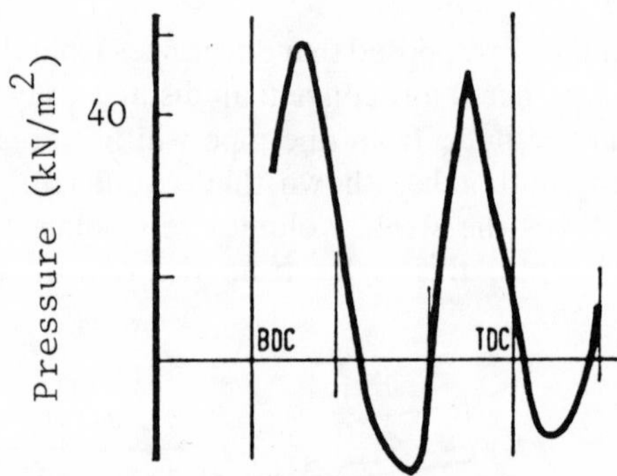

Fig. 3.52 Pressure fluctuations at blind end B in Fig. 3.50

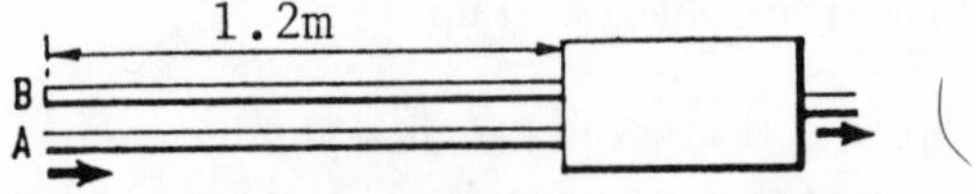

Fig. 3.53 Non-preferred downpipe layout

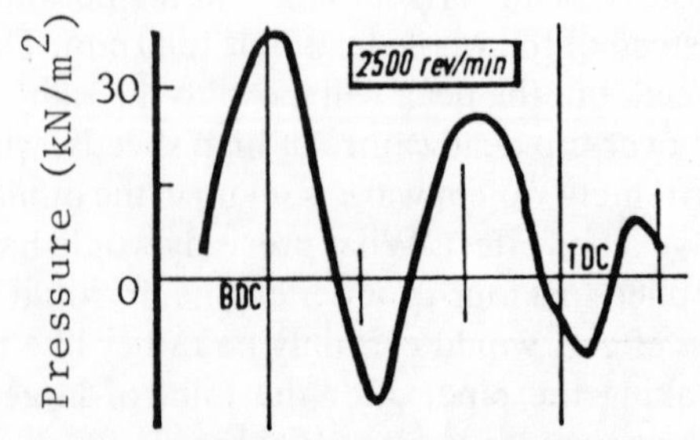

Fig. 3.54 Pressure fluctuation at A in Fig. 3.53

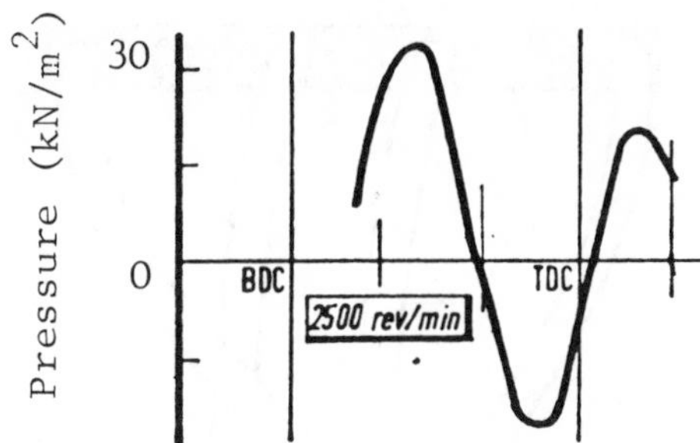

Fig. 3.55 Pressure fluctuation at B in Fig. 3.53

Suggested practice

Further examples of bad arrangement is shown in Figs 3.53, 3.54, and 3.55. The two downpipes are 1.25 m long and, instead of joining, go into a box which has a volume seven times that of the stroke volume. The interposition of the box, by shortening the period of vibration of the 1.25 m pipe, caused the second peak to be so early that even at 3500 r/min it was still over top dead centre, and the first peak up the other pipe arrived in time to add to the next primary wave. This would make matters worse if the arrangement were applied to a four-cylinder engine. The box would need to be double this size to allow the 1.25 m pipe to vibrate in its own period, and even then conditions would not be good till over 3000 r/min.

If the pipes were made shorter, say 0.45 m, the primary pulse would get round the join from one manifold to the other so quickly that there might as well be a single manifold. This seems strange, but the reason is that 0.45 m is well under the critical length for this speed, so that pressure builds up in both pipes at almost the same time. Another result of this short length is that the peak pressure is lower than it would otherwise be and the superimposing of the short period of the 0.45 m length on the period of the long pipe has a beneficial damping effect.

This curve is similar to the interference pipe in Fig. 3.49. The effect is not quite so good, because in the latter case the interference pipe was attached where it should be – near to the exhaust flange. However, the above experiment suggests an easy way of applying interference to the exhaust system of a four-cylinder engine. This is to take 0.45 m downpipes from the two outer ports and join them to another pipe (called the

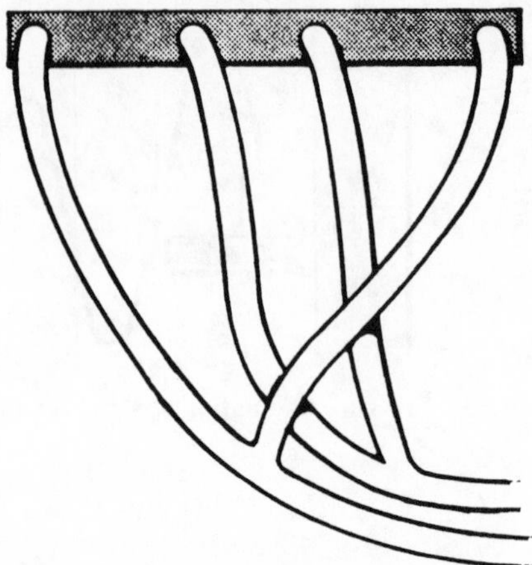

Fig. 3.56 Preferred downpipe layout

intermediate pipe) as shown diagrammatically in Fig. 3.56, and do the same for the centre pair. If the centre ports are siamesed so that there is a single pipe, an interference pipe 0.45 m long can be welded to this pipe near the flange.

The next question is what to do with the intermediate pipes. In the original interference experiment the total distance from the engine to the absorption silencer was 1.5 m for the diagrams shown. If the length were less than this, the effect would not be so good at low speeds, but would be good at very high speeds. Anything less than 1.25 m is not recommended.

Absorption silencers, it must be emphasized, are invariably much too small to give effective silencing. Two of them in series should be used, but it is possible for the second silencer to be of the baffle type, without altering the characteristics of the curves taken at the engine end, although the average pressure may be slightly raised. Two completely separate systems will give maximum power, but will be rather expensive. Joining the intermediate pipes before the first silencer is questionable unless the silencer is very big. A safer way is to put a moderate size absorption silencer on each pipe and merge the pipes a little way after the silencers and put a final silencer after that. The silencing effect of such a scheme is good. The interference method by itself reduces the snap of the exhaust note.

When schemes for the four-cylinder engine were being tried, the above

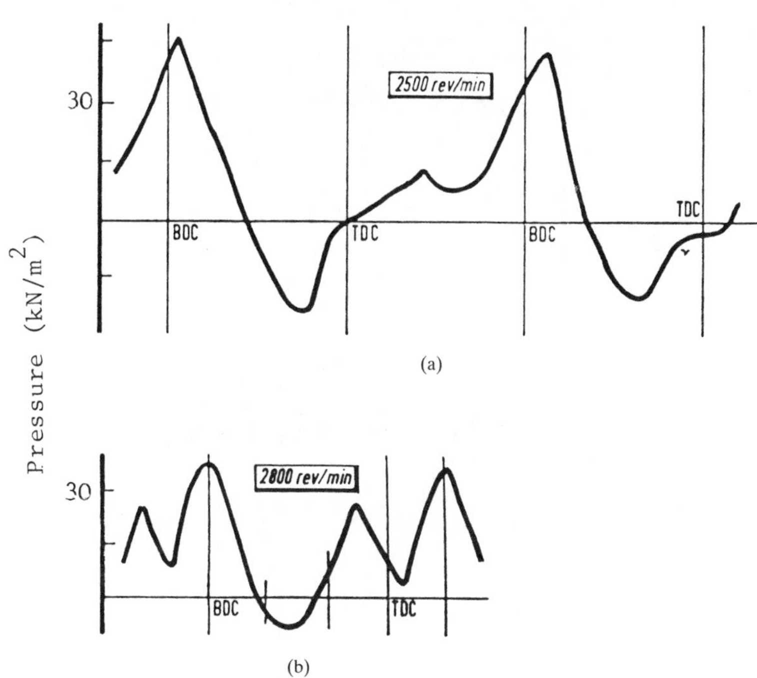

(a)

(b)

Fig. 3.57 Pressure fluctuation for 4-cylinder engine

application of interference had not been thought of, and the best solution for this engine was found to be a long pipe 1.85 m from each manifold going into a box 16 times the stroke volume, from which a pipe led to the original silencer. This gave a good diagram at 2500 r/min (Fig. 3.57(a)). Even with this large box it was evident that a wave was backing up from the next cylinder exhausting into the other pipe, but it was too late to do any harm.

Six-cylinder engines

The problems of the six-cylinder engine are well known to be more difficult than those of the four-cylinder and this certainly applies to the exhaust system, especially if separate manifolds are used and the downpipes are merged. Work done on a 2 litre engine showed how easily one could be worse off with separate systems than with the single manifold. For example, when the two downpipes were merged at approximately

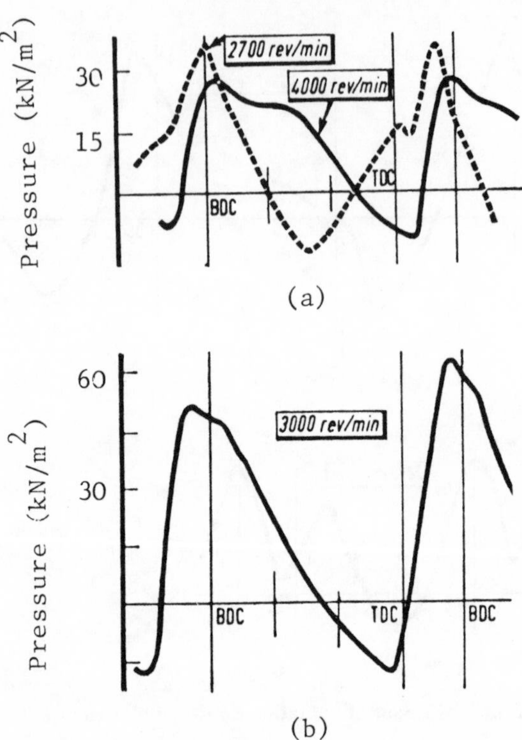

Fig. 3.58 Pressure fluctuation for 6-cylinder engine

0.45 m and the pipe then went another 0.45 m to an absorption silencer, the diagram at 2800 r/min (Fig. 3.57(b)) showed that the wave coming round from the next cylinder to exhaust arrived at exactly the wrong time. At a speed of 3400 r/min it was worse and would never have been late enough. The distance to the point of merging was then made 1.5 m, which moved the third peak so far to the right that at 3000 r/min it added exactly to the primary wave of the succeeding wave developing a high pressure – see Fig. 3.58(b). The pressure just before top dead centre is first low and then rises abruptly. At less than 3000 r/min this rise takes place before top dead centre, so the pressure at the important point changes from bad to good over a small speed range.

In the case of another engine which had a somewhat similar arrangement there was a change in the torque value of 18 per cent over this small range of speed, which is not a desirable feature. The author also ran a test using a 2.75 litre engine over a range of speeds with completely separate systems, consisting only of absorption silencers some feet away from the engine. Two curves are given in Fig. 3.58(a) for speeds of 2700 r/min and 4000 r/min. The curve at 2700 r/min is not unlike the one in Fig. 3.58(b), but the pressure at top dead centre dropped below atmospheric by 2900 r/min. So if one does not mind losing power below 2700 r/min the above simple scheme is sufficient. A second silencer would be necessary and would do no harm.

CHAPTER FOUR
Cooling system design

Efficiency of combustion depends upon high temperatures – and the suggested use of ceramic materials might one day bring the adiabatic engine into production. Meanwhile long engine life requirements cause the designer to pay careful attention to cooling and lubrication. Conventional materials, available inexpensively, in high volumes, require careful lubrication, and the lubricant in turn requires protection from unduly high temperatures.

This chapter opens with a section which proposes a systematic approach to cooling system design with discussion of the scope which still exists to improve cooling arrangements if the severe constraints of packaging and cost can be released. The next section deals with heat flows within the engine, while that following it examines the design of the coolant heat exchanger (the radiator); an approach to fan system design is next presented, followed by sections on analysis of the cylinder head gasket.

ENGINE COOLING – A SYSTEMATIC APPROACH

Dan Anderson[26] points out that lubrication, volumetric efficiency, tendency to 'knock', exhaust valve life, and corrosion are some of the main factors influenced by temperature distribution within the engine. It is also recognized that exhaust emission of hydrocarbons and carbon monoxide are measurably reduced by increasing the operating temperature of the coolant. The aim of the cooling system, therefore, is to control the engine at an optimum temperature distribution which is a compromise of considerations such as the above and the temperature limitations of the water jacket materials (≃260°C cast iron, 200°C aluminium). This is done, of course, by controlling the temperature of the coolant, typically at about 85°C. At 50 km/h it has been shown that a reduction of the coolant temperature from 88°C to 49°C resulted in a 3 per cent fuel consumption penalty.

Apart from the better known effects of 'hot or cold' engine-running, the influence on the temperatures of the so-called 'hot-zones' within the combustion chamber, including the piston, must be considered. Some Russian work has indicated the influence of coolant temperatures at these critical regions, and there seems little doubt that some scope exists for further refinement of the water-flow round the cylinder jacket and cylinder head. Also it should be noted that the coolant temperature should be reasonably uniform, to avoid thermal stresses, and the radiator designed to allow a temperature drop through it of the order of 6°C. The higher the coolant temperature, the smaller the radiator. The removal of heat from the coolant requires the expenditure of useful work to drive the water pump and fan. Aerodynamic drag associated with the cooling system has been estimated as high as 16 per cent of the overall drag and it is generally considered that cooling fans absorb 3–8 per cent of engine power.

Relevant heat transfer process

Within the engine, the heat is rejected to the coolant primarily by the mechanisms of conduction and convection heat transfer. Figure 4.1 shows

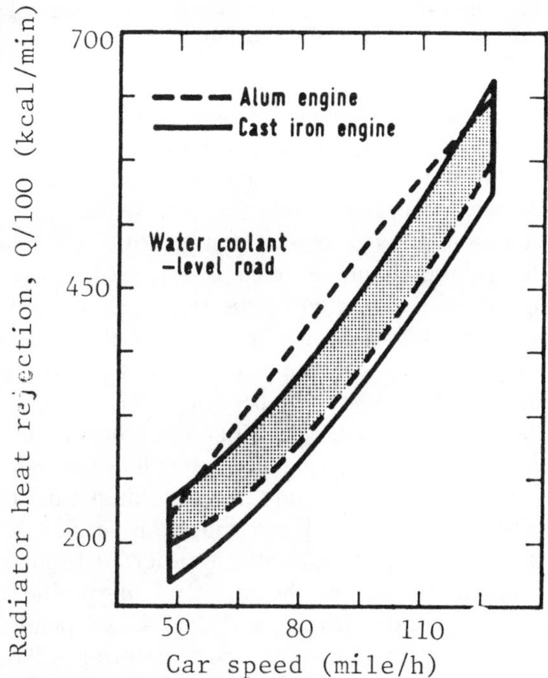

Fig. 4.1 Similarity in heat rejection to coolant for aluminium cast iron engines

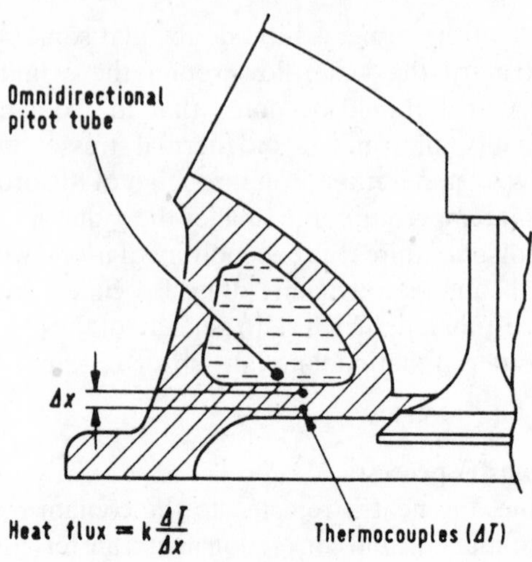

Fig. 4.2 System for measuring heat rejection to coolant involving coolant velocity and direction of flow

that the heat rejection to coolant is essentially the same for both aluminium and cast-iron engines. The conductivity of aluminium is approximately four times greater than that of cast-iron so that, for geometrically similar engines, the heat flux is the same, then the basic conductivity equation gives a smaller temperature gradient for the aluminium. That is, the aluminium engine runs cooler. The heat flux recorded in Fig. 4.1 was measured by the system shown in Fig. 4.2 in which coolant velocity and direction was obtained with a pitot probe.

The surface heat transfer coefficient from a hot surface to a liquid depends on the type of fluid flow, shape, and surface roughness of the passage. Satisfactory correlations exist for flow in tubes and well-defined shapes of passage, and by considering the coolant flow through an engine as analogous to turbulent flow in a tube. One researcher has obtained a simple non-dimensional correlation, experimentally verified, for predicting at any operating condition, the heat transferred from the engine cylinder to the coolant. He has also shown that cooling problems become more difficult the larger the cylinder size, Fig. 4.3, dependent on the piston speed. The use of aluminium instead of cast-iron will, of course, lower the cylinder head temperatures and, due to a smaller thermal capacity, also give a more rapid warm-up. An experimental comparison

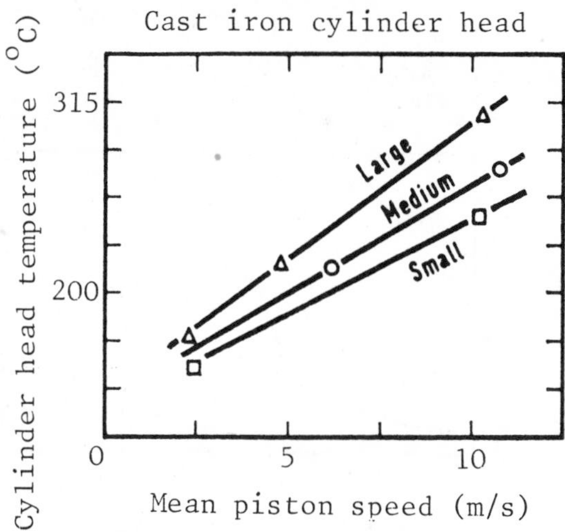

Fig. 4.3 Effect of engine cylinder size on cooling requirement

has shown that at the combustion chamber hot zones the temperature may be reduced under strenuous operating conditions by over 100°F, using aluminium.

To understand the influence the coolant may exert on the 'hot-spot' regions within the combustion chamber, other researchers have shown that the heat fluxes at extra-hot zones, such as exhaust valve seat area, are such that local boiling of the coolant is inevitable. Under such conditions the heat transfer coefficient depends primarily on the difference between the temperature of the metal surface in contact with the coolant and the saturation temperature (T_s) of the fluid. It varies approximately as shown in Fig. 4.4 when the flow velocity is zero. This was found from experiments with a special heat transfer chamber; it shows that nucleate boiling must exist.

Nucleate boiling is the familiar bubbling process which may be gentle (small bubbles) or extremely vigorous if the fluid-interface temperature is sufficiently high. The agitation and inrush of fluid caused by vapour bubbles rising to the surface, or simply moving downstream with the flow, can produce very high surface heat transfer coefficients. Film boiling occurs at high fluid-interface temperature differences and produces a continuous vapour sheet or film at the interface, with a progressively

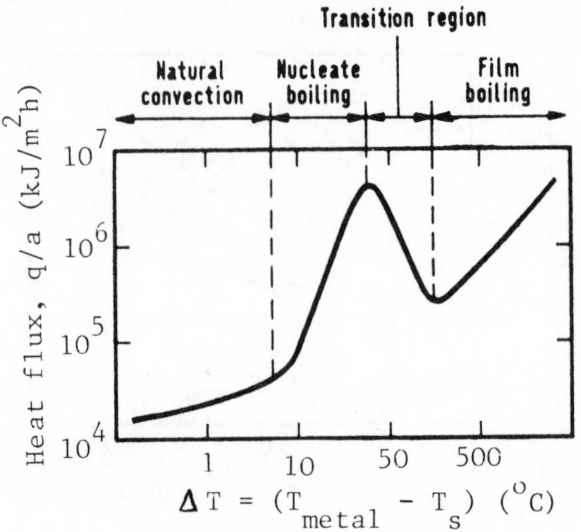

Fig. 4.4 Relationship of heat transfer with metal to fluid temperature difference

insulating effect. Obviously, film boiling and temperatures associated with it would be disastrous.

Results from work based upon a special heat transfer chamber to investigate the effects of coolant pressure and flow on the heat flux in the extra-hot regions, are shown in Fig. 4.5. The essential conclusions are that any increase in the boiling point (T_s) of the coolant (by pressurization, say) is reflected as a *pro rata* increase in the metal surface temperature; increasing the coolant velocity for a given heat flux suppresses the onset of nucleate boiling and results in lower metal temperatures. This is also true for non-aqueous coolants. A design objective should therefore be to have as high flow rates as possible through the cylinder head, especially if a high boiling point exists due to pressurization. Otherwise, the unfortunate exhaust valve may soon be a part to suffer.

Note that cast iron is basically an alloy of constituents having very different thermal conductivities, namely ferrite (Fe) with $k = 42$ in Imperial units, cementite, or iron carbide (Fe_3C), with $k = 6$, and pearlite, the eutectoid of ferrite, and cementite layers including 0.9 per cent carbon (C) with $k = 30$. Furthermore, Si, Mn, Ni, Cr, W, and Mo may be present in varyingly small quantities. Since heat treatment can change the distribution of the main constituents (ferrite, cementite, and

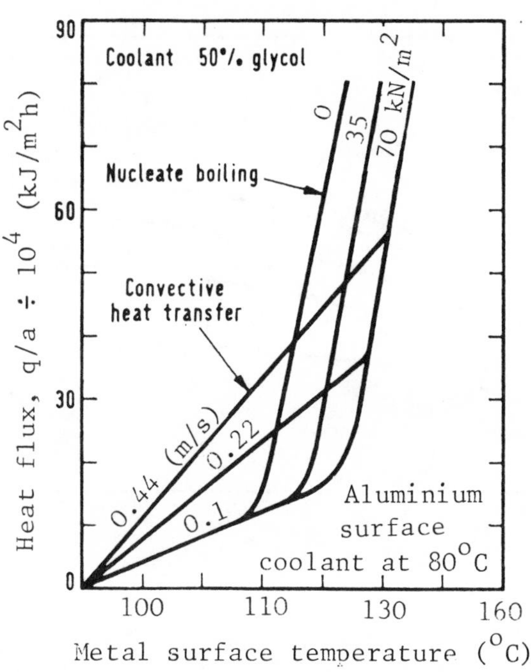

Fig. 4.5 Relationship of heat transfer with metal surface temperature in the very hot regions

pearlite) to a marked degree, it is very difficult to predict k for a cast iron, even though the chemical analysis is known. For example, at 100°C malleable cast-iron with about $2\frac{1}{2}$%C has $k = 37$ if 82 per cent of the iron is ferrite but $k = 27$ if 91 per cent is pearlite. For a non-treated cast-iron with 4.5%C, none of which was graphite, a conductivity of $k = 8$ was measured; by annealing for two hours at 1000°C so that eventually 4 per cent graphite was present, the conductivity increased by a factor of four, to $k = 32$.

It seems that too readily accepting a 'book value' for k in any analysis should be avoided, for in the hot-spot regions of some engines with cast-iron cylinder heads, the temperatures may be adversely affected if the conductivity is lower than that which may be readily obtainable by suitable heat treatment.

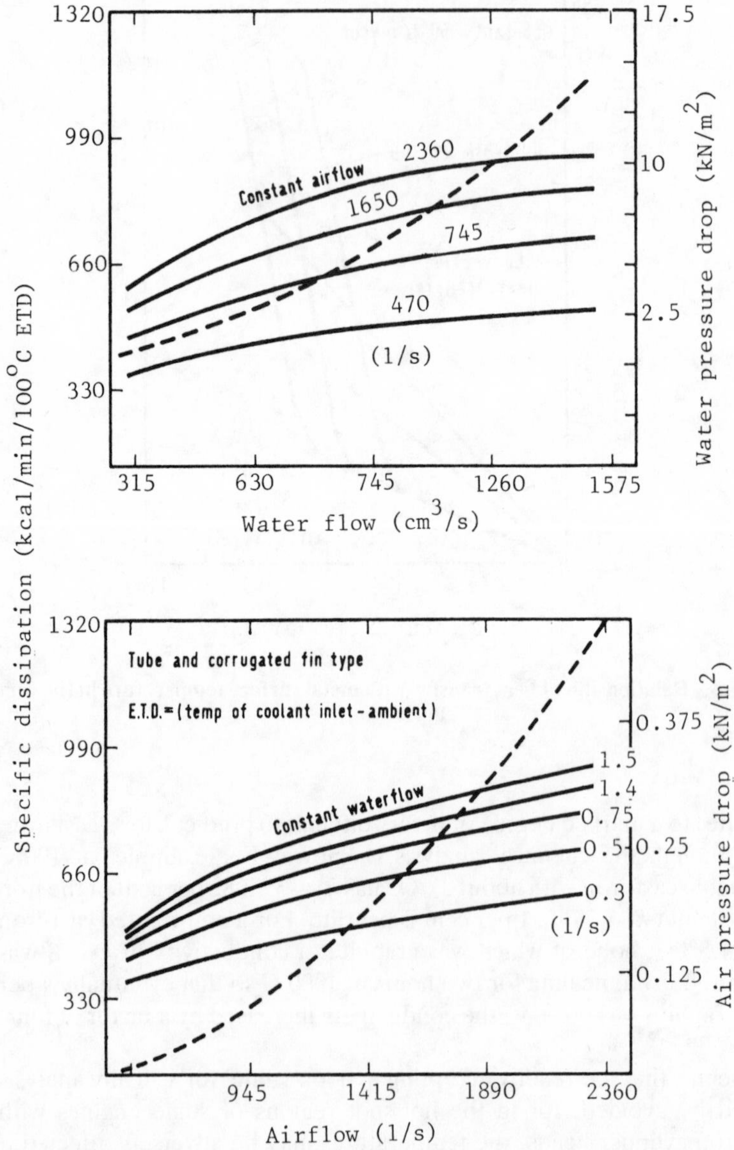

Fig. 4.6 Water flow and air flow characteristics for a coolant radiator (core width = 45 cm, core height = 35 cm, core depth = 3 cm; 3 fins/cm)

The coolant radiator

Heat dissipation from the radiator (cross-flow heat exchanger) is primarily dependent on the flow rates and temperatures of coolant and air passing through the radiator. Typical characteristics are shown in Fig. 4.6(a) and (b). The heat transfer rate equation must be combined with an energy equation relating the loss of enthalpy of the hot fluid (coolant) to the gain of enthalpy of the cold fluid (air) in order to relate the numerous heat exchanger variables. These variables may be grouped non-dimensionally, see Fig. 4.7, to simplify the treatment.

Exchanger heat transfer effectiveness

$$\begin{aligned} \zeta &= q/q_{max} \\ &= C_H(T_{H1} - T_{H2})/[C_{min}(T_{H1} - T_{C1})] \\ &= C_C(T_{C2} - T_{C1})/[C_{min}(T_{H1} - T_{C1})] \end{aligned}$$

where C_{min} is the smaller of C_H and C_O. For a radiator, $C_{min} = C_C$ so that

$$\zeta = (T_{C2} = T_{C1})/(T_{H1} - T_{C1})$$

Number of exchanger heat transfer unit (N.T.W.) $= AU/C_{min}$, where U, the average overall heat transfer coefficient, is considered constant for a given design. A is the surface heat transfer area.

$$\text{Capacity rate ratio} = C_{min}/C_{max}$$

In general, it is possible to express these non-dimensional groups as a functional relation dependent on the flow arrangement, that is

$$\zeta = \text{f (N.T.W., } C_{min}/C_{max}\text{, flow arrangement)}$$

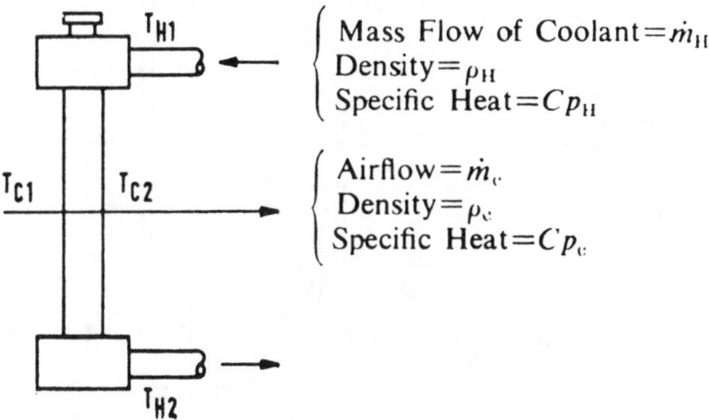

Fig. 4.7 Methodology of heat exchanger design

Since the radiator cannot dissipate any more heat from the coolant than the radiator airflow is capable of accepting ($C_c = C_{min}$), the airflow is particularly important for good dissipation (Fig. 4.7). Non-dimensional parameters in heat exchange are: $C_H = \dot{m}_H Cp_H$; $C_c = C_{min} = \dot{m}_c Cp_c$; h_ω = water to metal transfer coefficients; h_a = metal to air transfer coefficient; k = conductivity of metal. Average overall heat transfer coefficient is given by $1/\omega = (\alpha/h_\omega + \beta/k + \gamma/h_a)$ where α, β, and γ are factors depending on finning arrangement.

Coolant flow system

The main problems in maintaining the necessary circulation are in loss of fluid by expansion, aeration of fluid, poor water-pump inlet conditions, and possible cavitation reducing the pump output and perhaps the life of the pump itself. Sensitivity to loss of coolant can be greatly reduced by incorporating a separate expansion and make-up tank as shown in Fig. 4.8. This replaced a conventional radiator top tank and tests showed that a large volume of coolant could be taken from the system when full without any loss in circulation. Figure 4.9 shows an experimental system by Ford.

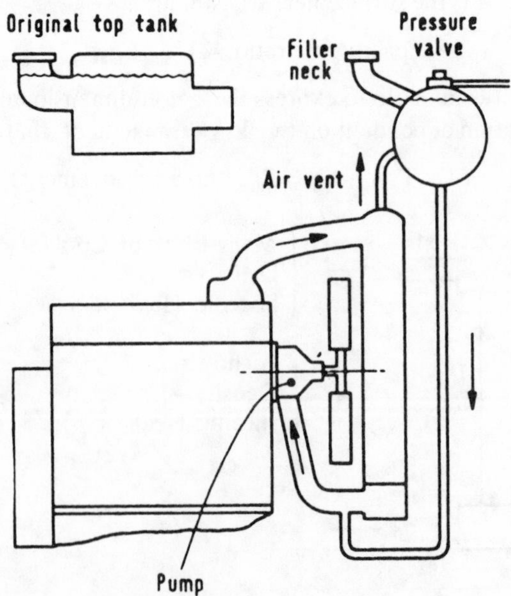

Fig. 4.8 Principle of separate expansion and make-up tank

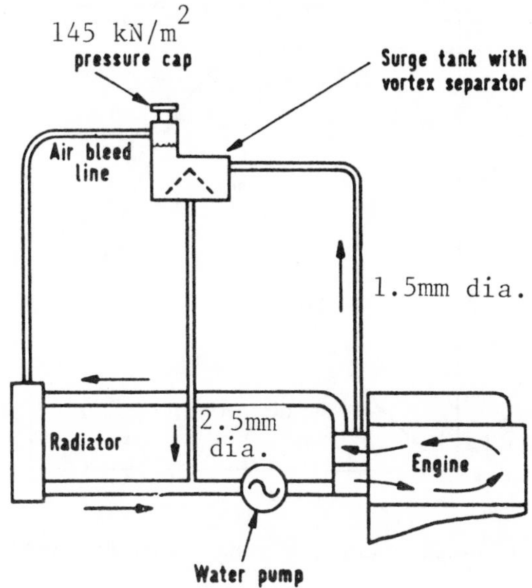

Fig. 4.9 Dimensions quoted for an experimental system

Conventional airflow system

The airflow system may be represented schematically as shown in Fig. 4.10. The basic design problem is to determine the airflow at any combination of fan speed and vehicle speed. The system resistance depends on the grille, inlet, outlet, and the pressure drop characteristics (or friction factor) of the radiator matrix. The resultant airflow depends

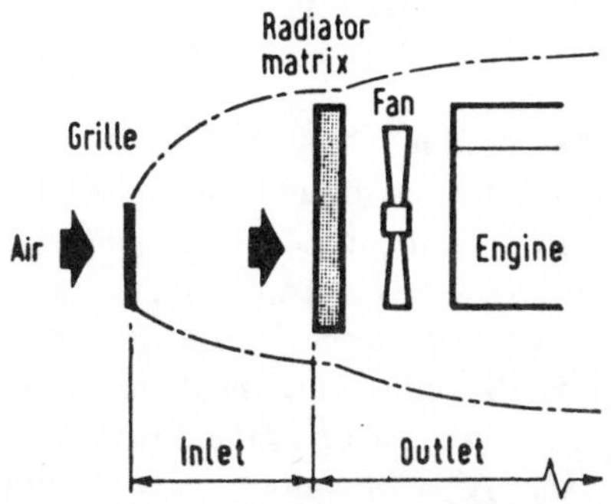

Fig. 4.10 Schematic representation of airflow system

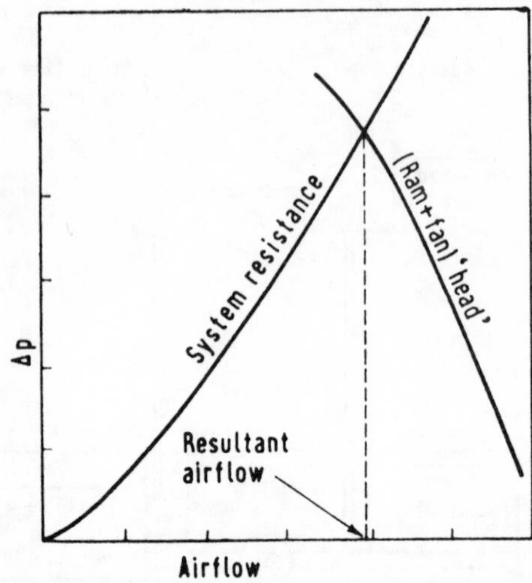

Fig. 4.11 Resultant airflow determination through a cooling system

on the extent to which the cooling air 'head' produced by ram effect and the fan can overcome the system resistance as shown diagrammatically in Fig. 4.11. Improvements can be made by shrouding the fan and optimizing its location between the radiator and the engine. Improved cooling performance may be readily obtained by locating the inlet at a region of high pressure and ducting it to the radiator. If an inlet grille is used then care should be taken that it does not inhibit the airflow. Generally the air is left to find its own way out of the engine compartment and the underbonnet pressures may increase the system resistance with a resultant reduction in airflow.

Air and coolant maldistribution

A radiator will dissipate the maximum heat when the airflow is uniform. Figure 4.12 purports to show the effect of air maldistribution on cooling. This representation is grossly over-simplified and merely gives an indication of the possible effect for example, if 40 per cent of the area gets 80 per cent of the total airflow, the reduction in heat dissipation is of the order of 20 per cent. An important feature of air maldistribution is whether or not areas of stagnation exist. Also, how does one quantitatively describe the degree of maldistribution? A brief exercise was done in which the radiator

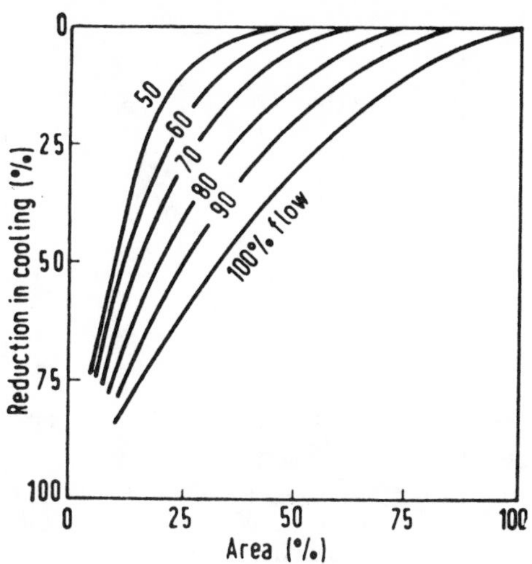

Fig. 4.12 Effect of maldistribution in airflow on cooling

face area was divided into eight equal parts with uniform but different, flows through each part. For a total nominal airflow of say 55 m³/minute the greatest deviation in flow was set at ±30 m³/minute. All possible combinations of flow with minimum difference between any two areas being 15 m³/minute were considered. The purpose was to assess, in a simple manner, the loss in heat dissipation depending on the maldistribution, a measure of which is given by the standard deviation. The exercise was repeated for airflows of 85 and 115 m³/minute, the waterflow being constant at 55 litres/minute. The results are shown in Fig. 4.13. The smaller the airflow the greater the effect of maldistribution. Areas of stagnation can obviously incur serious reductions in heat dissipation, and will also lower the total airflow due to increased system resistance.

Very little seems to be known about coolant distribution through the radiator tubes. Tests with circular tubes, connected across manifolds of equal diameter, as shown in Fig. 4.14 indicated an asymmetric pressure distribution tending to give higher than average flow through tubes at the entrance end of the manifold. If the outlet is as shown dotted, then the highest tube flow occurs at the dead end of the inlet manifold. Coolant distribution in a radiator is likely to be more of a problem the smaller the tube length/diameter ratios (less than 100) or wherever the bulk of the

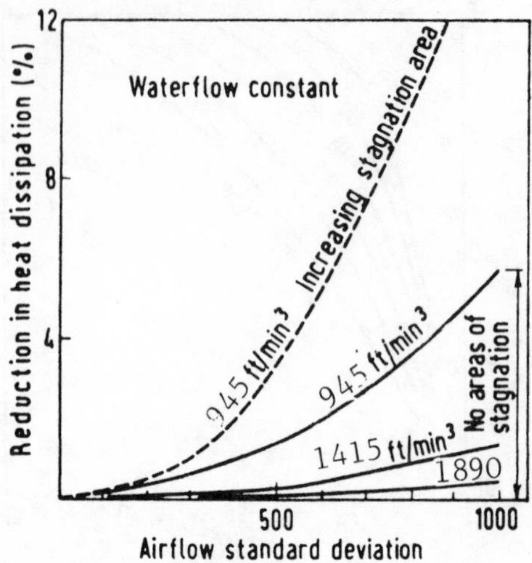

Fig. 4.13 Measurement of airflow maldistribution effect

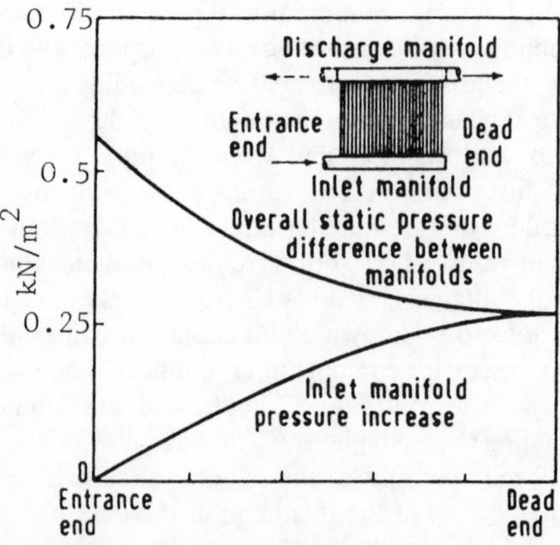

Fig. 4.14 Coolant distribution in simulated 'radiator' tubes

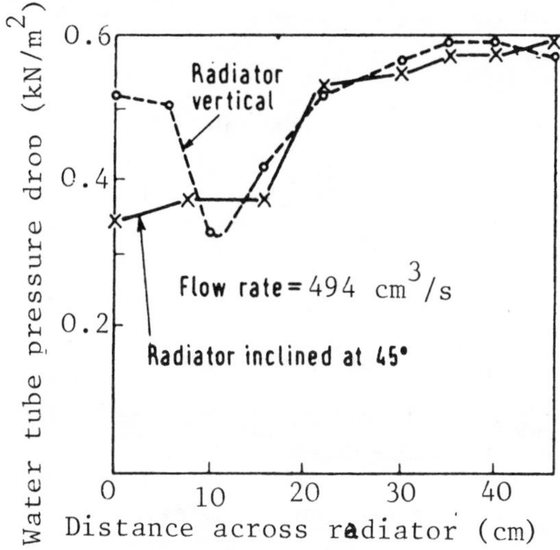

Fig. 4.15 Measured coolant distribution in an actual 'radiator'

pressure drop across the radiator occurs in the top and bottom tanks. The problem is essentially in the design of the shape of the top and bottom tanks related to the inlet and outlet location. Any maldistribution, or even flow reversal of the coolant in the tubes creates additional pressure losses which will result in lower flow rates. A conventional radiator was tested by the author on a water-flow rig and accurate measurements of tube flow were taken across the width of the radiator. Some degree of maldistribution existed, Fig. 4.15, but only a limited flow range was investigated and it is felt that with the trend towards wider and lower radiators in passenger cars it would in all probability be a mistake to dismiss radiator coolant distribution through the tubes as not being a problem.

Airflow measurement

To design a cooling system an essential requirement is a knowledge of the installed airflow. In a conventional system the vagaries of maldistribution, installed fan performance, and other factors, such as intake efficiency, preclude any reliable estimate of the airflow of the design stage. The problem is therefore one of development, which may be time-consuming

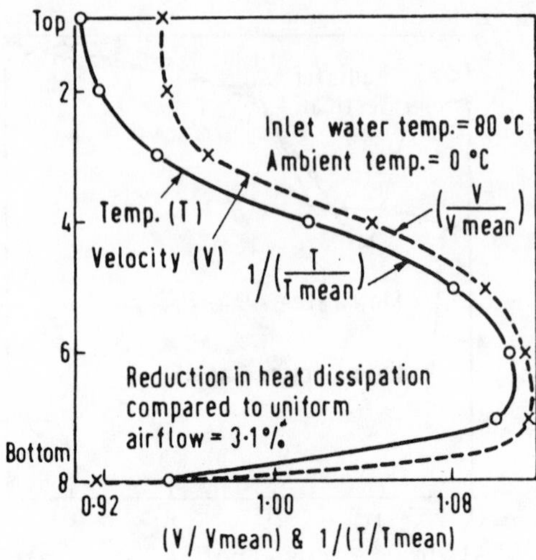

Rad. nom. airflow 1890 l/s
Rad. nom. waterflow 1510 cm^3/s

Fig. 4.16 Temperature distribution over the height of a 'radiator'

and costly. To directly measure the installed airflow (and maldistribution) previous researchers have employed 16 pitot and reverse pitot tubes located in the radiator to obtain pressure differences large enough to be read accurately on an inclined manometer, the probes being calibrated to give the velocity. This method is rather laborious but is the only one available (to the best of the author's knowledge) for measuring installed airflow and distribution through the radiator. Transforming this distribution onto a radiator area, a grid of 64 equal areas was constructed and an average velocity estimated for each area. For any given waterflow, airflow, inlet water temperature, and ambient temperature the temperature distribution of the air leaving the radiator can then be calculated. A sample calculation for a grid column gave a temperature distribution related to the velocity distribution as shown in Fig. 4.16. This encourages the thought that it may be possible to estimate the installed airflow and velocity distribution on the basis of temperature measurements.

The thermocouple grid of such a system would enable fast electrical print out of the results, a 49-point grid being currently considered.

Knowing the performance characteristics of the radiator, the velocity distribution could then be quickly estimated. Such a technique, if proven, has the advantage of being entirely contained within the car, with a read-out time of the order of 15 seconds being possible.

Choice of radiator matrix

Experience shows that a flat tube/fine-fin type of surface, with corrugations or louvres to promote turbulence, gives close to the optimum heat transfer performance obtainable, and is very suitable from cost and practical considerations. The heat dissipation increases with the number of fins/unit length, Fig. 4.17, but so also does the pressure drop and the tendency to fouling by dirt and insects. For any radiator the aim is to obtain the maximum heat dissipation for a given pressure drop (measured by friction factor). Ten to twelve fins/inch seems to represent a good compromise.

Based on experimental data, the ratio of the friction factor (f) divided by the Colburn factor ($J = \mathrm{NuSt}^{2/3}$) was plotted for three heat exchangers

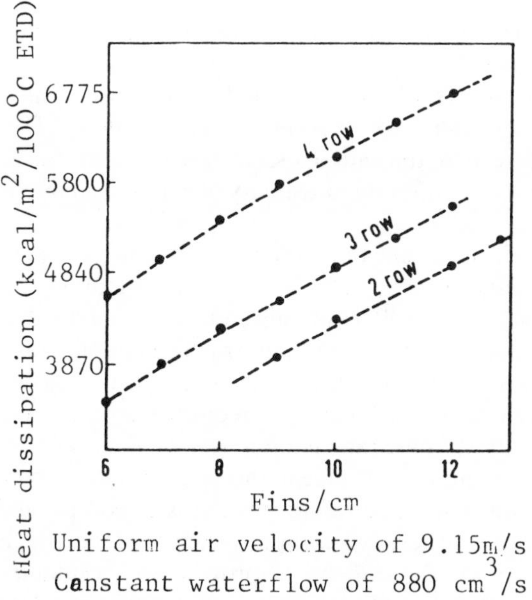

Fig. 4.17 Heat dissipation for different fin configurations

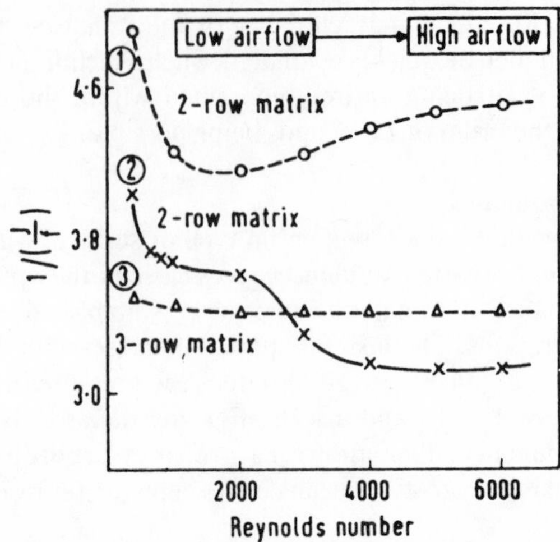

Fig. 4.18 Friction factor over Colburn factor, for these heat exchangers, related to Reynolds number

on a base of Reynolds number. Obviously, the larger the value of (J/f) the better the matrix. It is apparent from Fig. 4.18 that matrix number 3 is very efficient. Matrix 1 is not very good on this basis, whilst matrix 2 is much better and clearly very good at higher airflows. The interesting point is that such optimization plots, based on careful experimental measurements, may offer the possibility of a more enlightened choice of radiator matrix.

At lower vehicle speeds the airflow is increasingly dependent on the fan. Problems associated with the fan, air maldistribution, and prediction of installed airflow can all be minimized by using a fully ducted system, Fig. 4.19. Reductions in size, fan power consumption, and drag are the main incentives. The design and optimization of a ducted cooling system is not an easy task and will certainly require a careful choice of matrix efficiently matched to the ram and fan characteristics of the system. An optimization plot may be of value in this respect. The shape and location of the inlet and outlet and outlet ducting is important and should be related to the overall aerodynamic performance of the vehicle.

In practice it will be difficult to provide an acceptable outlet area greater than that of the radiator itself from the styling and ducting points

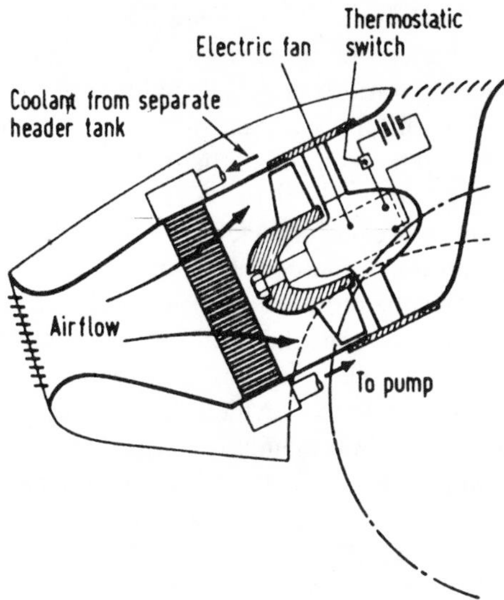

Fig. 4.19 Fully ducted system which could improve cooling system efficiency

of view. The air exit should preferably be in a region of low pressure, a condition which will be more readily obtained if a convergent outlet is used.

All practical ducted radiator systems used in piston-engined fighter aircraft were divergent–convergent. Such an arrangement is also likely to be the most practical for passenger cars. The philosophy of a free-wheeling fan for top gear operation, the system then being entirely ram dependent does not really give maximum economy since at lower ambient temperatures the fan may not be necessary at all. With the divergent–convergent duct the fan may be located in either the inlet or outlet.

HEAT FLOW WITHIN THE ENGINE

Cecil French[27] of Ricardo explains that temperatures of the cylinder head, barrel, and piston must be maintained at a sufficiently low level for the material to retain adequate mechanical strength under all conditions of operation. Furthermore, a reasonably low temperature on the gas face is essential for satisfactory volumetric efficiency, while the temperature of lubricated rubbing surfaces must be low enough for an effective lubricant film to be maintained without the risk of carbonization of the lubricant. Gas-side metal temperatures in petrol engines must be held to a reasonably low figure in order that excessive resistance to detonation and pre-ignition shall not be required of the fuels.

The main source of the heat to be removed is the hot gases in the cylinder. Friction, however, especially that of the piston and its rings, is an important additional source. Heat arising from the friction of the piston and rings must be removed either from the cylinder barrel or from the piston. Sufficient oil must be passed through other lubricated bearings, for example the crankshaft bearings, to carry away heat arising from friction, since it is difficult to provide other cooling. Heat transfer in the internal combustion engine mainly involves convection and, to a lesser extent, conduction. Radiant heat transfer is less than 5 per cent of the total engine heat loss, and it is convenient to include this relatively small proportion under convection.

Conductive flow

The problems encountered in applying the Fourier rate conduction equation in practice are primarily those due to the complex shapes involved. This applies particularly to the cylinder head and to the piston crown, a solution necessitating some form of mathematical analogue. It might be thought that the cyclic nature of the heat flow into the gas-side surface of the metal would make it difficult to calculate the temperature profiles. In fact, however, the transient temperatures only exist in a thin layer of metal at the surface, and the heat flow to the coolant is independent of the fluctuations, Fig. 4.20.

Perhaps the most complex conduction problems involved in engine design are those associated with the transfer of heat in the piston/ring/cylinder system. Some early workers showed that in large diesel engines, since the oil film between ring and cylinder is very thin, the rates of heat transmission are very high. The heat flow between piston and ring is less effective and the ring temperature is much closer to that of the liner than that of the piston. Piston rings actually play a very important part in transferring heat: on one large four-stroke diesel engine with an uncooled piston, it was shown that 40 per cent of the heat is transferred by the top ring.

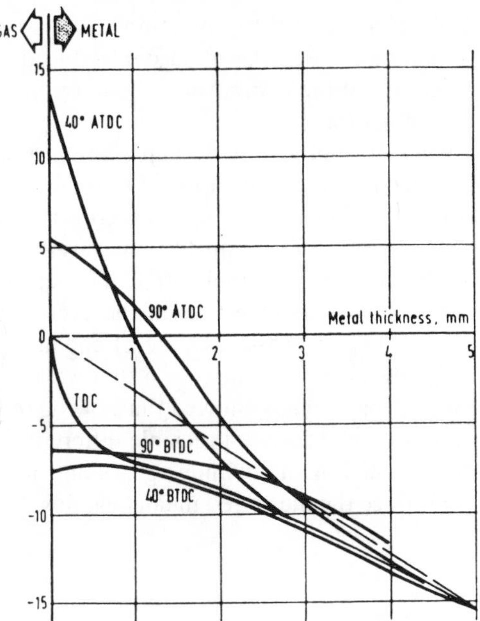

Fig. 4.20 Typical temperature fluctuations at a metal face, around a zero mean at 240°C, for different position in the engine cycle

Convective flow

This is the major heat transfer process in the internal combustion engine. It is unfortunate that so far as heat flow from the cylinder gases to the walls is concerned, it is also rather intangible, since the heat transfer at any one instant of time is some function of the temperature difference between the gas and the surface of the wall, the velocity of the gas, and its pressure. Clearly, all these vary considerably during the course of a single cycle. While it is possible, from an indicator diagram, to calculate the variations in the mean temperature of the gases in the cylinder, there are quite wide local variations.

Gas velocities also fluctuate during the cycle, ranging up to the sonic speeds that occur through the exhaust port during the early part of the exhaust process. Local variations are introduced by macroturbulence produced during the induction stroke, by squish and – in diesel engines –

by the air movements induced by the injection of fuel. In divided-chamber diesel engines, major gas velocities are also introduced by the movement of charge into and out of the combustion chamber during compression, combustion, and expansion.

For these reasons, it is impossible to apply any sort of rigid analytical approach. The procedure normally applied is to measure the heat flows experimentally and to apply a semi-empirical correlation. In most cases, experimenters have concentrated on the measurement of the average gross heat flows for the cycle. In their original experimental technique, however, the earlier workers investigated the cyclic variations in heat flow on a large slow-speed two-stroke diesel engine by measuring the corresponding variations in the length of a thin-walled tube introduced into the combustion space. They showed that, owing to variations in the heat transmission coefficient, an appreciable difference existed between mean cycle temperature of the cylinder gases and mean effective gas temperature to produce the measured mean heat flow.

Empirical formulae

The best known of the various formulae for heat flow from the gases to the cylinder walls are those due to Nusselt (4.1) and to Eichelberg (4.2)

$$h = 0.99\sqrt[3]{P^2T}(1 + 1.24C_m) \text{ Kcal/m}^3\text{/h/°C} \qquad (4.1)$$
$$h = 2.1\sqrt[3]{C_m} \times \sqrt{PT} \qquad (4.2)$$

where h is the heat transfer coefficient, P is the pressure of the gas in metric atmospheres, T is the absolute temperature, and C_m is the mean piston speed.

These equations were based on the results of experiments carried out in 'bombs' and on large slow-speed diesel engines with quiescent combustion systems. There are grave doubts as to whether they are applicable to automotive types of engines, although later workers have claimed good correlation in their cycle calculations when using the Eichelberg equation for an engine of 125 mm bore. However, others have shown that on small petrol and diesel engines, the heat flows follow laws which are closer to the $\frac{2}{3}$ power of the speed than the $\frac{1}{3}$ power suggested.

Another empirical approach has a form of equation which has proved very effective in correlating the results of a wide-range series of tests on heat flow in the case of fluids flowing through tubular passages. This form is

$$\frac{hL}{k} = c\left(\frac{GL}{\mu g_0}\right)^n\left(\frac{C_p\mu g_0}{k}\right)^m \tag{4.3}$$

where h is its heat transfer coefficient; L is a characteristic dimension (cylinder bore for an engine); k is the thermal conductivity of its fluid; G is the flow per unit area; μ is the fluid velocity; g_0 is the gravitational constant; C_p is the specific heat at constant pressure; and c, n, and m are constants derived from experiment.

Hence, for the gas-side heat flow

$$\frac{hL}{k} = C'\left(\frac{GL}{\mu g_0}\right)^n \tag{4.4}$$

where C′ is another constant. For the heat transfer to the coolant, be it air or water, forced convection is the process involved, and equation (4.3) is applicable.

With liquid cooling and the very high local rates of heat flow that are involved at high engine ratings, however, one enters the nucleate boiling regime in which steam bubbles are formed in the liquid at the metal face. Here, the metal surface temperature is almost independent of coolant temperature and velocity. To reduce the metal temperature, it is necessary to increase the coolant velocity until nucleate boiling is completely suppressed. Further increase in velocity will then reduce the metal temperature in the normal forced convection regime. The effects of nucleate boiling are shown in Fig. 4.21.

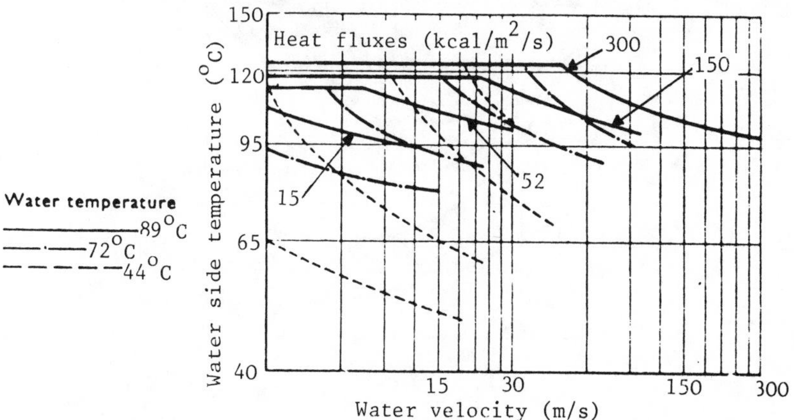

Fig. 4.21 Effects of nucleate boiling

Practical heat flows in engines

Considerable evidence is available nowadays concerning the gross heat flows that have to be removed from the engine. Typical jacket heat losses from a number of engines are given in Fig. 4.22; this information is largely taken from unpublished sources. For both petrol and diesel engines, researchers have shown that, where the speed and load are varied over a wide range of conditions, the gross heat is proportional to the fuel supply rate raised to the power 0.64–0.68, as shown in Fig. 4.23. So far as the engine designer is concerned, however, the total jacket heat is rarely a vital factor, since the heat loss does not seriously affect the efficiency of the cycle, and the final dissipation by means of the radiator is usually generously catered for. The real difficulty lies in the high local temperatures which exist due to either high local heat fluxes or local interruptions in the cooling.

A rather limited amount of information is available as to the quantitative values of local heat flows. One work gives comprehensive figures for

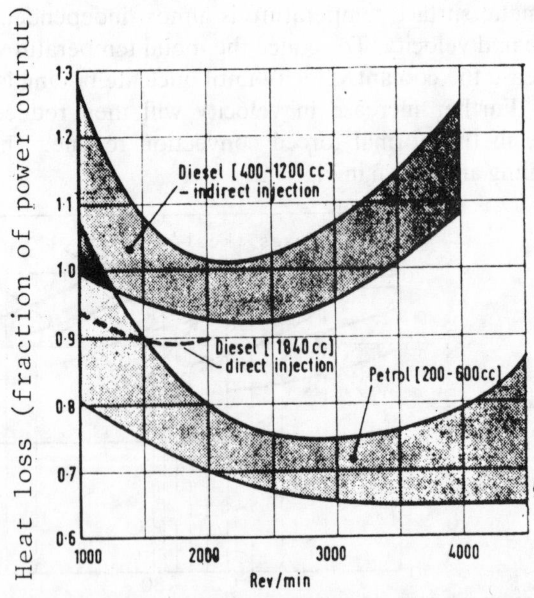

Fig. 4.22 Jacket heat losses as fraction of power output for cast-iron headed engines

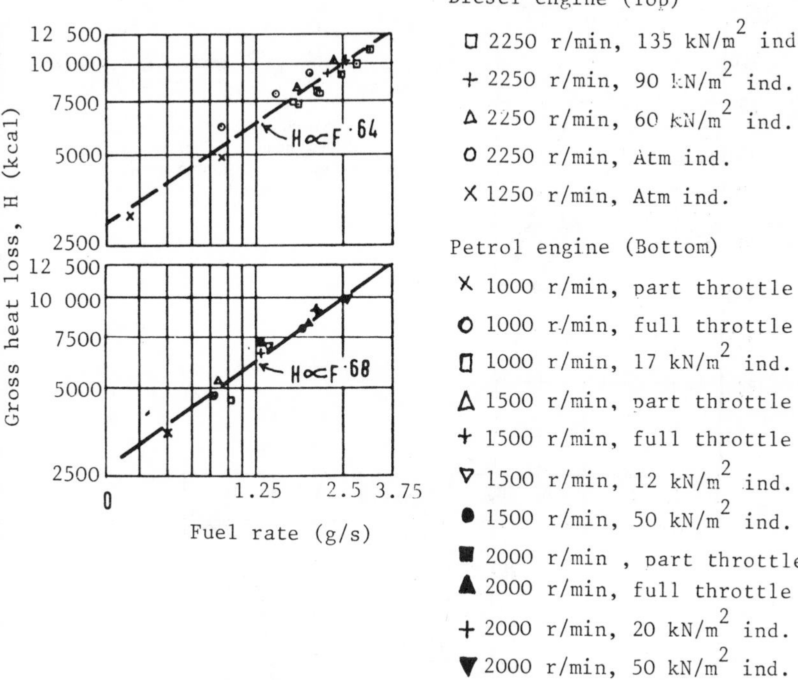

Fig. 4.23 Linear relation between gross heat and fuel supply for different engine types

two direct-injection diesel engines, together with spot values for a petrol engine (Fig. 4.24).

Use of basic data in engine design

Precautions have to be taken to increase the cooling rate in certain regions of high local heat flow, but in the smaller sizes of engines most design details intimately connected with the transfer of heat are decided by other considerations, such as foundry techniques. Since metal sections are a function of engine size, the gas-side temperatures increase with increasing cylinder size, assuming the coolant temperature to be constant. Consequently, heat transfer problems are much more severe in very large engines, and it is probably for this reason that the majority of the early work was carried out on large slow-speed diesel units.

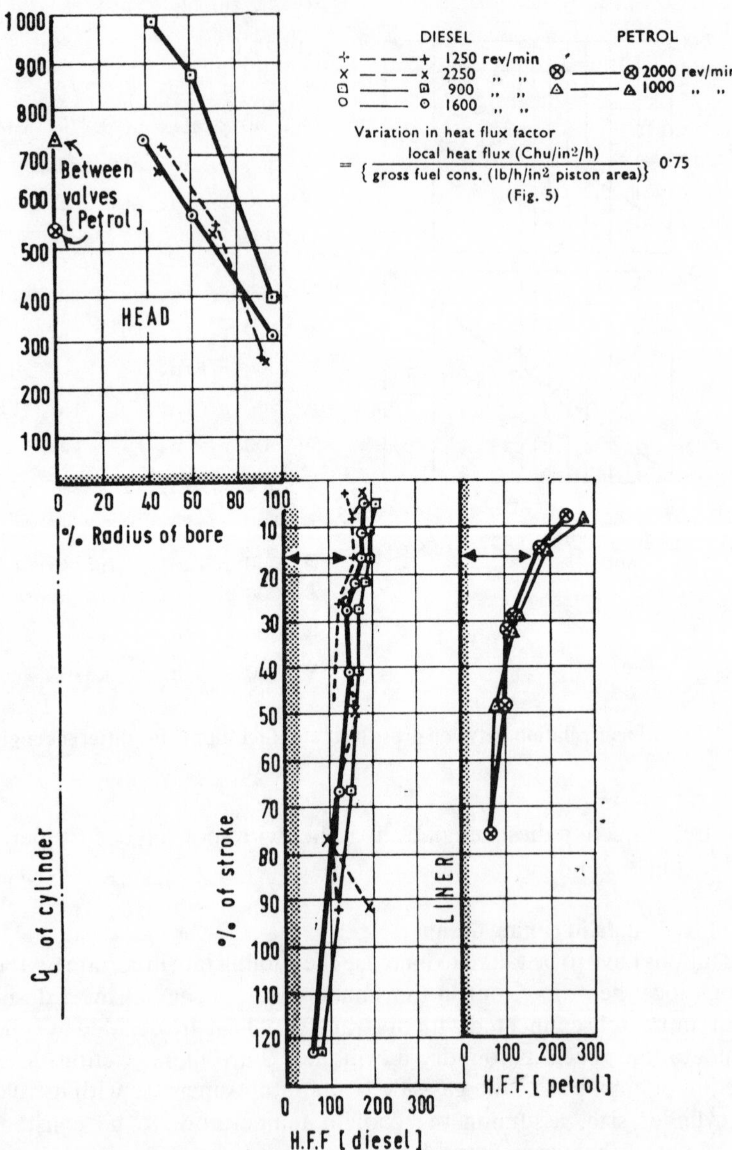

Fig. 4.24 Relative values of local heat flows for different engine types. Heat flux factor (HFF) = Heat flux/gross fuel consumption per piston area

In general today, however, the main thermal design problem in the vehicle petrol engine concerns the metal sections and coolant velocities in the immediate vicinity of the combustion chamber: the temperatures of the spark plug, exhaust valve and seat, and the gas-side metal must be prevented from becoming excessive. With increasing compression ratios and with the advent of turbocharging on petrol engines, these problems are likely to harden.

So far as the diesel is concerned, and especially the indirect-injection designs and supercharged units of all types, appreciably higher heat flows are involved than in the petrol engine, so difficulties are more serious. As thermal difficulties increase, it is more than ever necessary for the designer to consider how he can apply the available information in order to maintain metal temperatures at a reasonable level. Sufficient information has already been given to enable calculations to be made of the heat flow from the gas to the coolant and of that through the metal. The other required information is that concerning the heat from the metal to the cooling medium, and it is necessary to consider separately the cases of liquid and air cooling.

THE COOLANT HEAT EXCHANGER

S. F. Hurford[28] considers it necessary for forced convection to be usefully redefined as the process whereby heat is transferred from one fluid to another or from one fluid to a wall containing that fluid, the relative motion being due to a pressure difference applied to the fluid independent of the heat flow. The amount of heat transferred from one fluid to another – in the case of the car 'radiator', from coolant water to air – is dependent upon the mean temperature difference between the two fluids, the areas of metal in contact with each fluid, and the individual heat transfer coefficients. The areas involved are relatively simple to compute. In the case of the normal car radiator of the gill and tube type, they consist, on the air side, of the external surface area of the tubes and both sides of the gill or secondary surface; on the water side, they comprise the internal surfaces of the tube. The term secondary surface is used to describe any additional surface which is immersed in one of the fluids only.

Theory of the heat exchanger

For the normal theoretical analysis used for heat exchanger design, the mean temperature difference (MTD) depends upon the inlet and outlet temperatures of both hot and cold fluids.

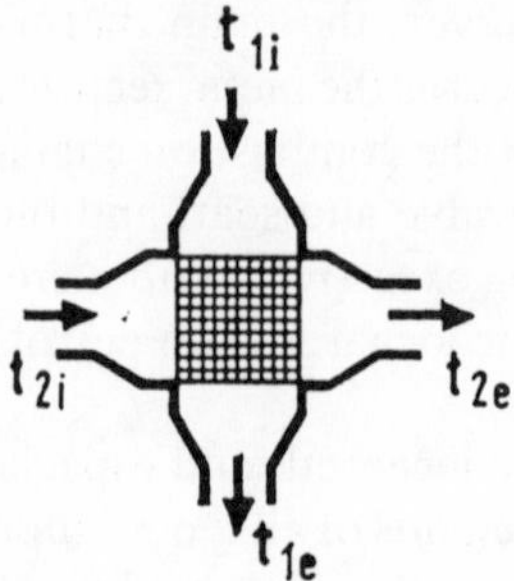

Fig. 4.25 Radiator considered as one-pass, cross-flow heat exchanger with both fluids unmixed

In the case of pure contra-flow

$$\text{MTD} = (A - B)/\log_e (A/B)$$

In this expression, which is sometimes referred to as log MTD, A is the largest of the two terminal temperature differences, t_{1e}-t_{2e} or t_{1e}-t_{2i}, B being the smaller difference. This is only exact for pure contra-flow (Fig. 4.25), and for other configurations correction factors must be applied, Fig. 4.26. Temperatures (t) are t_{1i} = inlet hot, t_{2i} = inlet cold, t_{1e} = outlet hot and t_{2e} = outlet cold, $P = (t_{2e}-t_{2i})/(t_{1e}-t_{2i})$ and $R = (t_{1i}-t_{1e})/(t_{2e}-t_{2i})$. Effective MTD = log MTD × F.

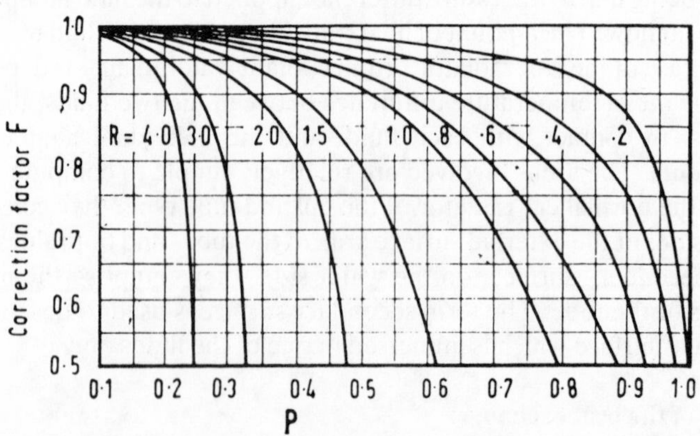

Fig. 4.26 Correction factors are needed in calculation for different heat exchanger configurations than that shown in Fig. 4.25

Heat transfer coefficients obtained from an analytical method can only be relied upon if the surfaces involved are simple ones, such as circular tubes or flat plates; the form of these equations is based upon dimensional analysis, the numerical values of the constants to be used being found empirically.

Choice of the equation to be used depends upon the flow regime existing. The value of the coefficient cannot be extrapolated between the limits of one equation and another, because violent deviations from a smooth transitional curve from streamline to turbulent flow are usually experienced. This applies to plain surfaces which do not give rise to any boundary layer breakaway. However, complex louvred, dimpled, and interrupted surfaces (or those where artificial turbulence is induced) produce a smooth transition curve. This is illustrated in Fig. 4.27, in which a matrix having staggered tube rows is compared with another having in-line rows. This shows that rate of increase of heat transfer (expressed in terms of Stanton and Prandtl number) and fluid friction (expressed by Fanning friction factor) changes as flow alters from laminar to turbulent. Transition is smoother in A than in B. For matrix A the slope of the curve for St $Pr^{2/3}$ for turbulent flow is similar to that of the orthodox relationship referred to earlier in this chapter. A sharp change in slope occurs at the critical Reynolds number, the curve becoming asymptotic to the usual curve for streamlined flow. The corresponding curve for matrix B shows a smooth change in slope through the transitional Reynolds number range. The best relationship between friction factor and heat transfer coefficient for surfaces producing induced turbulence is usually obtained at or near the critical Reynolds number.

Secondary surfaces

If the thermal resistance of the scale deposits is ignored and the coefficients for the laminar fluid layer and the turbulent film are combined, for the case of an actual water jacket metal/coolant interface, a simplified picture results from which it is clear that the heat transfer coefficient for water is many times that for the air side.

Since the product of overall coefficient and the heat transfer area $h_0A = (h_aA_a \times h_wA_w)/(h_aA_a + h_wA_w)$ (where subscripts a and w refer to air and water), then the best result is obtained when $h_aA_a = h_wA_w$ and $h_0A = (h_aA_a)/2 = (h_wA_w)/2$, so for balance $h_w/h_a = A_a/A_w$.

In other words, the air-side heat transfer area must be increased inversely as the ratio of the heat transfer coefficients. This is usually

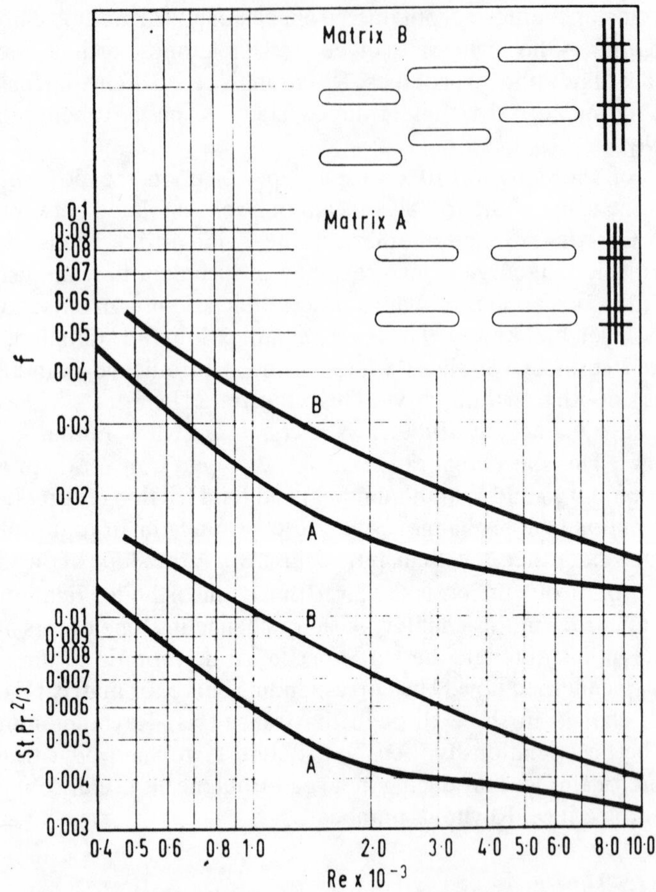

Fig. 4.27 Flow regime comparison for two types of heat exchanger matrix

achieved by adding secondary surfaces to the air side; their function is to conduct heat from the primary surface and transmit it by convection to the air stream.

Fouling factors

Standards of the Tubular Exchanger Manufacturers Association give factors of two times for the air side and three times for the water side (hard water) in the 'radiator', of the figure for the engine jacket on the water side. For a road vehicle cooling system, since the minerals in hard water are deposited in a fairly short time and the system is only rarely drained, quoted figures (reciprocal heat transfer coefficients or resistances to heat flow) should be treated with caution.

Taking the resistance of the tube material into consideration, with fouling factors for the air and water sides, the equation for overall thermal resistance becomes

$$\frac{1}{h_0 A} = \frac{1}{h_A(A_p + \eta_f A_s)} + \frac{r_A}{(A_p + A_s)} + \frac{1}{A_m k/t} + \frac{r_w}{A_w} + \frac{1}{h_w A_w}$$

where r_A = air side fouling factor; r_w = water side fouling factor; h_A = air side heat transfer coefficient; h_w = water side heat transfer coefficient; A_p = air side primary heat transfer area; A_s = air side secondary heat transfer area; A_m = mean tube heat transfer area; A_w = water side primary heat transfer area; k = tube material conductivity; t = thickness of tube material; and η_f = air side secondary surface fin efficiency. The value of $(h_0 A)$ obtained from the above is equal to heat dissipation/MTD.

Matrix design

The designer of the basic matrix is faced with the problem of transmitting the necessary amount of heat from one fluid to another in the most economical way; that is, with the minimum of material in a minimum space with the minimum power losses in terms of flow and pressure drop, always subject to material and process costs.

For a radiator, the cheapest matrix will, in general, be that which gives the highest air-side heat transfer coefficient. However, this statement must be qualified by considerations of fin efficiency and fouling: the commonly accepted ultimate heat transfer surface – that of the 'rose thorn' or even the 'plain pin' fin – would be very susceptible to fouling and far too costly to manufacture.

The practical approach is to manufacture a series of test samples based upon the most promising matrix forms suitable for mass production. These samples are mounted on a blower rig and tests conducted over a wide range of air and water flows. The mean temperatures are selected to be representative of those expected in the final full-scale radiator.

In the rig just mentioned, a sample matrix is mounted on a blower, and a grid of thermocouples is mounted at the air outlet face; these are connected in parallel to indicate the mean air outlet temperature. A weighing gear is used to measure the water flow, while air flow is measured by a standard form of venturi meter. Thermometer pockets are fitted in the water flow and return lines and the air inlet temperature is measured at the blower outlet. Static pressure tappings in the wall of the duct give a reading for the air inlet pressure, and hence air pressure drop across the matrix.

Results of tests carried out in this way are analysed and plotted in the form: overall specific heat dissipation over the facial area *vs* air velocity at the matrix face, Fig. 4.28. The mean temperature difference used for this is mean water temperature minus air inlet temperature. This method ignores the value of the air outlet temperature and is therefore not an exact one, but the error involved when the method is used in conjunction with the results shown in Fig. 4.29 is very small; the advantage is that the usual reiterative mathematical process is avoided when designing a matrix to fulfil a given set of conditions. Examples of standard design data sheets derived from such tests are shown in Fig. 4.30. This example is for a fin-and-tube type matrix having three rows, 60 mm deep in direction of air flow. Tubes are brass, 15 mm × 2.5 mm × 0.007 in seamed, having a through area of 0.04578 in^2 per tube, or 0.0321 ft^2 per foot width, and hydraulic diameter of 0.0118 ft. There are 101 tubes per foot width. Fins are 0.005 in thick B 4714 copper, dip-bonded to the tubes. Matrix weight = 12.25 lb/ft^2 facial area, corresponding to the Imperial units of the figure.

FAN TECHNOLOGY AND ENGINE COOLING

Recent development work undertaken by Airscrew Howden has covered many aspects. S. P. Hawes[29] considers that fan technology is now a major area of cooling-system design. Recent developments in system design include the improvement of radiator design to minimize the power required by the fan; reduction of the total noise by the development of new types of fan with higher pressure coefficients; techniques for eliminating discrete-frequency noise; thermostatic control of fan-speed to save fuel; and the use of self-contained cooling-system 'packages'.

In low-speed and off-road vehicles, the fan is needed under all operating conditions to transfer waste heat effectively from the cooling-water circuit to the atmosphere. Only at high speeds can the ram-effect provide all the necessary airflow, and even high-speed vehicles may need fan assistance when idling or moving slowly. The power consumed by the fan represents a parasitic demand on the engine and also causes a measure of

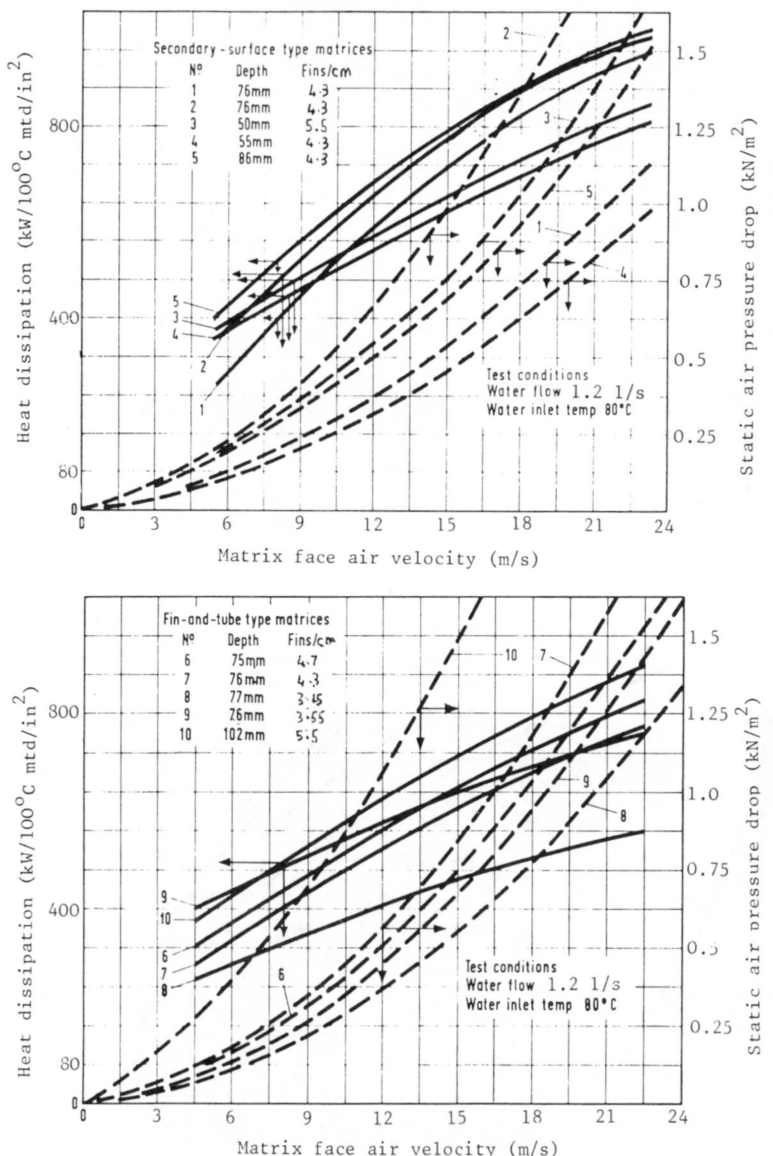

Fig. 4.28 Water-side performance curves for five configurations of extended surface type matrix (*left*) and five fin-and-tube (*right*)

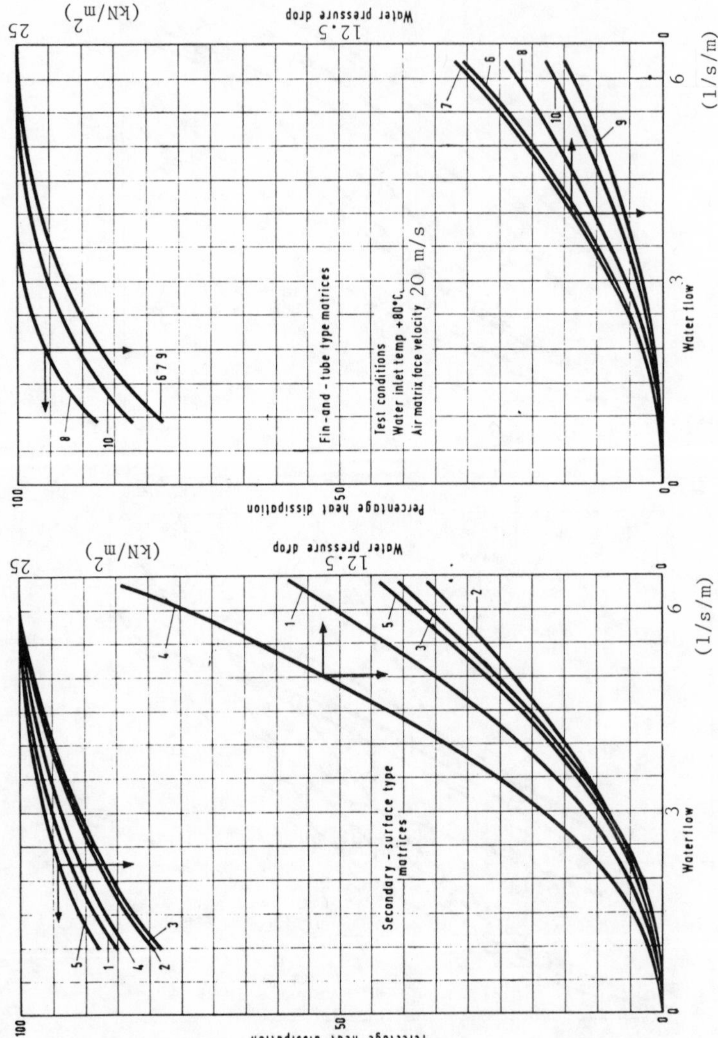

Fig. 4.29 Air-side performance curves corresponding to those in Fig. 4.28

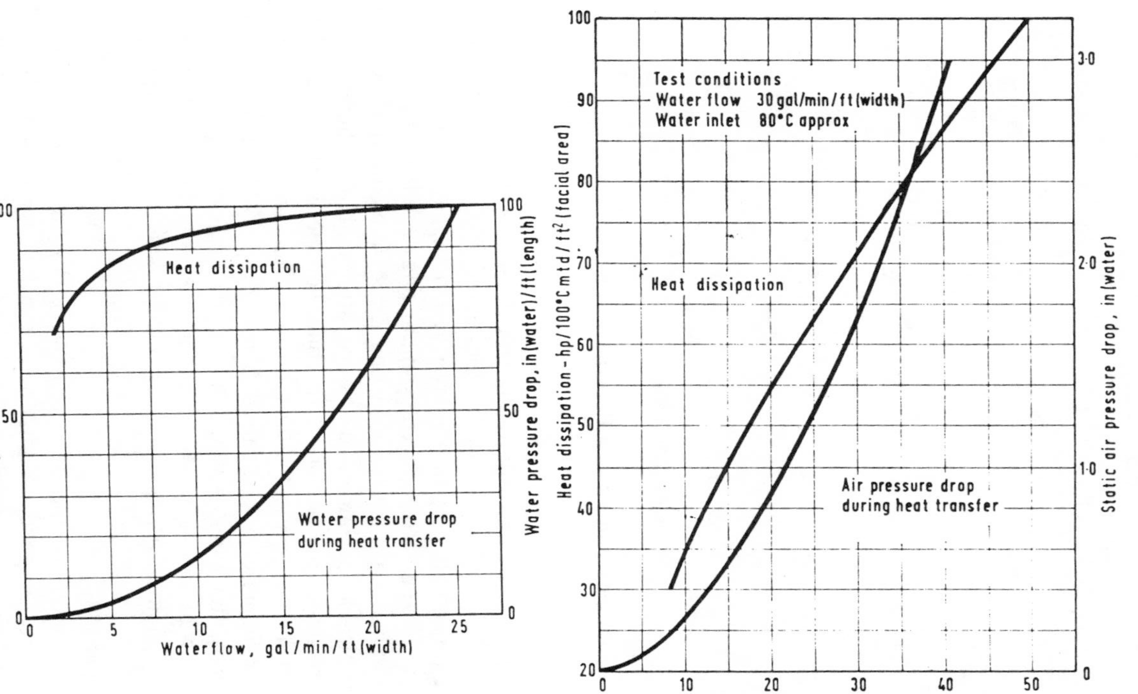

Fig. 4.30 Test data for use in matrix design

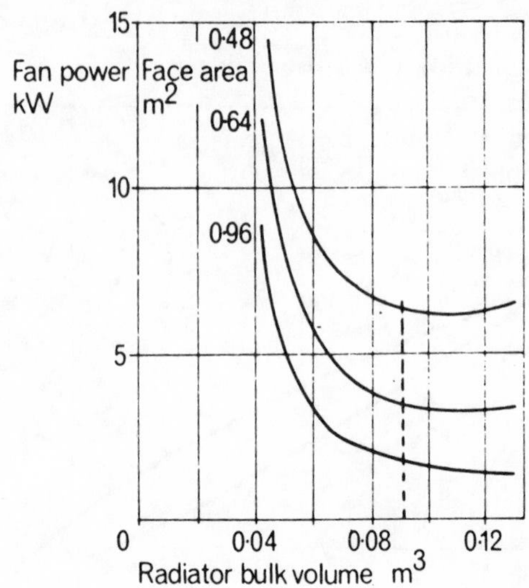

Fig. 4.31 Variation in fan power with radiator bulk volume for three face areas

the fan's noise output. On both counts it is desirable to keep the fan power low. However, the fan cannot be considered in isolation; its power requirement is affected by the radiator design and air-ducting arrangements, and can be greatly reduced by optimizing these features.

For the radiator, for example, two basic requirements are to use the largest practicable face area and an adequate total bulk volume. Figure 4.31 clearly shows the penalty if the bulk volume is insufficient. The curves apply to one commonly used type of radiator core and assume a fan efficiency of 50 per cent, a required dissipation of 210 kW, an air/water temperature differential of 50°C, and no air-circuit pressure losses external to the radiator. A bulk volume of 0.09 m^3 would give a reasonable fan power without incurring excessive radiator cost.

Design target

The corresponding specific dissipation of 2300 kW/m^3 can be considered a target figure for good design when pressure losses external to the radiator are low. As external pressure losses increase, this target falls to around 1900 kW/m^3, the optimum bulk becoming greater because of the beneficial effect of the smaller mass flow required. Notable savings of power have been obtained by this optimization technique.

Fin spacing has been analysed similarly. Minimum power requirements for the normal range of specific dissipations occur with spacings between 1.4 mm and 1.1 mm. Savings are greatest when external pressure losses are high. In practice, therefore, fins should be as closely spaced as possible. In one application, reducing the spacing from 3.6 mm to 2.1 mm lowered the power requirements by over 50 per cent.

Reducing noise

On many motor vehicles the fan is a major source of noise, but correct fan selection can avoid this. We have found that a fan's total sound output can be expressed with reasonable accuracy by making sound power level (dB) relative to 10^{-12} W equal to 47 + 9 log kW + 28 log tip-speed (m/s) + 2 log diameter (m).

Noise increases with both the power and the tip-speed of the fan. So, to reduce noise, the designer should: (a) minimize the fan's power requirement, as already discussed; (b) maximize fan efficiency; (c) select a fan with low tip-speed. And since the pressure coefficient = static pressure ÷ $(\text{tip-speed})^2$, reducing the tip-speed implies increasing the pressure coefficient (which may anyhow be necessary to overcome the higher pressure losses of optimized systems). The following indicates the possible savings. Halving the fan power by reducing the pressure losses, without changing the type of fan, will reduce the tip-speed by 1 : 1.41 and increase the diameter by 1.2 : 1. Noise level will accordingly drop by $9 \log 2 + 28 \log 1.41 - 2 \log 1.2 = 6.7$ dB. Halving the tip-speed for the same power, by changing the fan type, will scarcely affect the diameter, and the noise reduction will then be $28 \log 2 = 8.4$ dB. Noise reductions of this order normally suffice to bring the fan noise within the latest legal limits or below the level of the general vehicle noise.

New fan types

The pressure coefficient is a dimensionless quantity constant for all fans of one type, irrespective of size. A typical axial impeller for trucks has a value around 0.07, while centrifugal fans have values of 0.4 or more, but are ruled out for automotive applications by the bulk of their volute casings.

Two recent designs of fan have intermediate values, permitting higher pressure losses or lower noise levels, or a mixture of both. The *high-solidity axial fan* has numerous short blades around a relatively large hub; typically a 20-blade version of 710 mm diameter has speeds up to 3200 r/min and a pressure coefficient of 0.15. To reduce manufacturing costs, the aerofoil blades are mass-produced acetal-resin mouldings which can be bolted to steel hubs of various sizes. Adding stator vanes to remove the swirl generated by the fan can further raise the pressure coefficient where space and cost permit.

The *open-running mixed-flow fan*, used with open discharge and no casing, is a patented concept capable of still higher pressure coefficients. This design typically has a coefficient of 0.27 and is easily capable of 4000 N/m^2 pressure. Such fans are cast in aluminium alloy and can be mounted on a bulkhead or partition and driven mechanically or hydraulically. They are used mainly in military vehicles, earth-moving equipment, and heavy trucks.

Discrete frequencies

In constricted automotive layouts, fan inlet turbulence and interference from supporting structures are always present to some degree. Both can create discrete-frequency noise – corresponding to the blade-passing frequencies and their harmonics – which may well be more obtrusive than total fan noise. A new technique for eliminating such noise, and one that is particularly suitable for multi-blade fans, is to space the blades unevenly according to a planned sequence. This spreads the sound-power over a range of frequencies and usually eliminates any irritating whine. To prevent interaction tones between rotor blades and stator vanes, the best combination of numbers can be investigated at the design stage. Figure 4.32 shows noise spectra (bandwidth = one-third octave) for a small fan

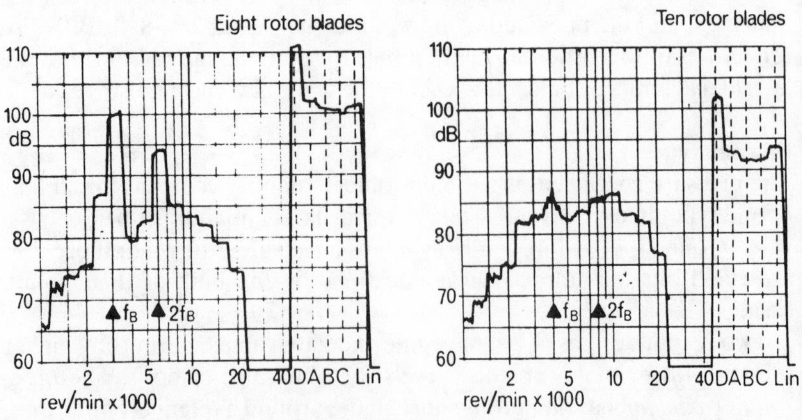

Fig. 4.32 Noise spectra for a small fan run in eight- and ten-blade form with seven upstream stator vanes

running at 22 000 r/min. There were seven stator vanes upstream of the rotor, which was tested with eight and ten blades. Adding two blades virtually eliminated discrete-frequency noise peaks (left), while overall noise level (right) fell by 9 dB. Subjectively the change was from a fan with an intensely loud scream, unapproachable without ear-muffs, to one completely acceptable.

Another technique that can sometimes be applied to existing rotor/stator fans is to cut back the leading edges of alternate stator vanes to create, in effect, non-uniform spacing. In one application recently a cut-back of 4 mm reduced the sound intensity at the blade-passing frequency by 7.5 dB.

Thermostatic control

An automotive cooling system must have the capacity to handle the highest possible rate of heat rejection by the engine at the highest likely ambient temperature – conditions generally occur infrequently in combination. At other times a much lower airflow would suffice; excess airflow causes the engine thermostat to close and represents wasted power at the fan. Conventional engine-driven fans may, therefore, be very inefficient in practice. Indeed, where ram air is available they work best when assisted airflow is needed least – at high road speeds. This problem is solved on some cars by the use of an electrically driven, thermostatically controlled fan which gives high airflow even at low engine speeds, and cuts out at high road speeds. The latest designs are unducted, round-tipped propeller-fans, usually one-piece polypropylene mouldings. For military and other heavy vehicles, hydraulic drive giving thermostatic control of fan speed is easy to arrange. Alternative methods are a fixed-displacement pump and motor with a temperature-controlled by-pass valve, and a temperature-controlled variable-displacement pump with a fixed displacement valve.

Figure 4.33 illustrates the great power savings possible for vehicles working at less than maximum performance. It is assumed that engine speed is constant at the maximum and that thermostatic control operates over the water-temperature range 85–93°C. The lower part of the figure shows how the engine's heat output (expressed as a percentage of maximum heat-rejection rate) and the ambient temperature affect the water temperature. The top part of the figure compares the power consumed by the two hydraulic systems and by an engine-driven fan at different water temperatures.

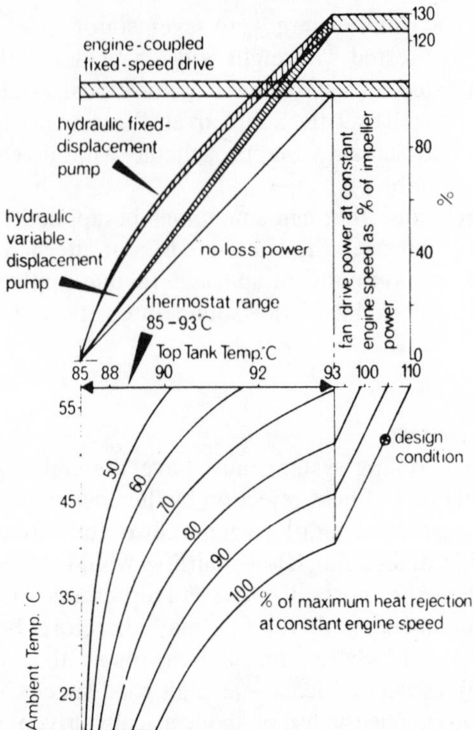

Fig. 4.33 Power savings possible with non-direct fan drive, and water temperature shown affected by engine output and ambient conditions

Up to a water temperature of around 92°C, which is reached only under severe conditions, both hydraulic systems save power continuously. For example, at *ambient* temperatures up to 36°C they save power even at the maximum heat-rejection rate. The variable-displacement pump has smaller internal losses (released as heat) and so saves more power.

Ducted systems

In an earlier article (*Automotive Design Engineering*, 1963, **2** (9), 56), the author discussed installation effects of fans.

Inlet loss

Many minor differences between test conditions and operating conditions can combine to have a pronounced effect on performance. The entry of

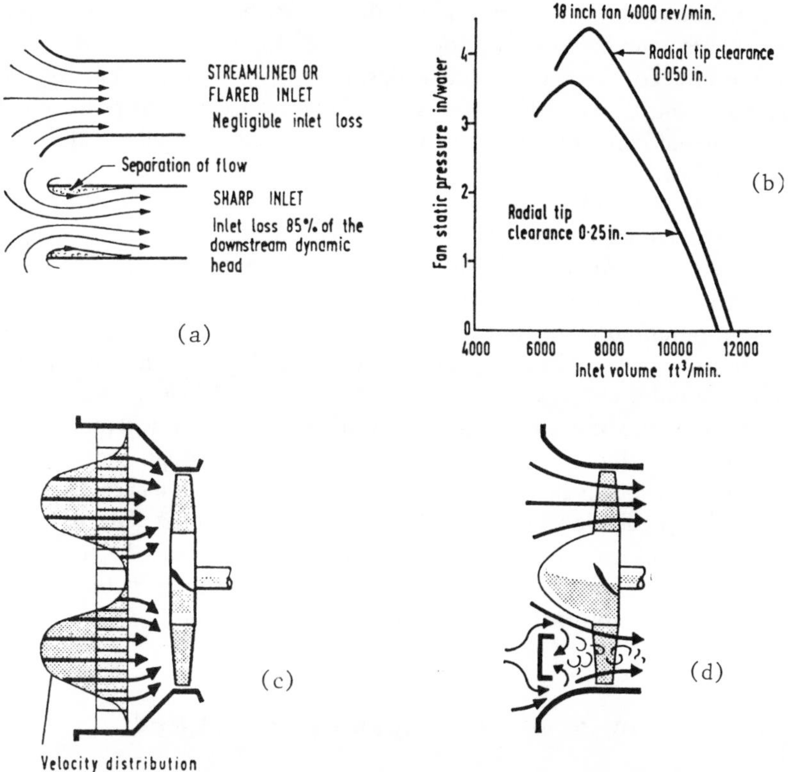

Fig. 4.34 Ducted systems – installation effects of fans

the air into the fan casing from an open space can be a source of pressure loss if a streamlined or flared inlet is not used. This is because an abrupt inlet causes the streamline of air to break away from the duct or fan casing, as shown in Fig. 4.34(a). Such losses also occur on entry to a duct system even if the fan is not located at the inlet, and should be allowed for. With a fan at the inlet, however, additional adverse effect is caused by the maldistribution of velocity over the fan blades, reducing the fan performance and efficiency.

Tip clearance

On axial flow fans excessive clearance between impeller and casing causes appreciable reduction in performance. Airflow near the tips of the blades is much reduced, with consequent reduction in the volume of air passed and in the peak pressure developed.

The effect of tip clearance on the performance of a 450 mm fan is shown in Fig. 4.34(b). To achieve the small tip clearance necessary for good performance and efficiency, a common mounting for impeller and casing is essential.

Interaction of system and fan

This is a very broad subject, and many things can happen to reduce the effectiveness of an installation. For example, in Fig. 4.34(c) an impeller is shown mounted close to a rectangular radiator. It is obvious that the cooling flow will not be uniform owing to the dead spots at the fan hub and at the radiator corners. The performance of the radiator will have been assessed on the basis of a uniform flow; non-uniform flow will reduce this performance and raise the resistance to airflow through the radiator owing to the locally high velocities. The installation shown in Fig. 4.34(d) reveals appreciable blockage in the region of the fan inlet. Owing to the high inlet velocity, pressure loss is excessive; moreover, the turbulent wake from the obstruction makes part of the fan annulus ineffective, as well as appreciably increasing the noise level of the fan.

DESIGNING CYLINDER HEAD GASKETS

Dr Ing. Gerd v. Bennisgen[30] considers the cylinder head gasket to be a component critical to cooling and lubrication systems. He points out that a controlled degree of differential compliance in the cylinder block structure around the liner has a major effect on minimizing liner distortion. Cylinder liner distortion is known to be caused predominantly by uneven distribution forces on the liner collar, or by an uneven compliance of the engine block due to its design. Because of these differences, axially disposed moments occur in the liner and these cause distortion transverse to the direction of the applied load. This has been effectively demonstrated using a simple paper model such as that shown in Fig. 4.35. Under the force system shown, the upper collar is deformed outwards at the compression points and inwards at the points of tension. A sort of ellipse is formed with its major axis passing through the compression points. The other end of the liner model forms an ellipse rotated through 90 degrees.

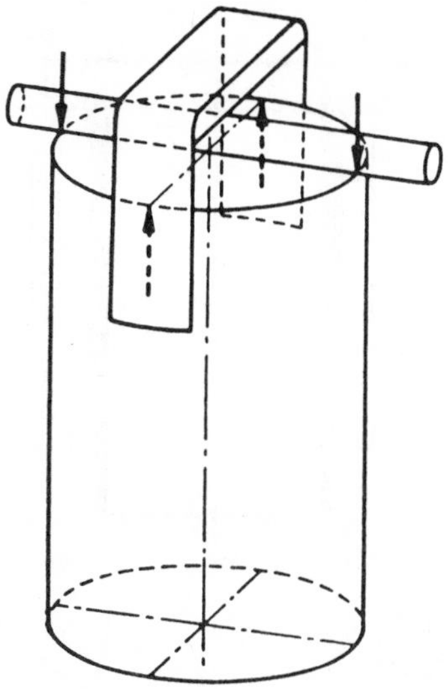

Fig. 4.35 Axial distortion of a cylinder liner

Transmission of cylinder head bolt forces

The engine designer has to cater for a variety of parameters: arrangement of the bolt holes, reinforcing ribs, and also the gas and liquid passages and walls; he has to ensure that the forces introduced by the cylinder head bolts are transmitted in such a way as to give as uniform as possible a pressure distribution on the edge of the cylinder liners. Cylinder heads and engine blocks are known to be statically indeterminate supporting systems full of openings; it is difficult, for example, to ensure completely uniform distribution of the pressure, even from the forces of four cylinder head bolds arranged symmetrically around a liner.

The forces introduced normal to the surface can only resolve themselves in an area transverse to the load direction, by forces and moments which are inevitably accompanied by elastic deformation. In addition to the suitable arrangement of the design elements lying in the plane of the applied force, the wall thickness of the top and bottom plates at right angles to the liner axis can also affect the amount of distortion. These wall thicknesses must be dimensioned so that the distortions, due to differences in the tightening pressure, remain within a controllable range. Needless to say, the bolt patterns have a decisive influence, and as far as

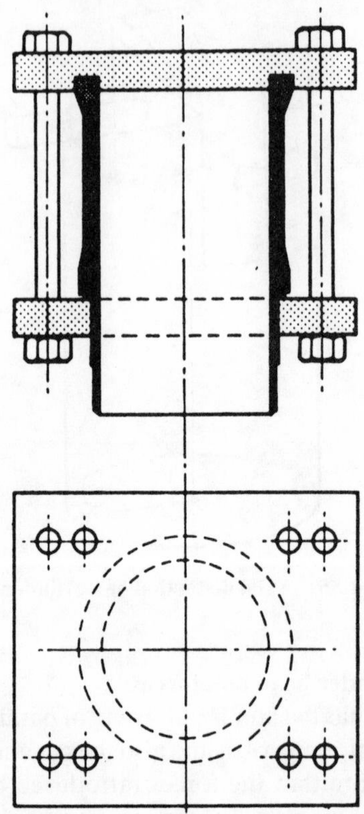

Fig. 4.36 Rig to study bolt hole patterns

possible should be arranged with a uniform but minimum radius, and at the same time be symmetrically disposed around the liner collar.

The effect of various bolt patterns has been studied on the idealized rig, seen in Fig. 4.36. Effects of tightening pressures on the circumference of the liner collar, reduced to a linear pressure on the developed cylinder, are shown in Fig. 4.37. The distortions are plotted as differential values with the 'level' unloaded liner as datum. As with the paper model, the diameter enlargements are approximately in the region of the pressure maxima, while the negative deformations appear roughly in the region of the pressure minima.

Even in an ideal embodiment of a radially symmetrical design, the pressure maxima would naturally always be in the close vicinity of the

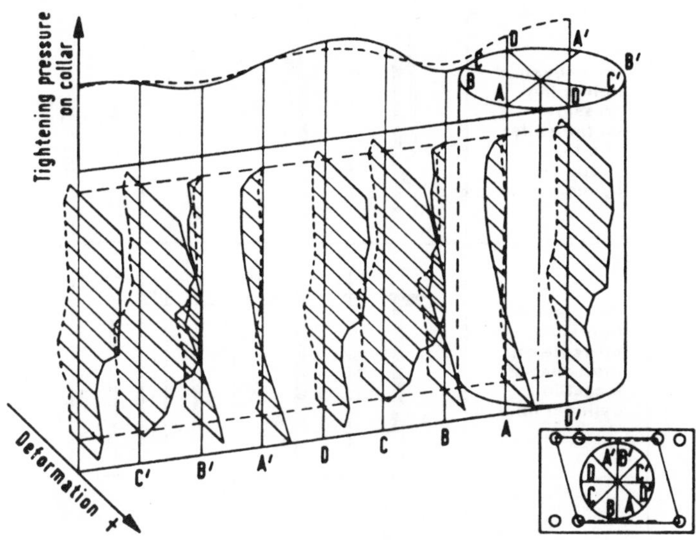

Fig. 4.37 Influence of bolt patterns and tightening pressure on liner distortion, using the rig shown in Fig. 4.36

bolts; the pressure minima between them can be raised to some extent, of course, by properly designed reinforcements. However, gas passages, spark plug arrangements, valve guides and passages for the coolant prevent the practical realization of an ideal design based solely on one function. Design measures for the functional development of simultaneous mechanical, physical, and chemical reactions in the limited region of the combustion chamber always require compromises to obtain an optimum at a reasonable cost. From this aspect, it appears almost impossible to control adequately the pressure distribution on the plane of separation, in particular in the liner collar region, by the design of the components alone. Figure 4.38 shows a number of possible factors in liner distortion.

Influence of the cylinder head gasket on pressure distribution

The necessary deformations resulting from the tightening force pattern must, of course, be contained in the force-locking connection between head and block. As it is hardly possible to do this at reasonable cost by exploiting the elasticity of the components, or even by suitably profiled

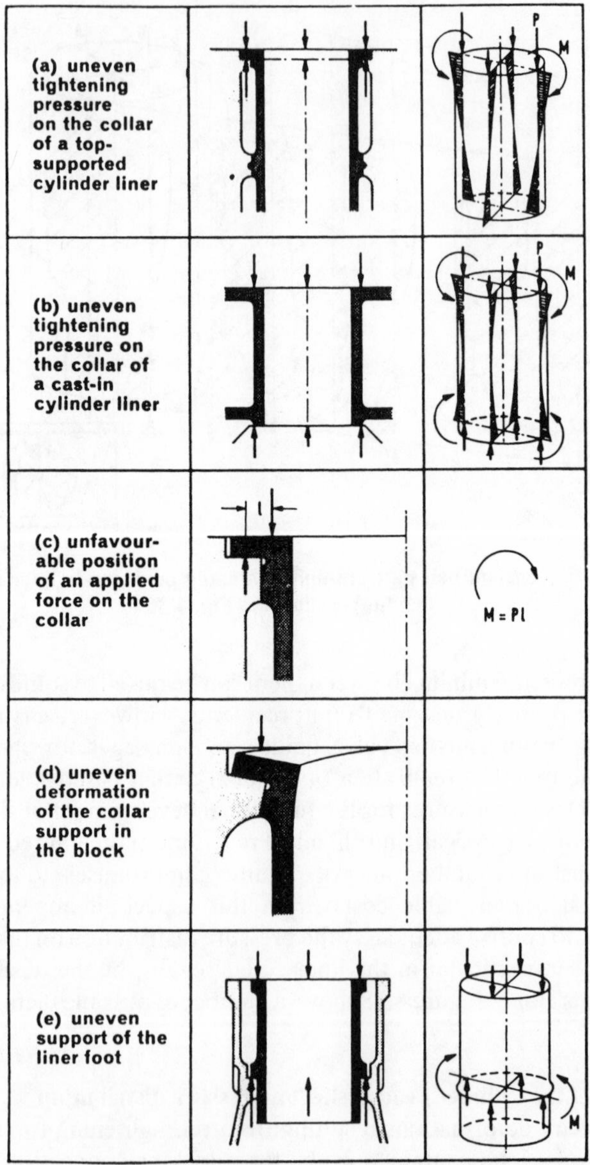

Fig. 4.38 Possible influences leading to liner distortion

separating surfaces, the compressibility of a gasket between them plays an important role in pressure distribution. This has led to the recent development, for small combustion engines, of cylinder head gaskets for which the maximum tightening pressure is concentrated on the combustion chamber periphery by means of edge-binding flanges. Conditions over the remaining area can be ignored, since on this area the specific surface pressure is only a fraction of the circular tightening load around the chamber.

Type and design of the edge-binding flange of combustion chamber apertures on cylinder head gaskets offer a number of possibilities for improving the required unform pressure distribution, and of compensating for liner distortion. Important factors are: (1) Forces tending to twist open the liner can be avoided, or at least reduced, by concentrating them into a narrow ring. The force resultant acting within the flange width and arising from the specific tightening pressure should as far as possible correspond to the position of the liner axial supporting forces. Annular gaps, which generally have an unfavourable effect on combustion, should be avoided. These requirements can be met by the use of folded flanges as shown in Fig. 4.39. (2) A local displacment of the resultant force radially can be obtained by flanges provided with lugs, Fig. 4.40. (3) An increase in the specific tightening forces on the circumference can be obtained by individual flange-folds. The increase in total thickness by the sum of layers must be allowed for, Fig. 4.41. The point where these additional thicknesses are required can be determined by pressure distribution measurements and by experimental determination of the liner distortion. To minimize cylinder head movement, it is necessary that cylinder head gaskets should be as thin as possible.

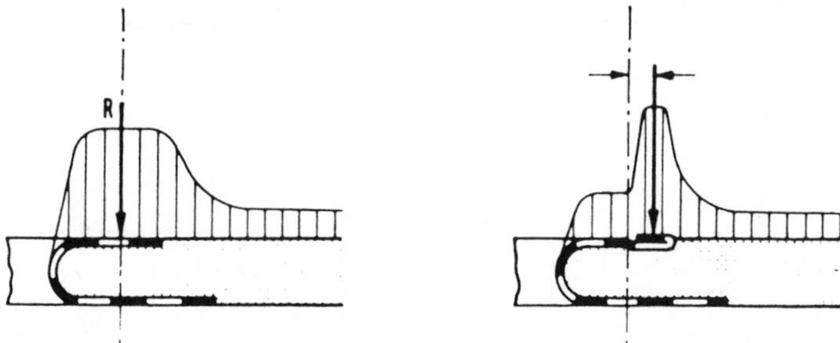

Fig. 4.39 Displacement of force-resultant by additional flange turn-over

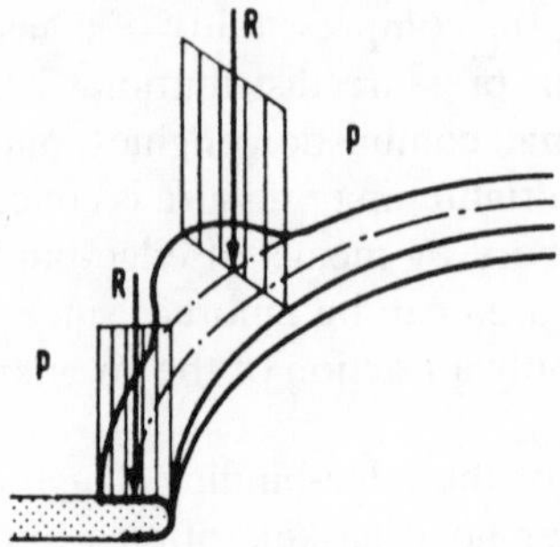

Fig. 4.40 Displacement of force-resultant by flange lugs

Dynamic properties of flat gaskets

G. Stahl[31] has established that the static elasticity coefficients of the gasket material may be determined relatively easily, but the dynamic properties are of greater interest to engine designers. To determine these properties an hydraulic simulator (Fig. 4.42) has been developed. This imposes an almost sinusoidal load, $F_i(t)$, on the gasket material, causing a dynamic variation in thickness (a sealing gap variation). It can also be used on joints compressed by bolts. Internal force, $F_i(t)$, and bolt elongation, $y_B(t)$, are measured by strain gauges which are linked up to an oscilloscope enabling the bolt elongation/internal force diagram to be read directly.

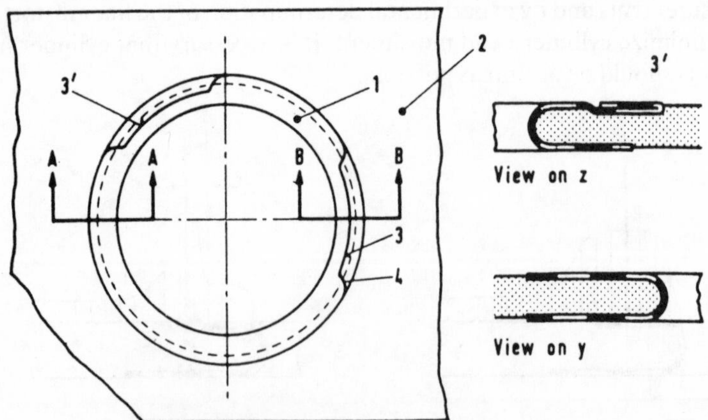

Fig. 4.41 Local increase in tightening pressure effect due to circumferential folds

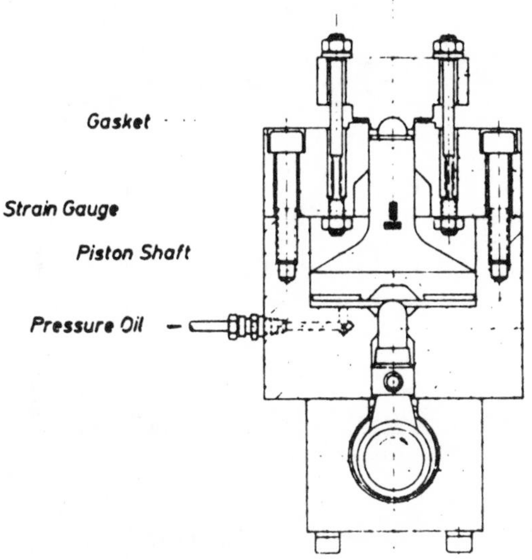

Fig. 4.42 Hydraulic simulator for studying dynamic properties

The diagram can then be used to determine the elasticity coefficients of the elements in the gasket, but the simulator must be subjected to dynamic loads without the gasket in place so that these can be found from

$$G_G = \frac{b}{(y_B - y_B')} = b/y_G = \tan\beta$$

where

$$y_G = (y_B - y_B')$$

represents the increase in sealing gap.

As the point of bolt action is not in the centre of the sealing surface, an additional measurement of gap variation is made using a very small flat spring element fitted with strain gauges. This is fitted into the sealing gap and indicates changes of gap measurement y_G directly. Figure 4.43 shows F_i against y_G for a reinforced soft gasket. Note the hysteresis effect due to internal damping, W_G, of the material. Tests have shown simulator components to have negligible influence on damping. It has also been noted that the origin of the damping loop is displaced by y^* from the initial static case $y_G = 0$.

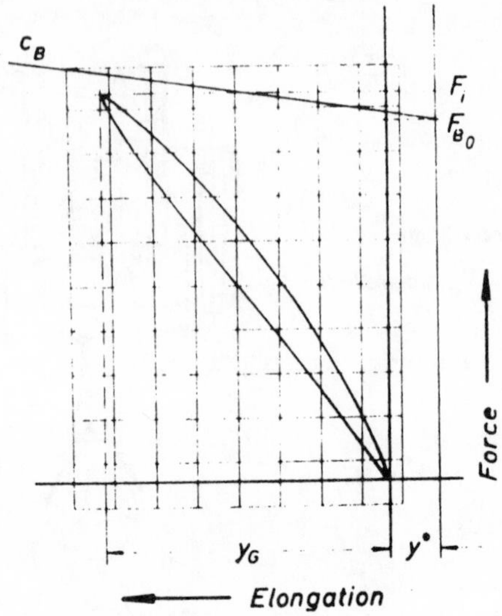

Fig. 4.43 Damping loop of a flat gasket

Numerous measurements have shown that y^* increases almost linearly with increasing y_G. This is obviously due to additional internal friction in the gasket material. On relieving the dynamic stress, the sealing gap returns very slowly to the original static dimensions, $y^* = 0$. Under dynamic stress the cover appeared to be raised above the initial static state so that the bolt no longer returns to F_{B_0}.

A correction factor is therefore needed as the bolt is displaced in practice by y^*c_B for every damping loop measurement with maximum bolt elongation y_B (Fig. 4.44). As tests have shown that y^* increases linearly with y_G, a modified bolt characteristic can be designed c_B^* with a somewhat steeper slope than characteristic c_B.

The following equation therefore applies for dynamic force

$$F_B = F_{B_0} + (y_B + y^*)c_B = F_i + F_G$$

where F_G is sealing force and F_{B_0} is bolt tension. At the origin of the damping loop the bolt force is

$$F_{B_0}\,\text{dyn} = F_{B_0} + y^*c_B$$

F_B has thus increased by y^*c_B. Since F_i is a measured value, y^*c_B causes an increase of effective sealing force F_G. In dynamic tests on a reinforced

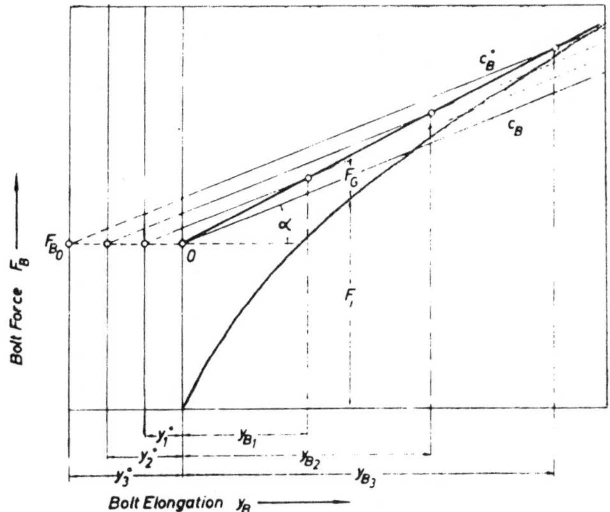

Fig. 4.44 Dynamic displacement of origin

soft material neither the size nor shape of the damping loops varied in the range 3000 to 12 000 cycles/min of the simulator.

Before measuring the dynamic properties of a material, a fatigue strain must be applied until no further plastic deformation can be recorded. Figure 4.45 shows the damping loops on an F_i/y_G diagram. As sealing gap increases (as gasket 'relaxation' becomes greater) the curve gradually becomes that of the bolt characteristic with elasticity coefficient, c_B^*.

The dynamic elasticity coefficients of the gasket can be deduced directly from Fig. 4.45 using the formula

$$\overline{c_G} = b/y_G = (F_i - c_B y_B)/y_G = \tan \beta$$

Figure 4.46 shows the dynamic elasticity coefficients of a material for initial unit load $p_{G_0} = 300$ and 500 kgf/cm^2, as a function of sealing gap y_G. The coefficients, as can be seen, decrease considerably with increasing sealing gap. Damping loops of the gasket material used in Fig. 4.43 were taken for different sealing gap expansions and the enclosed areas measured by planimeter. Results for initial loads of $p_{G_0} = 300$ and 500 kgf/cm^2 have been plotted (Fig. 4.47) and show that the damping work, W_G, increases approximately by the 2nd power of sealing gap expansion, whereas the initial load for the range tested is of minor importance.

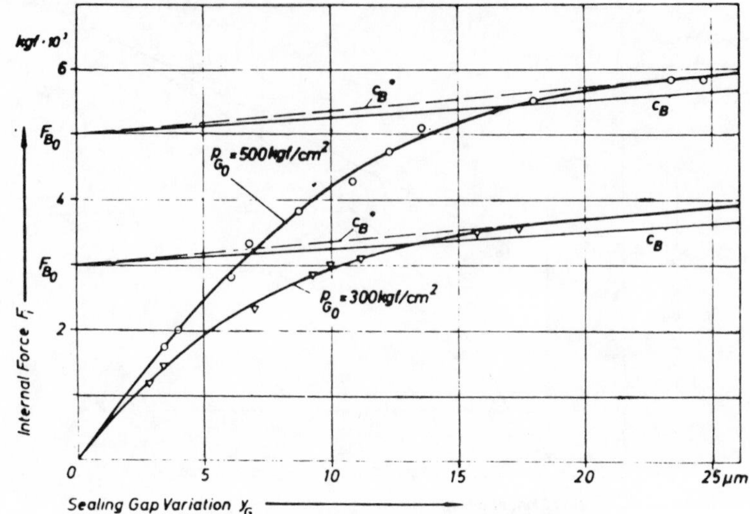

Fig. 4.45 Dynamic characteristics of a flat gasket

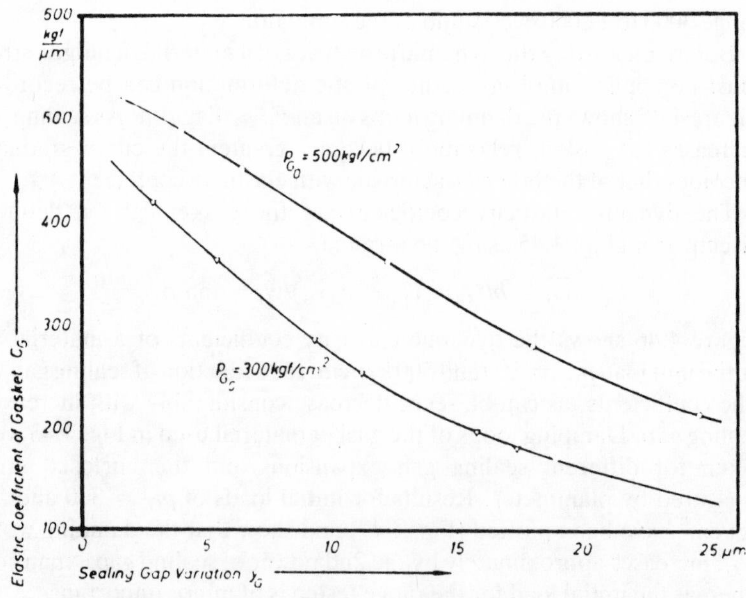

Fig. 4.46 Dynamic elasticity coefficients of a flat gasket

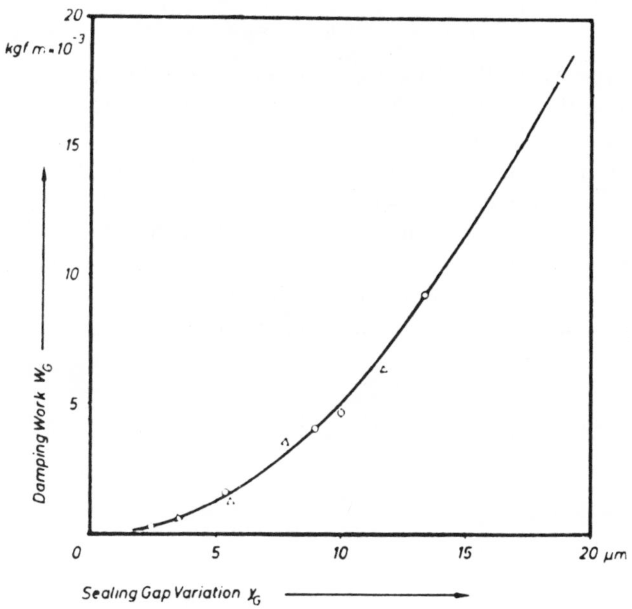

Fig. 4.47 Damping effect of a flat gasket

The dynamic elasticity coefficient and damping work of the gasket material are material properties and may be used to predict the rate of sealing gap expansion under varying internal force if all other characteristics of the joint are known.

Sealing gap variation in a dynamically stressed joint

The previous section showed how the dynamic properties of a gasket – elasticity coefficient, c_G, and damping work, W_G – may be determined as a function of initial load, p_{G_0}, and sealing gap extension, y_G. From these properties the change of sealing gap under pulsating stress (i.e., with a pump or combustion cycle) can be found to first approximation. The joint represents a vibratory structure stressed by a pulsating internal force $F_i(t)$, which – in this case – is assumed to be sinusoidal. The resulting 'sealing gap variation' generally depends on the following factors: mass of cover, m; elasticity coefficient of bolt, c_B; elasticity coefficient of gasket, c_G; and damping work of gasket, W_G. Assuming that the sealing gap 'vibrates' between $-\eta_{max.}$ and $+\eta_{max.}$, where $2\eta_{max.}$ is y_G (Fig. 4.47), a differential equation may be applied to the forced vibration

$$m\ddot{\eta} + D\dot{\eta} + C\eta = F_{\text{Emax.}} \sin \omega \cdot \tau$$

where D represents the damping coefficient of speed-proportional damping and C is the elasticity coefficient. The sinusoidal force is the internal force corresponding to $F_{\text{Emax.}} = F_i/2$.

The vibration process can also be seen from Fig. 4.48. For the sake of simplicity it has been assumed that the gasket is the only compressed component – that sealing gap movement y_G is the same as bolt elongation y_B. In the first approximation, the sinusoidal exciting force $F_{\text{Emax.}} \sin \omega\tau$ produces a sealing gap vibration $\eta_{\text{max.}} \sin \omega\tau$.

In Fig. 4.48 the total elasticity coefficient $\tan \gamma = c_B + \overline{c_G}$ has been

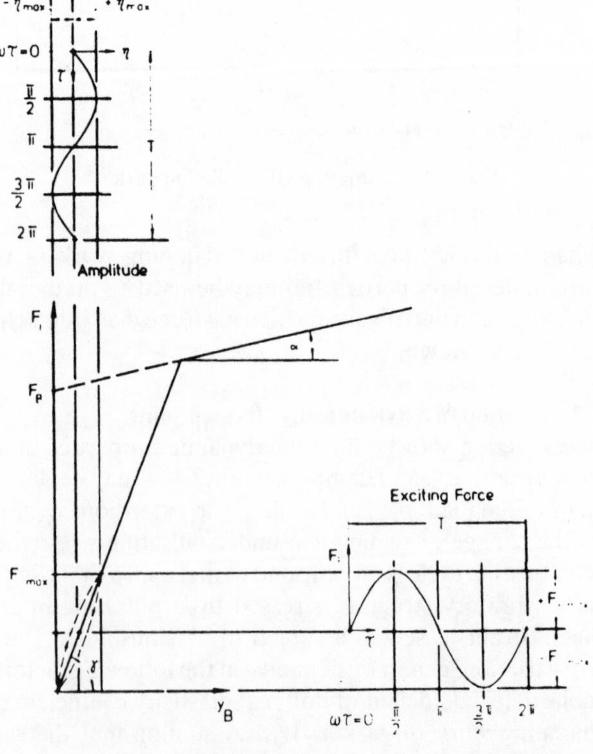

Fig. 4.48 Sealing gap vibration in the F_i, Y_B diagram

introduced into the differential equation for the elasticity coefficient, C. The damping coefficient, D, of the gasket material may be derived from the damping work (Fig. 4.47). The 'equivalent damping coefficient' is derived from

$$D = 4W_G/(\pi \omega y_G^2)$$

Since W_G increases almost by the 2nd power of y_G, the following applies approximately: $D\omega \approx$ constant. In this expression, ω is the 'circular' frequency of the exciting force and $y_G = 2\eta_{max.}$ is the total sealing gap 'variation'.

The solution of the differential equation gives the maximum amplitude of the sealing gap vibration to the first approximation

$$\eta_{max.} = \frac{F_{Emax.}}{\sqrt{\{(D\omega)^2 + (C - m\omega^2)^2\}}}$$

where $\omega = \tau\eta/30$ = frequency of exciting force $F_i/2 = F_{Emax.}$; $C = \overline{c_B \omega c_G}$ = sum of elasticity coefficients of bolt and gasket; D = equivalent damping coefficients; and $\eta_{max.} = y_G/2$ = max. amplitude of sealing gap vibration, or half of sealing gap 'variation', respectively.

A direct solution of this equation is not possible, since C as well as D, depend on the sealing gap variation, y_G, so it is necessary to introduce estimated values for $y_G = 2\eta_{max.}$ for the gasket material. The correct solution for $\eta_{max.}$ is then easily found. The equation is particularly valuable, as it enables gasket materials with optimum C and D values to be selected; that is, materials which show the smallest possible sealing gap 'vibrations' under dynamic stress. Many applications to date of this principle have shown that a reduction of sealing gap vibration amplitude of a few thousandths of a millimeter can be decisive in ensuring a reliable high performance seal.

CHAPTER FIVE
CAD for gasoline engines

The way in which Computer Aided Design (CAD) is replacing traditional approaches to evolutionary design and development of s.i. engines is first considered, alongside the integration of component testing and manufacture into Computer Aided Engineering (CAE). The section immediately following it then shows application of these techniques to analysis of piston, cylinder head, block and gasket, alongside a detailed case study of engine noise reduction using computer techniques. Then follow case studies on connecting rod life evaluation, engine lubrication system, and crankshaft bearing design. A method of predicting fuel consumption precedes the description of CAD being put to use in a major international engine design office. The chapter is concluded by an account of an engine claimed to have been designed almost exclusively by computer: the Fiat FIRE 1000.

COMPUTER AIDED ENGINEERING: THE INTEGRATED APPROACH TO DESIGN, TEST AND MANUFACTURE

Ian Dabney[32] points out that while computer aided design/drafting (CAD/D) and computer aided manufacturing (CAM) technologies are rapidly gaining acceptance, far greater time savings and productivity gains can be obtained by fully integrating the computer in the design, development, production and test stages (Fig. 5.1). CAE emphasizes the use of a common, continually-evolving data base and the flow of information between each stage in the process. Implementation of CAE over and above CAD/D–CAM will ensure improvements in efficiency, productivity, and product quality to a far greater extent than that currently being attained by implementing CAD/D–CAM over and above the traditional design process.

In the traditional (now outdated) design approach, the product concept is manually committed to drawings, then built and performance tested. The build-and-test cycle is repeated until the product satisfactorily meets its functional specifications. Usually, one concept is selected early in the design programme and that single concept is iterated upon in successive prototypes until the requirements are met. This traditional approach to

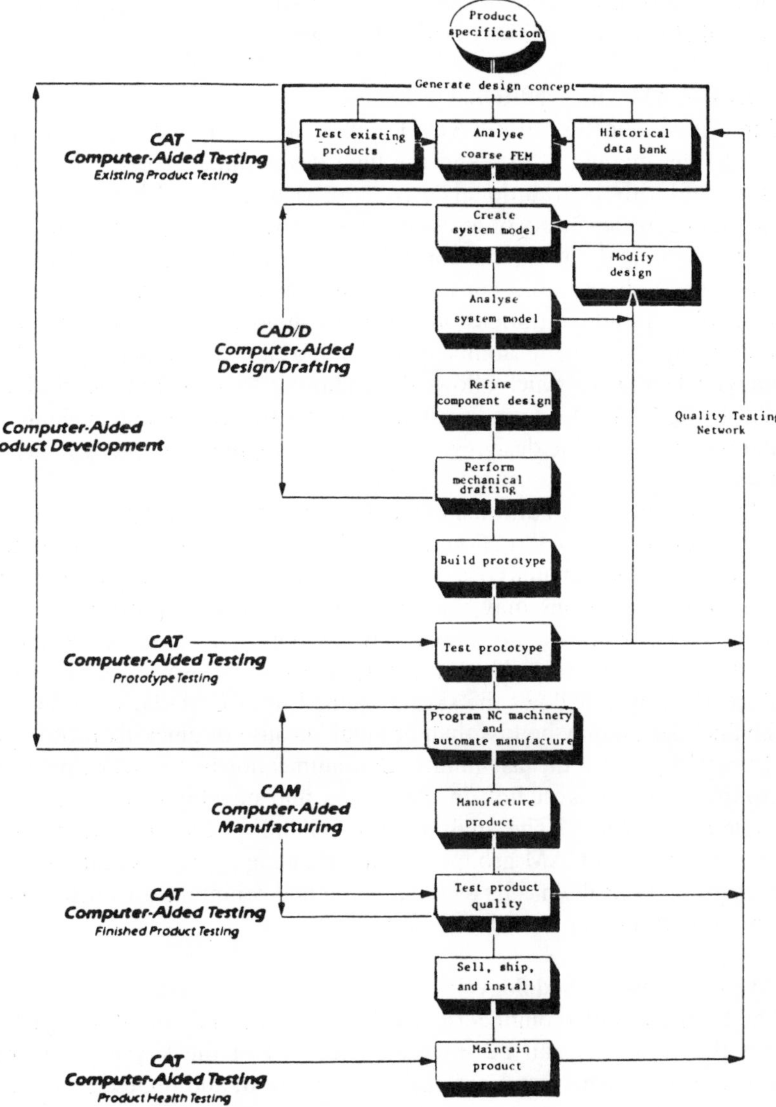

Fig. 5.1 Flow diagram to describe the CAE process

the creation of a new product is characterized by the use of predominantly manual efforts in all phases of its development.

CAD/D–CAM – the traditional technology

There is no question that CAD/D and CAM systems can substantially reduce the manual effort, and the high development costs, which are associated with the traditional design approach. CAD/CAM systems can automate a wide variety of individual tasks, such as engineering drawing preparation and numerical control (NC) machine tape preparation, which are tedious, time consuming, and error-prone, when executed manually. An added dimension available on advanced computer aided drafting systems is the ability to use finite element modelling (FEM) analytical software which allows the computer to assist product design. Within the CAD/D system, analytical models are created and evaluated, and the most suitable design goes on for development as an engineering prototype.

In common with traditional manual design, the CAD/D process relies heavily upon the construction and testing of one or more prototypes. While finite element analytical tools can shorten the design cycle, certain structural deficiencies may remain undetected until a prototype fails. Several build-and-test iterations may be required before a design is produced satisfactorily. Unfortunately, the cost of prototype development often overshadows the savings gained from CAD/D. In addition, the final design may be less than optimal because of quick-fix prototype changes inserted at the last minute. Communication between engineering groups is the weakest link in the design and development process as implemented today. The analysis, test, reliability, drafting/documentation, and related CAM groups may all use computer workstations, but each group typically has its own data base and communicates with other departments using plots, charts, and memos.

CAE – the new methodology

The computer aided engineering (CAE) approach first of all recognizes that the computer can be integrated into all of the key engineering functions of design, test, drafting/documentation, and related manufacturing. CAE then goes beyond the individual process elements to focus design and development efforts on the computer 'system' model rather than have individual department data bases. The CAE philosophy, shown in Fig. 5.1, encompasses the segmented CAD/D, computer aided testing (CAT), and related CAM activities implemented today. To develop the design concept of a new product, computer aided testing is performed on any available comparable products and any existing components that may be used in the new design. CAT is also used to establish functional characteristics such as vibration, noise, or other loadings. This test information, and coarse finite element analysis work performed with the product specifications in mind, form the beginnings of the system model taking shape within the computer.

Software available today allows the system model to be exercised under simulated loadings so that the design deficiencies can be identified. In this manner, many different design concepts can be evaluated inexpensively. Design deficiencies that are encountered are solved in the analytic design modification step depicted in the illustration. As successful concepts are determined, more extensive simulation will allow the model to represent more concisely the optimum design. While it may be necessary to build and test certain components of the final design to update the system model, the expensive system prototype is not needed until product specifications and performance are achieved. A computer aided product development philosophy, in contrast to the traditional and CAD/D–CAM approach, defers detailed drafting description and prototype construction until the design is well advanced in terms of its performance and, therefore, is relatively proven. After construction, the CAE system prototype is tested to verify the quality of the system model. Any unexpected problems are again simulated and solved within the computer, using the CAT experimental data integrated into the system model. Accordingly, using the CAE method to arrive at the optimum design, prototype requirements are substantially reduced.

In the production phase, the CAE system model is used both in several product-related CAM functions, and in the dynamic quality control tests of the finished product. CAM activities such as operation planning, fixture design, and NC programming can all benefit from the engineering data base captured earlier in the development cycle. To assure the quality of the customer's product, the finished assembly is dynamically tested and its responses are compared with those defined by the system model and, therefore, the product specifications. The CAT data, collected at critical points in the product design, development, and manufacturing stages, forms a 'quality network' to check and verify that each stage in the process has been successfully completed. Quite often, dynamic testing and monitoring will be continued for product maintenance purposes even after the product has been delivered to the customer. This provides an on-going reliability and service life data base. The test information from the quality network will therefore be available for use in a historical 'data bank' for the next generation of the new product cycle as shown in the illustration.

Automotive applications
With the traditional demands in the motor industry for fully tested designs in the minimum possible time, CAE has shown itself to be especially suited and, in many cases, of extreme value. In recent years, the introduction of sophisticated finite element programs, and the development of computers capable of running them, has substantially extended the scope of CAE. The preparation of large matrices to define detailed structural characteristics, has made it possible to address and solve problems previously considered too complex to be practically approached from either a time or cost standpoint.

ENGINE COMPONENTS BY CAE

Bob Southall[33] outlines an approach to automotive design in which a method 'based largely on experience' is replaced by innovative design based on 'image analysis' and 'concept analysis'. In these cases, design experience is synthesized by accelerated test and analysis routines and extensive mathematical modelling is carried out. This approach to design by SDRC is shown in Fig. 5.2.

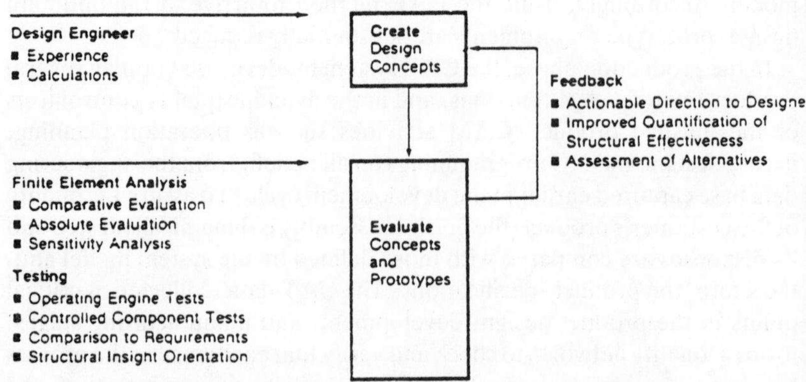

Fig. 5.2 Flow diagram for engine component design

The main consequences of the traditional approach is that design innovation tends to be suppressed because the lack of previous experience leads to high risk. Innovative designs often require a longer development process in which major design changes sometimes have to be made to achieve performance or durability.

Image analysis allows the designer to rapidly synthesize design experience in areas where his knowledge is limited. The process uses *test* and *analysis* of similar or competitor products to identify design targets and to identify and understand the significance of features that contribute to the desired (or undesired) performance. In essence we are now taking the time-honoured process of evaluating competitor products one step further than normal. Instead of just evaluating performance we are using computer analysis to really understand the key features of a design so that they can be incorporated in a new configuration.

In concept analysis, simple mathematical models are used to assess, develop, or reject a variety of concept ideas. The process allows the designer to assess a whole variety of options where normally he would have had to make a judgement based on his experience alone. One of the major problems in concept analysis is that the models have to be simple so that the analysis can keep pace with the rapid progress of projects at this stage. These models, whilst they appear simple, can only be built by very experienced engineers because of the many assumptions that have to be made.

The importance of making good early design decisions can be visualized in the relationships within a 'decision tree' (Fig. 5.3). Making the early decisions right has a controlling influence on the final outcome. For example, in the 1970s vehicle makers were beginning new engine development programs based on alternative technologies such as rotary engines, turbines, diesels, or controlled combustion schemes. Those early decisions were followed by years of expert work to improve the details of engineering and manufacturing. Yet it was that early choice of a basic engine design approach which determined, more than anything else, the success of the program.

CAE system simulations can give companies an opportunity to predict the consequences of early design decisions at a fraction of the cost of building and evaluating prototypes. Outstanding designers say that one secret of their success is that they try to step back from the problem as it is originally stated and consider it from a broader perspective. Unfortunately, the increasing specialization of engineering means that most

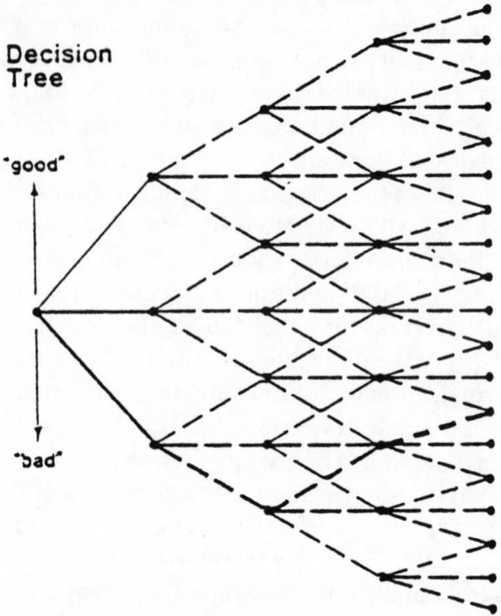

Fig. 5.3 Graphic illustration of the importance of early decision making

engineers end up designing components. The responsibility for product design is often so widely dispersed among component designers, that no one truly understands overall system performance. In order to manage the complexity of engineering today, engineers need tools to try out alternative concepts and select the best one before spending money and time on detail design.

Piston analysis

The structural design and evaluation approach to 'engine development' is shown in Fig. 5.2. In piston analysis, for example, 2-D axisymmetric or full 3-D modelling is carried out. Capability includes studying fatigue, design load evaluation, concept design, design modification to eliminate crown cracking, thermal stress related to piston ring location, and two-piece piston designs.

Recently, models have been created by first building a solid geometric representation in GEOMOD than transferring geometric data to the pre-processing program SUPERTAB. The mesh is created using a mix-

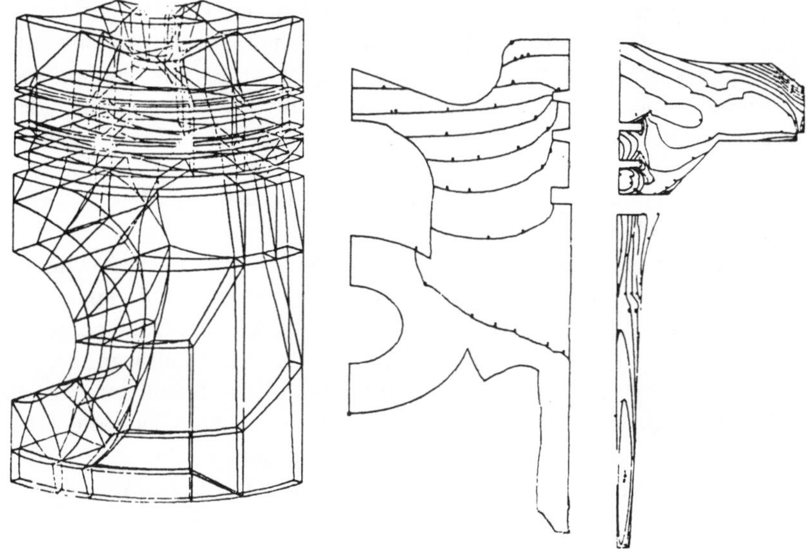

Fig. 5.4 FEM mesh for one-quarter of piston (*left*), with stress gradients (*centre*), and those for temperature (*right*)

ture of manual and automatic generation techniques; it permits analysis of distortion, stress, temperature distribution, fatigue life, weight reduction, and ring stability for both concept design and development purposes (Fig. 5.4).

Cylinder head, block, and gasket

While 23 previous SDRC projects concern the block and head individually, there is particular interest in a combined analysis of head, block, and gasket; perhaps a first step towards the structural design of the complete engine assembly.

Since the performance of both the cylinder head and the cylinder block (in the upper region) are influenced by their interaction with each other through the head gasket, in order to obtain realistic stress and distortion analysis it is necessary to have a realistic model of the whole head, block, and gasket assembly. Similarly, the performance of the gasket is very dependent on the head and block characteristics. Finite element modelling can be used to represent the basic head and block structures without major problem, except that of complexity and model size. The gasket is a non-linear plastic problem and, therefore, much more difficult to model. Techniques are, however, now being used by SDRC to simulate the problem economically. Analyses possible with this technique include: stress and fatigue life, temperature distribution, gasket performance, and bore distortion. The technique is ideally suited to concept design as it allows the analyst to trade off between structural versus gasket performance requirements.

Finite element models of cylinder heads and, in many cases, cylinder blocks are normally constructed from 3-D (brick) elements. The resulting models are very complex, being difficult to build and visualize. Geometric modelling can also be used here to assist the modelling by allowing colour visualization of the complex structures. While the head/block/gasket assembly model is used primarily to study characteristics which affect the top region of the block, other studies on the block, particularly the durability of the main bearing region, can be carried out in the block model alone. Previous studies cover aspects such as bearing saddle stresses, bearing cup wear, noise, and vibration. An important new development here is the use of these engineering models for manufacturing studies to investigate, for example, machining errors due to component distortion under the cutting loads.

Noise prediction

Models for individual components and interfacing components have been discussed above. Models which simulate a number of components in an assembly are needed to study noise or vibration, particularly where the vibration mode of a crankcase, say, can be affected by a fuel pump bolted to its side or a sump which closes the open end. The basic process for engine noise control is shown in Fig. 5.5. Operating loads such as bearing forces and piston slap are typical of the factors contributory to the forced response analysis.

SDRC have developed an analytical process which utilizes finite element models in a calculation of radiated noise. This process can be used both as a development and a concept design tool. Initially several iterations on the model were made to achieve an agreement of 10 per cent between the predicted and measured model frequencies. Later models (with experience) were found to correlate to roughly 3 per cent for most modes at the first attempt. The same models can be used for weight reduction studies – worst case loads from both assembly and operation are

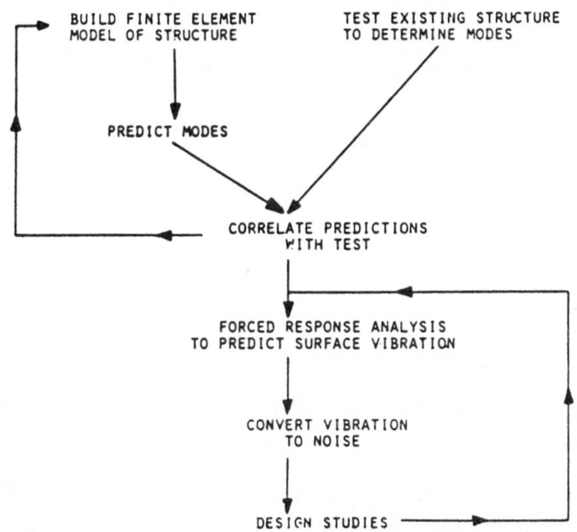

Fig. 5.5 Design and analysis concept for engine noise control

applied to give a coarse indication of stress distribution in the structure. Within the constraints of minimum casting thickness weight is then reduced by removing material from the low stressed areas. (However, as the analysis is only global, and does not include residual stresses, prototype durability trials are necessary following any modifications.)

The noise prediction process can be used for concept design now that confidence has been built up in modelling ability. This allows basic design features (such as ladder frame crankcases, bed plate crankcases, and various sump designs) to be assessed rapidly during the concept phase. This is very important because many noise reducing features have been found in the past not to successfully read over from one engine to another. Prototype tests on the options would on the other hand take many years. For the future SDRC propose to develop an optimization procedure. The idea is to set a noise target and allow the computer to identify the necessary structural changes to achieve the noise level. A work programme is fully planned out and the basic mathematical approach has been devised.

A CAE approach to engine noise reduction for the Ford York 4-cylinder IDI-2.3 litre unit was described by R. W. Johnson in Conference

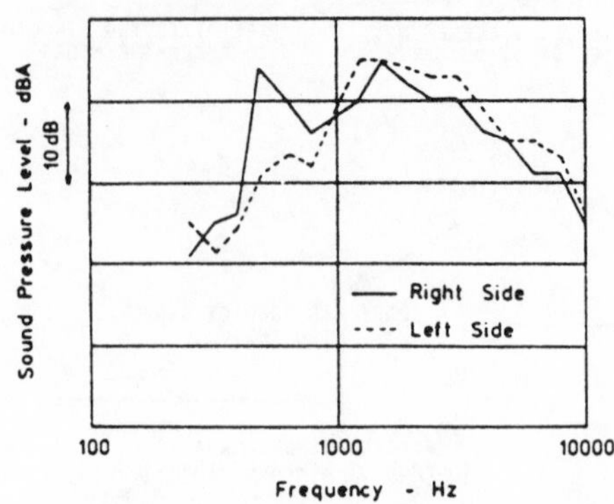

Fig. 5.6 Bare engine noise characteristics prior to redesign

Paper C123/82 published by the Institution of Mechanical Engineers. Noise testing on the engine has, in this case, been carried out and typical bare engine noise characteristics are shown in Fig. 5.6. The overall engine noise level at rated speed is controlled by three peaks in the $\frac{1}{3}$ octave bands with centre frequencies of 500/630 Hz, 1250/1600 Hz, and 3150 Hz. Using the lead cladding technique a noise source ranking table of the dominant noise radiating components was compiled: Fig. 5.7 identified the block sides, sump, and rocker cover as the major noise sources. Prior to the creation of a finite element model of the cylinder block, an experimental modal analysis exercise was carried out to provide a correlation basis for

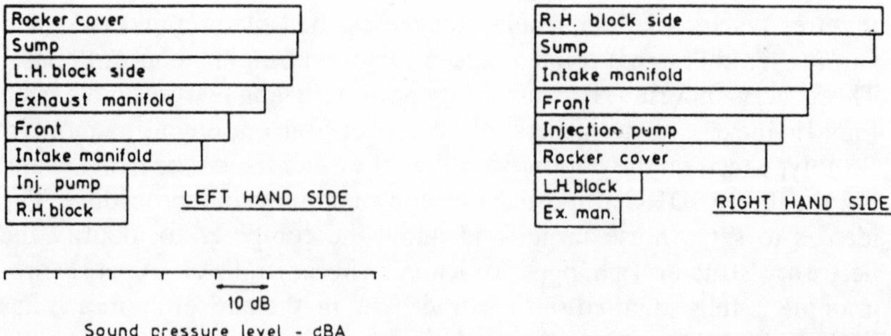

Fig. 5.7 Ranking of component noise sources obtained by the lead-cladding technique

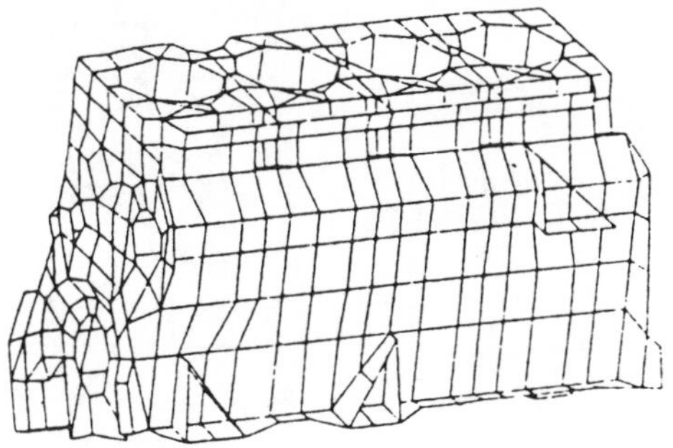

Fig. 5.8 FEM mesh for cylinder block

the model. Using a combination of impulse and forced vibration inputs, frequency response functions were measured at a mesh of points on the block, identifying approximately 30 normal modes. Mode shapes were generated from this data and the individual shape descriptions identified by inspection of the animated mode shapes.

A finite element model of the cylinder block, head, and main bearing caps was generated using mainly thin shell elements with a total of 1181 modes and 1462 elements, Fig. 5.8. Using the FEM program NASTRAN, a dynamic analysis was performed to determine all natural modes occurring in the frequency range 500–3000 Hz. Comparison of the mode shape characteristics was made by visual inspection of the animated modes. Generally good agreement was found between the experimental and analytical modes in order, frequency, and shape compared. Excitation of the main bearing caps on the crankcase was carried out using an impulse hammer to excite the structure at locations representing the point of input of piston slap and main bearing forces. Using a measured cylinder pressure diagram and a Ford computer program, main bearing force time histories were computed for each main bearing. Transforming these time histories into the frequency domain provided the forcing function for the model which simulated running engine conditions. Figure 5.9 compares the predicted $\frac{1}{3}$ octave band noise levels for the left-hand side of the cylinder block with that measured from the running engine.

Proposed design changes were limited to those capable of being manufactured on the existing production line and consistent with an objective of minimum noise and weight. As no internal modifications were feasible, the effect on the overall modes was restricted to that

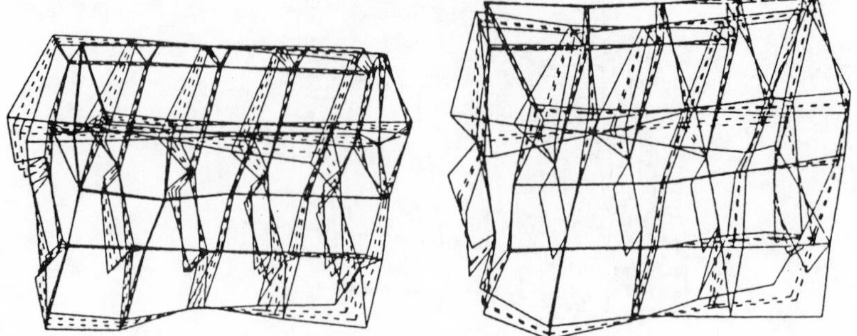

Fig. 5.9 Experimental and analytical mode shapes compared

achievable by the sump rail changes. Panel resonances were treated by attaching horizontal ribs on the left-hand side. Due to natural discontinuities on the right-hand side wall, panel resonances on this side were already outside the frequency range of interest. For ribbing, an 'optimum' design (Fig. 5.10), identified by the noise prediction program was cast and a running engine noise test carried out in a semi-anechoic test cell. This resulted in an overall bare engine noise reduction of 1 dBA against a predicted benefit of 2 dBA (from cylinder block radiated noise). The lower measured attenuation was due to the contribution from other noise sources remaining at a dominant level, for example, sump noise.

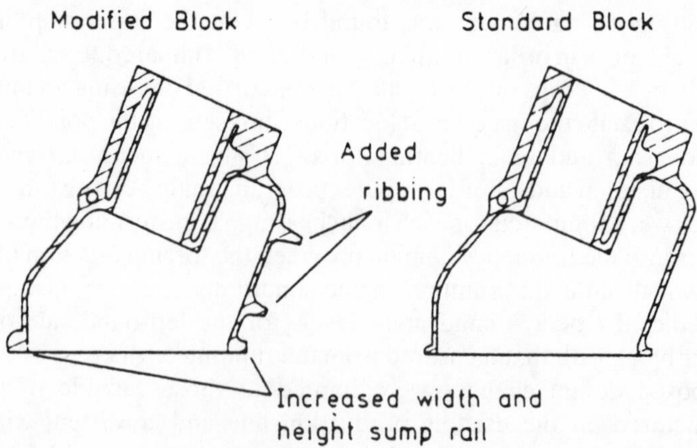

Fig. 5.10 Section through crankcase before and after modification

One of the initial objectives of the project was to ensure that design modifications did not result in a significant increase in engine weight. The modified cylinder block design combined the addition of metal in the form of ribs for noise reduction purposes, with the deletion of non-essential bosses, for weight reduction. The resulting modified block was predicted from the finite element model to give a 0.8 kg weight reduction over the standard block. This compares well with the actual 0.6 kg measured weight reduction.

Proposals for crankshaft/bearing/structure analysis

SDRC has identified a need for a more sophisticated method of analysing the 'bottom end'; to identify bearing loads, bearing performance, block stresses, and crankshaft stresses. Traditionally, bearing loads have been calculated in a statistically determinate process. The total load on a main bearing being taken as the vectorial sum of the loads from the bays either side of that bearing. No account is taken of cylinder block flexibility or oil film effects, and the crankshaft is usually pinjointed. Similarly, much of the bearing analysis carried out assumes rigid components and uses the statically determinate loads as boundary conditions. It is of great credit to the engineers involved that this approximate data has been successfully applied by careful comparison with extensive test data to produce an empirical design process.

The problem is clearly quite complex because of the non-linear nature of the oil film and also because the rotation of the crankshaft presents a time varying stiffness with respect to the bearing housing. Advances have been made recently by manufacturers, and researchers; programs now exist which go someway towards a full model of the problem but only in a static sense. SDRC are now underway with a research programme that is planned to develop a complete dynamic simulation of the block/bearing/crankshaft system.

Flow analysis

The company are also currently developing FE-based software for incompressible fluid flow analysis. A two dimensional capability already exists and a three year programme has been set up to extend this to 3-D. The project is funded on a club basis by SDRC, vehicle and engine manufacturers and the National Computing Council. The resulting program will have application to water flow in engines, air flow in manifolds, and flow around vehicles. The program will handle laminar, turbulent, and transitionary flow.

To date most numerical fluid flow studies have been carried out using finite difference techniques. This process is much less flexible because the elements used in models cannot be distorted from true rectangular shapes with 90 degree corners. While this is overcome on smooth flowing aircraft structures using transformation mapping techniques, it is not really a feasible option for the less regular structures found in vehicles and engines. Using the finite element process overcomes these problems and will also allow the new processes to be easily integrated with current analysis and with current pre- and post-processing and graphics programs.

CONNECTING ROD LIFE EVALUATION BY CAE

Major forging manufacturers GKN are reported[34] to be greatly involved in producing engine components to given performance requirements rather than adhering to a strict drawing and material specification laid down by the vehicle or engine builder. The result has been a considerable move down-line in design activity towards the component maker. Larger suppliers have, in some cases, invested heavily in hardware and software for CAE.

The ability to measure and statistically describe in-service loading is now available, and the refinement of strain-life fatigue analysis techniques makes for a useful interpretation of design load histories for design calculation. This is backed by improved methods for cyclic measurement of material properties coupled with finite-element and related methods of studying internal stresses in components. A continuous design and evaluation 'loop' is linked to rig testing and in-service monitoring, and a system approach to the interaction of components like pistons, connecting rods, and crankshafts has now been realized. Strain gauges at the critical areas provide data for building accurate FEM models. Subsequent analysis reveals areas of redundant metal which, when removed without referring additional stress to the critical areas, can lead to equivalent strain distributions before and after modification. A CAD system can then describe the new shape, from which NC machining of forging dies can be arranged with an associated CAM program. Service loads are held in a data bank where they are digitized in amplitude *vs* time form as an event corresponding to, say, an engine condition. The solid models of forgings are built up from brick-shaped elements which are analysed from a point

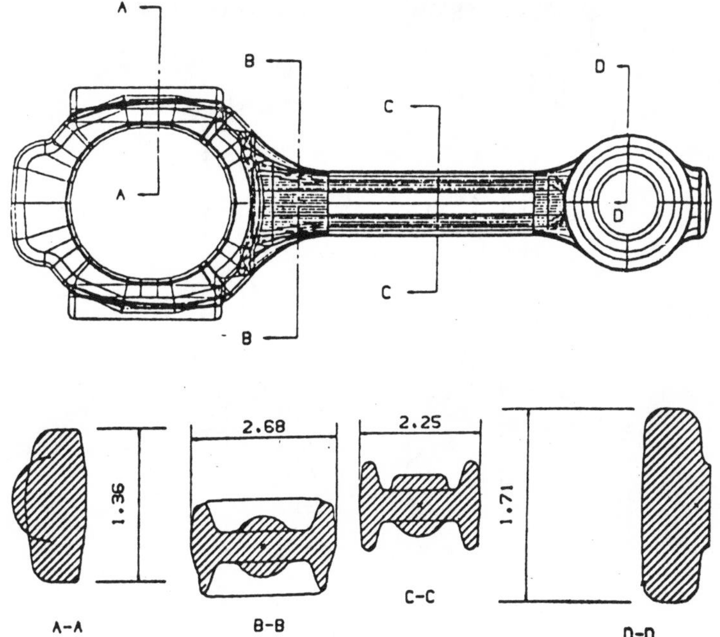

Fig. 5.11 Refined cross sections resulting from both stress and buckling analysis

of maximum stress in the model that will lead to a fatigue failure; low stress areas point to opportunities for weight saving.

The FEM analysis can also be used to determine overall deformation (Fig. 5.11), as well as stress level, and even buckling failures can be examined. The diagram shows a con-rod, the conventional form of column for which, withkeep H-section, is traditionally designed for buckling resistance. With the modern trend to shorter con-rods, this factor is of lesser significance and an easier-to-forge more rounded section becomes feasible. In this design material has actually been added to the big-end region, for improved bearing support. However, overall life is as much a function of surface condition and, for example, suitably shaped sinter-forged rods produce a longer life. The computer reads an in-house developed contour-plotting program to generate the FEM mesh for these analyses. A comparison of the complete con-rod before and

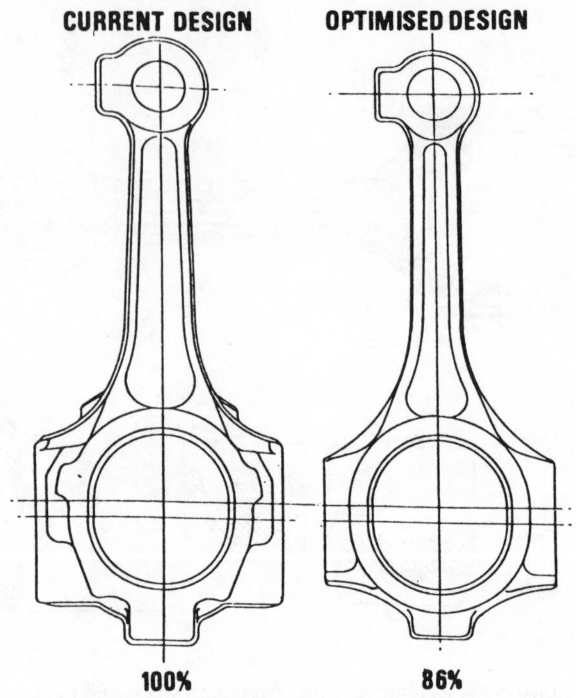

Fig. 5.12 Weight reduction achieved in redesign

after design optimization is shown in Fig. 5.12, to give a 1 per cent weight reduction and Fig. 5.13 illustrates a safety factor plot alongside the FEM mesh.

Specimens are sinusoidally stroked in cyclic test machines which monitor strain and load. Both strain/life and dynamic stress/strain (Fig. 5.14) characteristics are obtained for different materials. Total strain is derived from a combination of plastic and elastic strains, and results taken when evaluating materials are in the form of scattered points which must lie in a strain band for acceptance. In the eventual design analysis a stress–strain product rather than a modulus is used; another parameter needed is the stress concentration factor.

The importance of surface on fatigue life has led to a classification of surface forms and treatments, according to conditions likely or not to initiate fatigue cracks. Photomicroscopic techniques are used to deter-

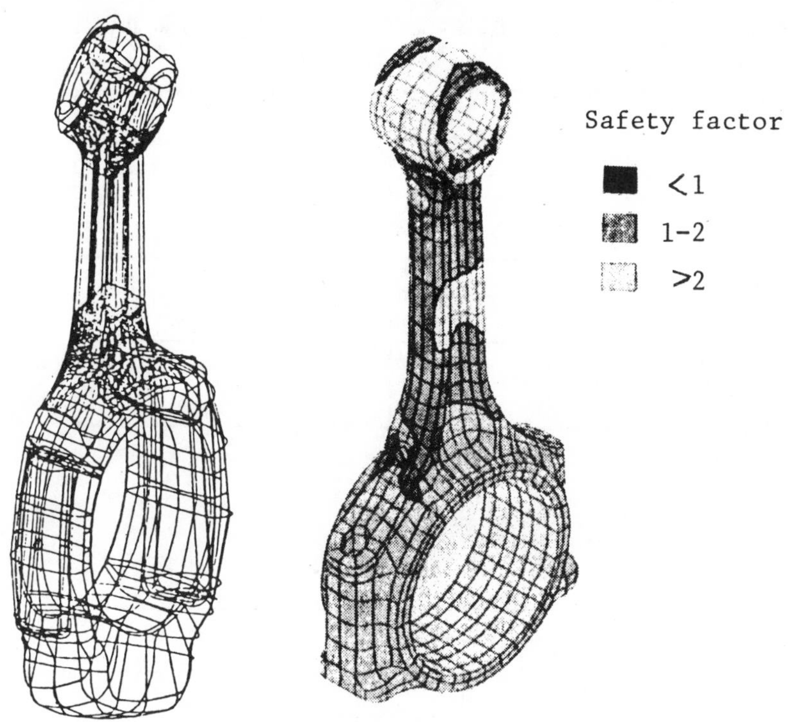

Fig. 5.13 Safety factor plot alongside FEM mesh used

mine, for example, decarburizing loss due to pre-forge heat treatments, as well as comparing as-forged, shot-peened, as-cast, and machined finishes. Fine-surfaced sintered materials show flat strain/life curves indicating low ductility, but the fatigue limit is well defined. Composite materials are also being examined, with the emphasis on improving consistency of properties. Material fatigue characteristics and cyclic stress/strain curves are used in the computation. Other inputs are the effects of changing sections or overall component shape on loads and the geometry effects on stress concentration. A Neuber analysis, used here to relate normal and local conditions, and the technique of rainflow cycle counting are used to break down the loading history, identifying cyclic events from the local response and estimating their damage contribution by comparison with relevant conventional constant-amplitude fatigue

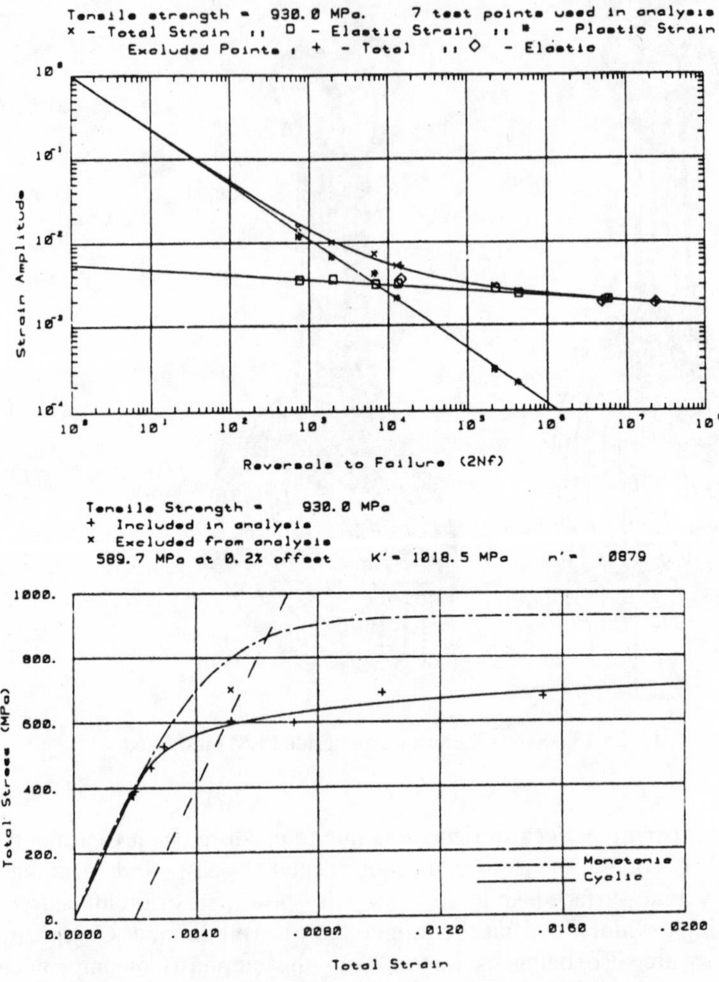

Fig. 5.14 **Strain/life (*top*) and dynamic stress/strain (*bottom*) curves**

data, damage being summed linearly. The reciprocal of the calculated damage (accumulated by one application of the service history) defines the predicted service life in terms of 'blocks' of history. The computer converts this into numerical road mileages, or a typical life summary can be displayed.

ENGINE LUBRICATION SYSTEMS

A program is reported[35] which has been developed by Perkins to model the entire oil circulation system of a high speed diesel engine and which can predict flow in both steady-state *and* transient conditions. In the latter respect, the company claim to be comfortably ahead of any other volume diesel manufacturer in the UK. The pie-chart of steady-state oil flow in Fig. 5.15 is of particular significance. It shows not just that the engine oil pump has a high reserve of flow capacity, dumped (or routed to piston spray-cooling) by the relief valve above a certain speed, to cater for increased flow as clearances increase throughout engine life – but also shows the dominant effect of the crankshaft bearings. An understanding of crankshaft bearing behaviour is thus of key importance in the design of the engine lubrication system. The ability to apply computer techniques in this area has also been described by Perkins engineers Dr Brian Law and Mr A. K. Haddock in the paper 'Prediction of main bearing and crankshaft loading in reciprocating engines' presented at the 1983 CIMAC conference in Paris. The work is complementary to that which is on-going by bearing manufacturers (see the next section of this chapter). From the standpoint of lubrication flow, the work establishes that textbook methods for calculating oil-flow through the idealized 'hydrodynamically lubricated' bearing are now inappropriate.

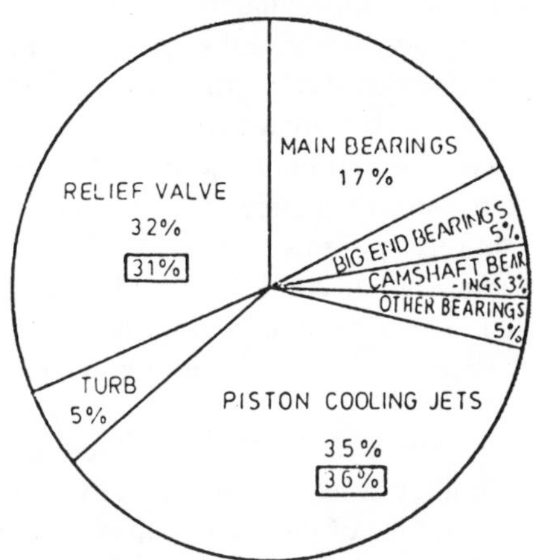

Fig. 5.15 Steady state oil flow distribution in an engine

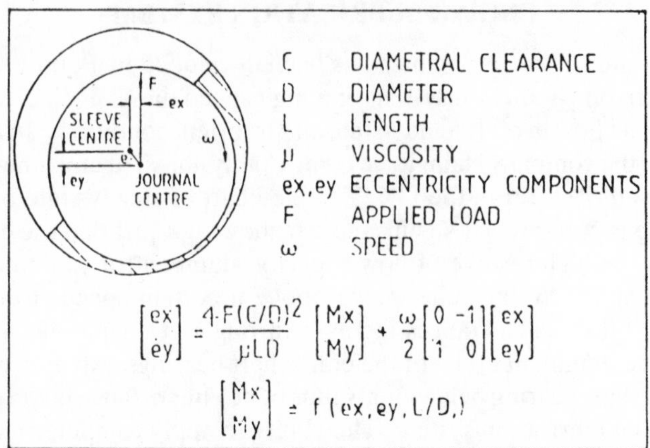

Fig. 5.16 Parameters used in dynamic analysis of bearing

Dynamic changes in loading within a single crankshaft cycle destroy the concept of the journal rotating on a fixed wedge of oil; cyclic displacement of the journal within the bearing actually causes oil to be 'pumped' in and out of the bearing at a higher rate than suggested by equating prevailing line oil-pressure with journal clearance. In fact peak oil pressure within the bearing clearance volume may be up to one thousand times the supply pressure. In the procedure for predicting crankshaft bearing loading, sequential solution of the hydrodynamic and structural equations has been found to correctly describe a real engine situation. Sophisticated FEM techniques, for crankshaft and crankcase, are used to obtain the structural equations while the hydrodynamic operation of the main bearings is modelled by the 'mobility' method. A pair of equations has been derived to obtain rates of change of eccentricity (Fig. 5.16), based on the Reynolds relationship, in matrix, form. Non-linear oil-film stiffnesses are elements of an overall crankshaft model (Fig. 5.17). Compared with the determinate model, shown, used in earlier analysis, a remarkable corelation exists between predicted behaviour and measured strains on a running engine.

The derivation of oil flow from the data obtained in the above prediction is found by a technique similar to one already published by Glacier Metals; it determines the size of the main gallery and feed pipes for the crankshaft bearings. The distributed oil flow prediction program can then be brought into use to determine the relative feed rates to other parts of the engine (Fig. 5.18). The camshaft requirement is likely to be only 5 per

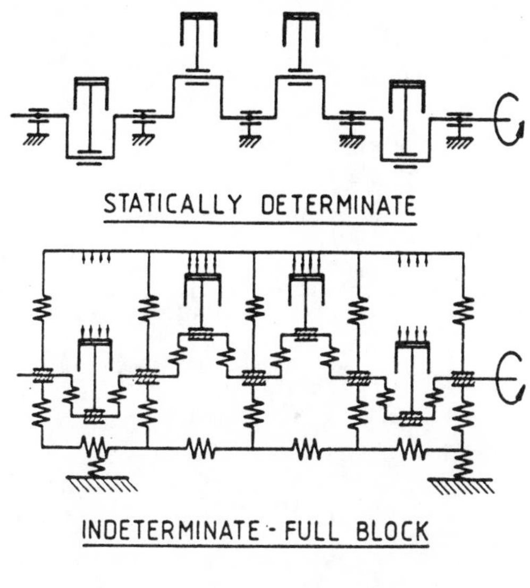

Fig. 5.17 Overall lubrication system for an engine

cent that of the crankshaft, using hole rather than groove-fed bearings. For the valve rocker gear, non-hydrodynamic conditions apply in the bearings and much development know-how goes into designing the overall shape of the rockers in order to route oil along them for drip feeding the valve guides. Feed to the rockers is usually achieved by a flat on one of the camshaft journals, allowing an intermittent supply once every revolution.

The oil pump, as a positive-displacement device with linear relationship between speed and output, thus requires a relief valve to dump increased flow above the idle rate. In the case of turbocharged engines with spray cooling to the piston undersides, a two-stage relief valve is fitted. This is another area where considerable development experience is applied in designing, in the Perkins case, a plunger and slot arrangement. The company also use a special design of oil pump with an in-built weir to ensure that the pump is self priming after engine oil changes. The pump is

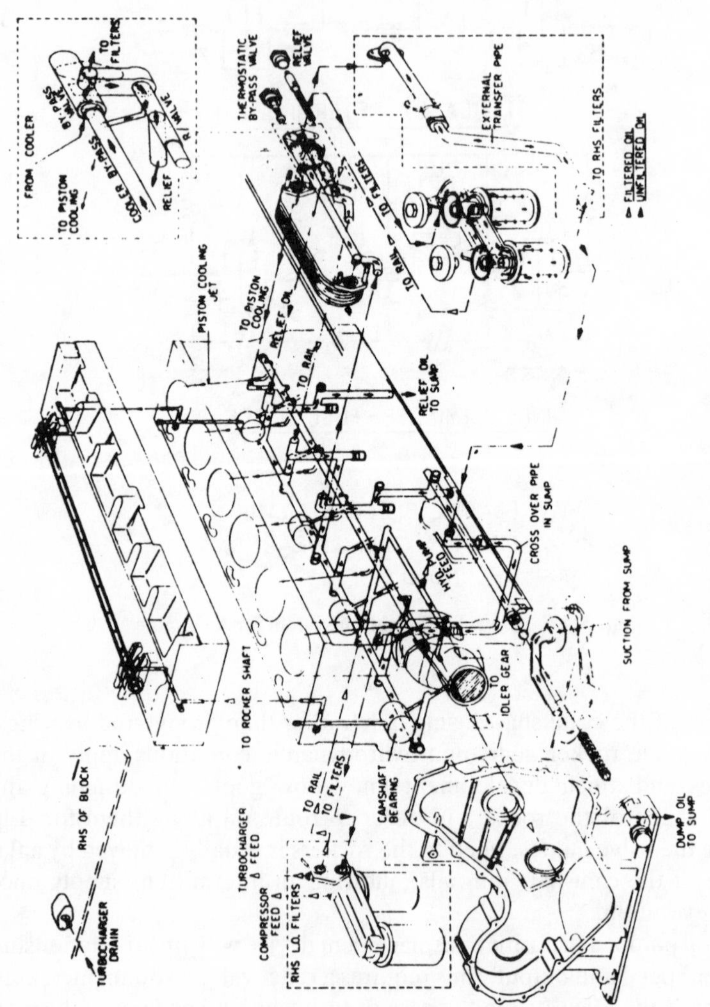

Fig. 5.18 Oil pump characteristic

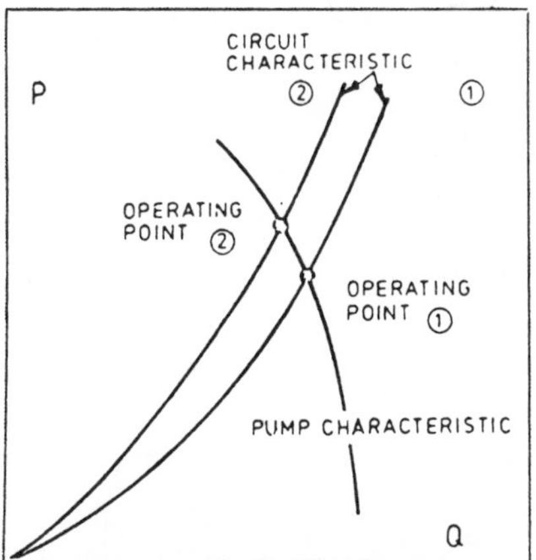

Fig. 5.19 Analytical models compared for former and current calculation methods

matched to the requirements of the overall circuit as indicated in Fig. 5.19, which gives pressure flow at a given speed. Both it and the relief valve become elements in the overall circuit network built in to the prediction program. Generally, the flow elements describe a section of pipe, its pressure/flow relationship being determined by diameter, length, and shape of bends; each element is analogous to a resistance in an electrical circuit. Some elements are switches (time-dependent) and, in the case of oil feeds formed by crankshaft drillings, careful modelling is required to allow for centrifugal effects as the crankshaft rotates.

CRANKSHAFT BEARING WEAR PREDICTION

F. A. Davis[36] illustrates the methods available in controlling the wear of plain bearings. The first requirement is to optimize design so that during operation, hydrodynamic conditions prevail and no misplaced oil holes or grooves lessen the oil film thickness to one that becomes critical. Secondly, the wear of the bearing alloy must be carefully assessed and the consequences of increasing its abrasion resistance must be realized. Thirdly, an understanding of the bearing wear mechanism is necessary.

Of prime importance in the selection of bearing materials for internal combustion engine applications is the alloy's ability to support the very high loads that originate from the combination of gas and inertia forces. This analysis is simplified and assumptions are made to enable the design engineer to easily compute the plain bearing loading. A typical example

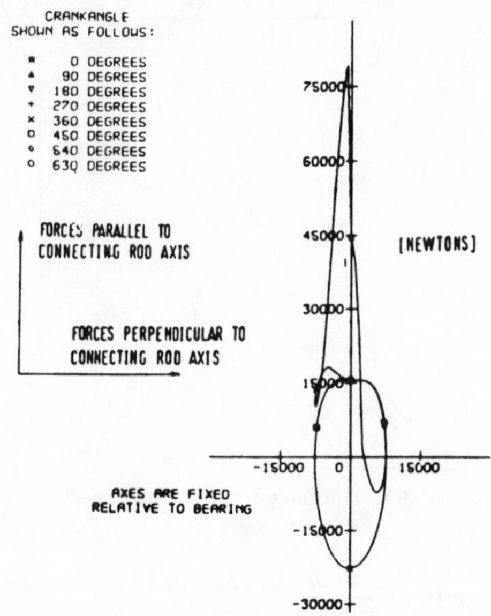

Fig. 5.20 Polar load diagram of a big-end bearing for a four-stroke engine

is shown diagrammatically in the form of a polar load diagram for a big-end bearing of a four stroke internal combustion engine in Fig. 5.20. Comparison of these loadings with pressure design ratings of bearing materials enables the correct material, based on its load carrying capacity to be chosen.

Regrettably, in practice, hydrodynamic lubrication conditions are never wholly realized, and the factors that favour the breakdown of the oil film resulting in surface interactions and bearing wear must be fully appreciated. These factors will be of even greater importance as the design criteria for automotive internal combustion engines necessitate small, high power units.

Apart from wear of the bearing which will lead to excessive running clearance accompanied by a drop in oil pressure resulting ultimately in failure, limited wear will result in increased engine noise, which has, in certain cases, been suggested to be in excess of the European noise regulations. Although the degree of wear observed visually on the bearing was thought to be acceptable to the severity of the test, the noise emission rendered the test a failure. Thus, it is apparent that the subject of plain bearing wear involves firstly its design to ensure that hydrodynamically lubricated conditions predominates, and secondly the selection of the material to minimize the consequence of wear when non-hydrodynamic condition (boundary lubrication) prevails. This section is therefore concerned with both parameters.

Effects of design on plain bearing wear

Based on the information provided by the load-vector diagram it is possible to input data to a lubrication analysis program which will calculate the resultant motion of the crankshaft journal within the bearing clearance. To enable rapid data return, assumptions are incorporated into this analysis. For instance utilization of the short bearing approximation which suggests that the circumferential pressure gradient is negligible compared with the axial pressure gradient, and that the bearing and shaft surfaces are: perfectly round, microscopically smooth, rigid, and axially parallel, whilst the lubricating oil is: a Newtonian fluid, isoviscous throughout the film, and incompressible.

Figure 5.21 shows the shape of the journal orbit diagram corresponding

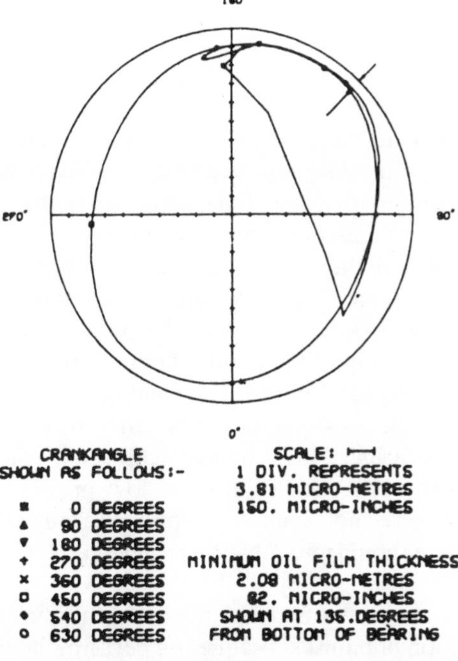

Fig. 5.21 Orbit diagram for a crankshaft bearing

to the polar load diagram shown in Fig. 5.20. The radial distance between each orbital point and the outer base circle represents the instantaneous minimum film thickness on the same scale whereby the radius of the circle represents the radial runnning clearance. All the film thicknesses, but in particular the overall minimum, are used to judge the likelihood of unacceptable bearing wear. Throughout this analysis an oil viscosity of 7 cP is used and this provides the basis for any wear comparison. In this approach the bearing circumference is divided into 36 ten-degree segments and the reciprocals of the minimum film thicknesses of the orbital points within each segment are summed, these being the parameter values. A high value indicates a great probability of wear because it arises from closely spaced orbital points and/or, low film thicknesses. Hence, the most prone areas can be easily assessed. The ability to theoretically predict is illustrated in Fig. 5.22 which compares the wear parameter histograms at two speeds, for a plain lower half and one with a 360 degree oil groove. The parameter values differ by a factor of four and the histogram indicates a very substantial change for the better in the probability of wear if a plain lower half bearing is installed.

The supply of oil to the connecting-rod bearings from the main bearings is via drillings in the crankshaft. The location of the crank pin oil supply hole, or holes, can critically affect the big-end bearing performance. An oil film extent diagram (Fig. 5.23) aids the selection of the correct positions, by showing the position of the pressurized oil film on the bearing at 5 degree crank angle intervals throughout the cycle. The lighter shaded areas represent the edge of the oil film where the general pressure is less than the supply pressure, whilst the darker areas show the pressure central regions. Within these areas the hydrodynamic oil pressure can be several magnitudes of the supply pressure. Superimposed on this diagram are the loci of two crank-pin oil supply holes positioned at 80 degrees and 280 degrees from the top of the crank-pin measuring around its circumference in the direction of rotation. The ideal supply oil hole is one which completely avoids the pressurized regions, thereby ensuring continuous oil supply to the bearing. This is almost achieved by the 80 degree hole which only slightly encroaches into the low pressure region briefly, whereas the 280 degree hole continually passes through the high pressure region. Such a hole should undoubtedly be deleted as it would act as a sink removing oil.

Often the oil at the big end has not only to lubricate the bearing/crankshaft junction but must subsequently perform the task of small-end

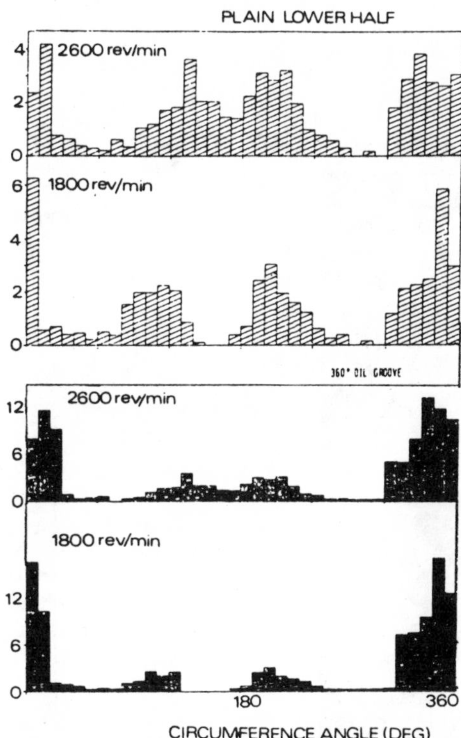

Fig. 5.22 Wear parameter histograms compared for main bearing with a plain half (*top*) and with 360 degree oil groove (*bottom*)

lubrication, piston cooling, or liner lubrication. This requirement often has a detrimental effect on the oil film thickness necessitating the drilling of a secondary big-end hole which will almost inevitably be poorly located. Small-end lubrication is sometimes taken from a drilling down the connecting rod to the bearing crown, while some designs contain slots in the bearing to supply oil squirt holes. Such a reduction of the bearing area can have deleterious effects on the oil film, as holes in the crown can reduce the film thickness by as much as 70 per cent. The movement of the oil holes (or slots) in a direction opposite to the engine rotation can substantially restore the oil film, as shown in Fig. 5.24.

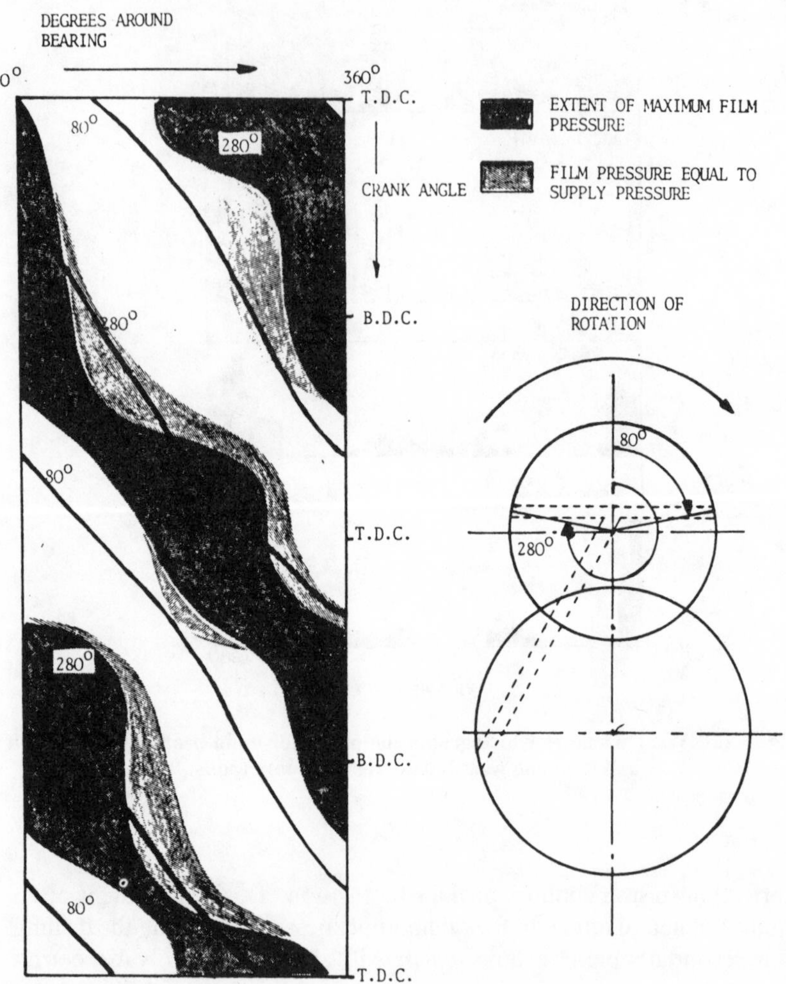

Fig. 5.23 Oil film extent diagram

HOLE TYPE	MINIMUM FILM THICKNESS AT 1200 RPM	MINIMUM FILM THICKNESS AT 2800 RPM
NO HOLE	2.67 (105)	3.08 (121)
¼ IN. HOLE AT CROWN	0.85 (34)	1.64 (65)
IN. HOLE AT CROWN	0.93 (36)	1.95 (77)
¼ IN. HOLE OFFSET 15°	2.20 (87)	1.76 (69)
¼ IN. HOLE OFFSET 25°	2.67 (105)	2.33 (92)
¼ IN. HOLE OFFSET 45°	2.67 (105)	2.66 (105)
¼ IN. HOLE OFFSET 65°	2.67 (105)	2.85 (112)
¼ IN. HOLE OFFSET 75°	2.67 (105)	2.91 (115)

UNITS ARE MICRON (MICROINCH)

Fig. 5.24 Effect of oil hole position on oil film thickness

PREDICTING FUEL CONSUMPTION

R. T. C. Harman[37] has shown how the energy used in a test driving cycle can be predicted, and thus fuel comsumption figures obtained, in advance of vehicle manufacture, using the concept of driving cycle intrinsic energy. He points out that the major source of energy loss in petrol cars is the engine. Whereas one standard rule of thumb has been that equal thirds of the fuel energy go to the radiator, the exhaust and the driveshaft, this analysis shows that less than a sixth of the energy gets to the driveshaft in an urban driving cycle.

The cycles distribute the energy losses differently, as shown later in Table 5.3. There is no reason why a simple cycle should not give the same answer as a complex one, although it would only cover a narrow spectrum of speed and load combinations. The ECE15 cycle clearly places less emphasis on wind effects than the EPA Urban cycle while all urban cycles lose more energy to the brakes than the other cycles. The braking energy can be reduced to a considerable extent by coasting to gentle stops, thus saving a substantial amount of fuel, but that would be a deviation from the cycle specifications. It would, however, account in part for the measured difference between drivers.

The potential for reducing fuel consumption by reducing these sources of loss may be predicted from these intrinsic energy calculations. In changing the vehicle characteristics from the 1985 levels to the ECV3 levels, for example, the percentage fuel savings when driving to the European average method would be 11.6 from the 24 reduction of loss to the chassis and tyres alone, 22.8 from the 28.6 reduction of aerodynamic loss alone, or 11.2 from the mass reduction of 18.5. The effect of all these changes together is a fuel saving of 29.6 per cent. Similar predictions may be made for any other changes of the characteristic values and driving pattern or cycle. It is also possible to predict the effect of converting a vehicle to electric operation, with its increased mass offset partially by regenerative braking.

While the intrinsic energy concept may reduce the need for testing to obtain competitive fuel comsumption data, it could only be extended to indicate emissions levels if used in conjunction with fully mapped engine speed versus load data.

The intrinsic energy is defined as the quantity of energy required to propel a vehicle of known drag characteristics and mass through a driving cycle having a prescribed variation of velocity relative to time. A simple technique for calculating intrinsic energy reveals the relative significance of the vehicle's mass, aerodynamics, tyre, and chassis losses while comparison with its measured fuel comsumption reveals the power plant efficiency.

The calculation of intrinsic energy starts with the energy consumption at constant speeds. The energy used per metre is, in fact, the tractive effort, which may be defined as

$$\mathrm{d}E/\mathrm{d}x = F = C_c M + C_t M + \tfrac{1}{2}\varrho C_d A V^2 \quad \text{(Nm/m)} \qquad (5.1)$$

where M = vehicle mass, laden (kg); V = velocity (m/s); C_c = chassis rolling resistance coefficient (N/kg, normal range = 0.05 to 0.20); C_t = tyre rolling resistance coefficient (N/kg normal range = 0.10 to 0.20); C_d = aerodynamic drag coefficient (normal range = 0.30 to 0.70); A = vehicle frontal area (m^2); ϱ = air density taken as 1.23 kg/m^3.

This expression could be made more general by including a term for the gradient of the road in the form Mg/s, where the gradient has a slope of 1 in s, but this is not used in the driving cycles.

The second term of the equation is used to represent operation with radial ply tyres. It takes the form $C_t MV$ for cross ply tyres, whose greater sidewall stiffness requires considerably greater flexing energy, and may also apply to radial tyres at higher speeds. For this exercise the velocity term is ignored and the combined coefficient C_{ct} is used, where

$$C_{ct} = C_c + C_t \tag{5.2}$$

When the vehicle accelerates at rate a (m/s^2), the tractive force is increased by an Ma term and the aerodynamic drag increases as the velocity rises. Taking V_1 to be the velocity at the beginning of the acceleration (or deceleration) segment, the rate of usage of energy on wind drag at any time, relative to the distance travelled by that time, is

$$dE_w/dx = \tfrac{1}{2}\varrho C_d A V^2 = \tfrac{1}{2}\varrho C_d A(V_1^2 + 2ax) \tag{5.3}$$

for which the total energy used over a distance x is

$$E_T = (C_{ct} + a)Mx + \tfrac{1}{2}\varrho C_d A(V_1^2 + ax^2) \tag{5.4}$$

When comparing the driving cycles listed in Table 5.1, the V_1, a and x terms which apply to each segment may be combined into coefficients X, Y, and Z which uniquely describe the accumulated events of the cycle

$$E = XC_{ct}M + YM + ZC_d A \tag{5.5}$$

These three terms represent the rolling resistance, inertia effects, and wind drag, X is simply the distance travelled in metres if the energy is expressed in Joules but, in the specific form of Wh/km, X takes the value of 0.278 for all cycles. The inertia term YM proves to be the energy dissipated by the brakes, as the energy expended in acceleration is recovered during deceleration by overcoming the resistance losses if no

Table 5.1 Driving cycle data

Driving cycle	Origin	Distance (m)	Time (s)	Idle (s)	No. of stops	Max. V (km/h)
ECE 15	EEC	1018	195	60	3	50.0
EPA Urban	USA	12050	1371	245	18	91.2
EPA Highway	USA	16480	765	4	1	96.3
Urban	NZ	1382	221.8	60	3	65.0
Suburban	NZ	4086	337	31.3	2	96.0
SAE J227(a) C	USA	336*	80	25	1	30.0
EEC Average†			—	—	—	120.0

* Slightly variable due to a coasting segment.
† Cycle consists of 40% ECE 15, 50% constant 90 km/h, 10% constant 120 km/h.

braking effort is used. Z takes a different value for each cycle, appropriate to the velocity levels used, and is constant except in the electric vehicle cycles to SAE J227(a) which include a coasting segment. For instance, a heavy, low-drag vehicle would retain a higher velocity than a light, high-drag vehicle at the end of the time prescribed for coasting. Z, therefore, varies slightly for these cycles, but an average, constant value incurs little error in calculating E.

Values of Y were determined for the cycles in Table 5.1 using computed energy requirements for a wide range of vehicle characteristics. Y was found to be expressible in terms of

$$Y = (f_0 F_r - f_1 C_{ct} - f_2 C_d A/M)(1 - F_b) \tag{5.6}$$

where f_0, f_1, and f_2 are factors having specific values for each cycle, as listed in Table 5.2. F_r is a factor by which M is multiplied to include the effective inertia of all rotating parts, while F_t is the fraction of the braking energy which is recovered by regenerative braking in the case of electric vehicles. F_r is taken here as 1.05 for fuelled vehicles and 1.035 for electric vehicles, while F_b is zero for the former. If equation (5.6) is substituted back into equation (5.5) and the terms are rearranged, the final expression for the intrinsic energy of the driving cycle becomes

$$E = f_0 F_r M + (0.278 - f_1) C_c M + (Z - f_2) C_d A \quad \text{(Wh/km)} \tag{5.7}$$

The numerical values of the factors in Table 5.2 are valid only for a limited range of vehicle characteristic values. Considering any particular

Table 5.2 Coefficients *Y* and *Z* for several driving cycles, excluding idling

Driving cycle	Factors for *Y* f_0	f_1	f_2	*Y* values within ±	Coefficient *Z*
ECE 15	0.03944	0.0557	3.0	0.00001	17.17
EPA Urban	0.0395	0.0450	3.7	0.00025	37.58
EPA Highway	0.00964	0.0140	3.0	0.00035	88.25
Urban	0.05725	0.0875	8.1	0.00001	26.02
Suburban	0.0156	0.030	4.5	0.002*	50.44
SAE J227(a) C	0.0252	0.0560	1.4	0.00020	9.40†
EEC Average*	0.0158	0.0223	1.2	0.00001	79.30

*Error within ±0.00010 if Y is between 0.0050 and 0.0085.

†Z varies from 9.18 for a light, high drag vehicle to 9.66 for a heavy, low drag vehicle within the specified range of characteristics.

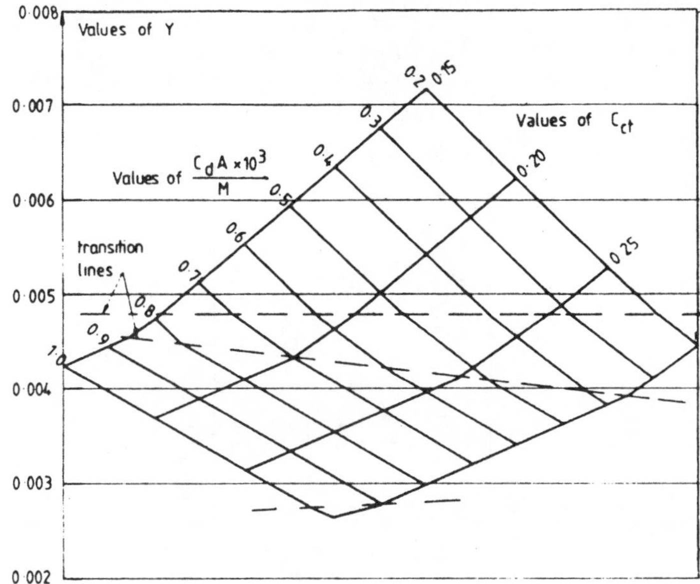

Fig. 5.25 Variation of coefficient *Y* with values of vehicle characteristics

deceleration segment, progressive reduction of the vehicle drag relative to its inertia will at some stage require a transition from the use of engine power to the use of the brakes. Some cycles show the effect of such transitions within the normal range of vehicle characteristics, as illustrated by the triple mesh of Fig. 5.25 for the EPA Highway cycle. These transitions are less significant in the EPA Urban cycle and absent in the ECE 15 cycle, but are severe in the much simpler New Zealand Suburban cycle. In practical terms, the effect of these transitions is small over the range for which the factors in Table 5.2 apply: $1000C_dA/M$ from 0.2 to 1.0, and C_{ct} from 0.15 to 0.30. The factors quoted describe a single, compromise mesh which departs from the multi-mesh reality only to the extent of the tolerance listed.

Drive cycle and vehicle comparisons

The values of Z and the factors for Y in Table 5.2 differ noticeably for the different cycles, which may be expected to show quite different intrinsic energies for any given vehicle. In Table 5.3, the cycles are compared on

Table 5.3 Energy comsumption of different cycles, 1985 car

Driving cycle	Intrinsic (Wh/km*)	Percentage contributions from Chassis	Tyres	Brakes	Wind	Travel (Wh/km*)	Idle (Wh)	Overall (Wh/km*)
ECE 15	106.4 (1.14)	34.0	31.3	24.0	10.7	709 (7.63)	155	862 (9.3)
EPA Urban	122.3 (1.31)	29.6	27.3	22.7	20.4	8.5 (8.76)	632	868 (9.3)
EPA Highway	132.8 (1.43)	27.2	25.1	3.5	44.2	885 (9.52)	10	886 (9.5)
Urban	119.7 (1.23)	30.2	27.8	27.5	14.5	797 (8.58)	155	909 (9.8)
Suburban	108.9 (1.17)	33.2	30.6	5.4	30.8	726 (7.80)	81	746 (8.0)
SAE J227(a) C	87.3 (0.94)	41.4	38.2	13.2	7.1	582 (6.26)	65	775 (8.3)
EEC Average	132.5 (1.42)	27.3	25.2	7.7	39.8	883 (9.50)	62	944 (10.1)

* Bracketed terms show /100 km figures based on conversion rate 100 Wh/km = 1.075/100 km.

the basis of a good 1985 car, weighing 1000 kg including a payload of 150 kg and having $C_c = 0.13$, $C_t = 0.12$, $C_d = 0.35$, $A = 1.9$, and $F_r = 1.05$. The percentage contributions of the various losses highlight the differences between the cycles.

The second column shows how little energy is intrinsically required to overcome the vehicle losses incurred by following any of these driving cycles. The discrepancy between this and typical measured fuel consumption is accounted for by the engine and transmission losses, but account must also be taken of the idling time in the driving cycles. In the absence of better information, a power plant efficiency of 15 per cent was used to convert the intrinsic energy into the energy required for the travel segments. When the petrol consumed during the idle segment is included, at 1 litre per hour (9.3 kW or 2.58 Wh/s), the overall calculated fuel consumption appears realistic. It can be seen that, in some cycles, the effect of idling is very significant, but it would be less significant with diesel engines and no cause of loss with electric vehicles.

This technique now provides a basis for determining typical levels of power plant efficiency. Table 5.4 compares five fuelled vehicles and two electric vehicles operating to appropriate driving cycles. Their characteristics are tabulated as realistically as possible, with a query where a figure is estimated. The fuelled vehicle masses include an arbitrary passenger load of 150 kg and the engine efficiency, which is very low in the petrol vehicles, includes transmission loss and the loss due to idling. The electric vehicle masses include the known payloads and F_r is given the value of 1.035. A braking allowance of $F_b = 0.2$ is included, covering a battery charge/discharge efficiency of 60 and a generation efficiency of 33.3 per cent.

The experimental fuelled cars might represents common practice in 1995. BL's ECV3 was designed to save energy in every possible way, yet is a five seater capable of 185 km/h. Its measured fuel consumption data covers several conditions, of which two are used here. The Citroen SA 109 is a potential replacement for the 2CV, gaining its very low C_D value from a single box form and, like the ECV3, using a three cylinder petrol engine. It is believed that the fuel consumption quoted applies to the European average method, but it is shown also as applying to the ECE 15 test. The Aero 2000 is reported to use the body planned for GM's 1984 electric car, which is now fitted with a diesel engine. It is assumed that its quoted fuel consumption was measured to the EPA Urban driving cycle.

The electric vehicles considered are an AC motor powered car made at

Table 5.4 Energy comsumption and efficiency of different vehicles

Vehicle type	Characteristic values C_c	C_t	C_d	A	M	Cycle No.	Energy usage Wh/km or intrinsic	(/100 km) tested	Engine efficiency (%)
1985 car	0.13	0.12	0.35	1.90	1000	1	(1.144)	(9.5?)	12.0
BL EVC3	0.1	0.09	0.25	1.9?	815	1	(0.805)	(5.76)	14.0
BL ECV3						7	(0.970)	(4.50)	21.5
Citroen SA109	0.08?	0.08?	0.21	1.53	590	1?	(0.540)	(3.0)	17.9?
Citroen SA 109						7?	(0.634)	(3.0)	21.1?
GM Aero 2000	0.1?	0.1?	0.23	1.9?	950?	2?	(1.055)	(3.3)	32.0?
U of C Mk II	0.08	0.12	0.35	1.85	1250	3	124.0	340	37.5
Lucas Bedford	0.1?	0.15?	0.45	3.5	3495	6	297.7	557	53.5

the University of Canterbury and the Lucas Bedford van. The efficiency of the car is based on recorded use over 725 km on flat roads, in comparison with the New Zealand Urban cycle. The Lucas Bedford was tested through 140 cycles though with a quoted distance of 1.8 times that listed in Table 5.1, which casts doubts on the energy figures. The efficiency of both vehicles is based on power supply from the mains, so includes the battery with the motor: it is considerably higher than for the petrol vehicles.

ENGINE SYSTEMS ANALYSIS, STRUCTURAL DESIGN, AND DRAUGHTING BY CAD

A recent report[38] has shown how a major consultancy has put CAD to work in engine design. At Ricardo manual draughting is becoming the preserve of the more experienced designer who is released from most 'routine' layout work. Design staff working on terminals are also skilled in both interactive CAD (adding one line at a time) of utilizing software, developed in-house, for automated draughting of 'standard' components like valves or tooth-belt drive assemblies. The latter involve designing within parametric rules, built into the programs, devised from many years' design practice within the company or by regularized calculation procedures. There is also skill in the *use* of 'automated' programs as these continually ask questions of the designer who must enter the appropriate replies via menu boards of design data.

In producing revised cylinder head designs, the interactive approach is partly necessary, but a line need only be called up once and thereafter can be repeated as necessary throughout the design. And as 80 per cent of the company's design work involves new layouts, the application of CAD is challenging. CAD has brought a 45 per cent improvement in productivity; some would consider this a relatively low figure, but it is important to realize that this improvement has meant the release of scarce high-grade design engineers for concept work.

The CAD system was chosen on the basis of a variety of criteria, not least of which was ability to link in with parallel FE 'structural' and Fortran CA – analysis programs. CAM links were also possible as were the easier application of company *and* client standards, multi-user access to drawings, simultaneously, easy reproduction at any scale, and automatic accounting of design time. The system chosen allows comprehen-

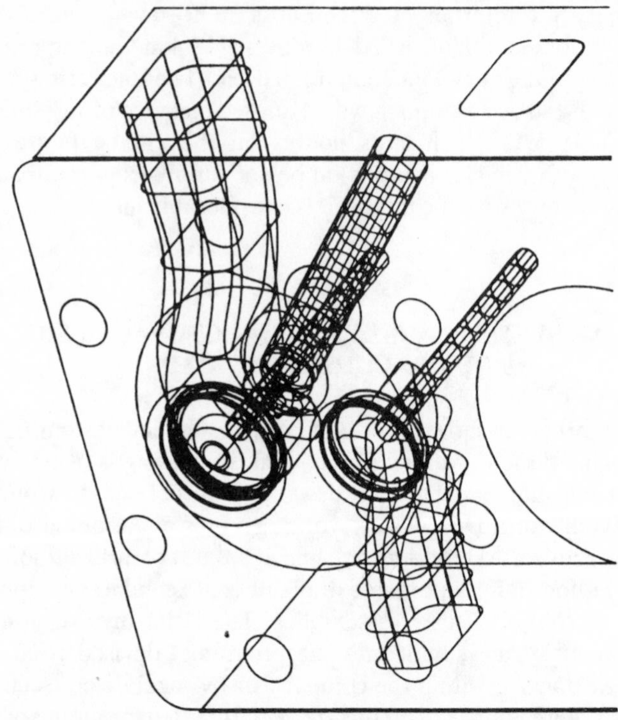

Fig. 5.26 3-D wire frame model from which a section as in Fig. 5.27 can be obtained

sive 2-D draughting and a 3-D mesh generation (Fig. 5.26) facility for surface visualization which allows the taking of oblique sections for examining clearances (Fig. 5.27), and wall thickness in complex castings, for example, interactive programming language GRIP is used for automatic generation of geometrically similar parts such as gears, nuts, bolts, valves, and seat inserts. The graphics software allows rapid geometry creation by storing shapes, text, and dimensions as 'patterns'. Typical creation times in seconds are as follows, for the parts in Fig. 5.28. Hydraulic tappet (A) 30; Injector (B) 40; Heater plug (C) 30; Valve and seat (D) 180 and Ricardo Comet chamber (E) 300.

Design-analysis facilities available

The company's service, for a new range of engines, might include; concept studies, determining basic configurations, preliminary layout

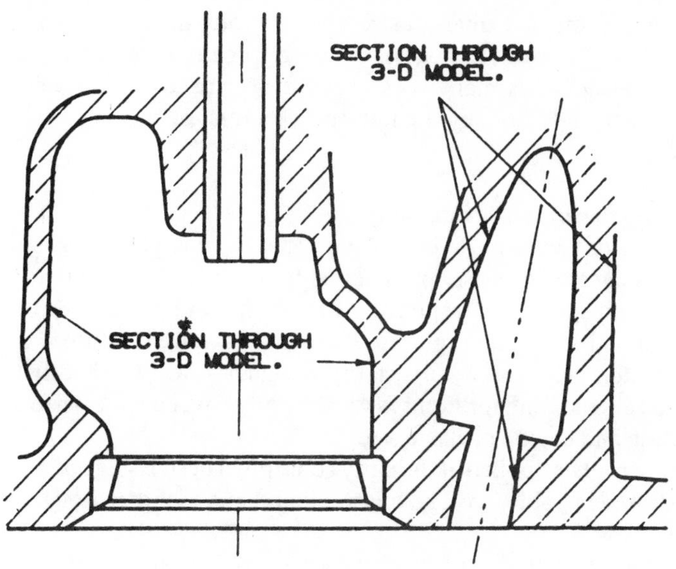

Fig. 5.27 Cylinder head section from 3-D plot

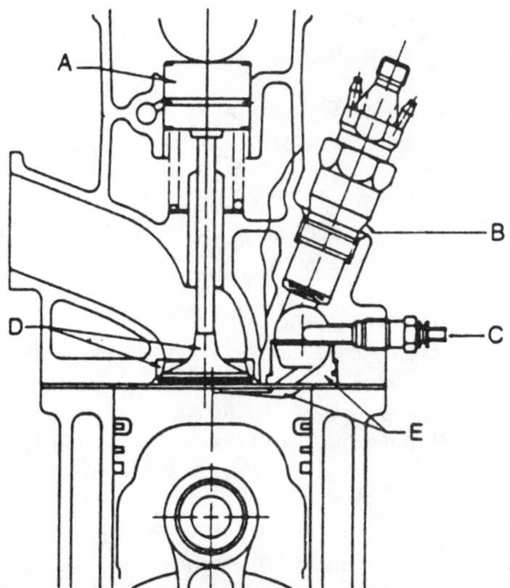

Fig. 5.28 Rapid geometry creation for standard parts annotated in text

and optimization of auxiliaries, stress calculations, definitive layouts, and detail drawings with, if necessary, parts procurement for prototype engines. Design assessment work may also be carried out, on completely new concepts; additionally the uprating potential of existing models can be studied.

Design and performance interaction

A key part of Computer Aided Analysis at the company surrounds the concept stage: fixing basic parameters like number of cylinders, bore, stroke, rated speed, and aspiration with relation to performance, economy, and emission requirements. The company have developed a range of computer programs to assist the designer towards identifying the optimum engine configuration; these programs cover engine performance prediction and engine package size.

They can all be applied at the early conceptual design stage of a project, as illustrated in Table 5.5. The relative fuel economies of various combus-

Table 5.5 Interaction of performance and design: the concept phase

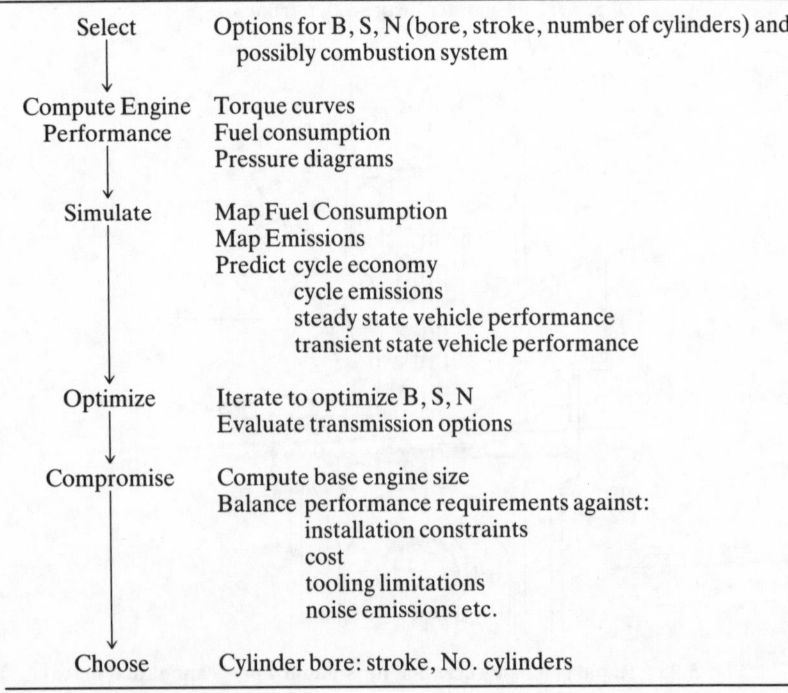

Select ↓	Options for B, S, N (bore, stroke, number of cylinders) and possibly combustion system
Compute Engine Performance ↓	Torque curves Fuel consumption Pressure diagrams
Simulate ↓	Map Fuel Consumption Map Emissions Predict cycle economy cycle emissions steady state vehicle performance transient state vehicle performance
Optimize ↓	Iterate to optimize B, S, N Evaluate transmission options
Compromise ↓	Compute base engine size Balance performance requirements against: installation constraints cost tooling limitations noise emissions etc.
Choose	Cylinder bore: stroke, No. cylinders

tion systems can be determined and the options for cylinder bore, stroke, and number of cylinders studied. The computer greatly assists in this decision making process by quantifying many of the performance parameters.

For larger projects, particularly those involving medium-speed engines, it is common to extend the computer analysis much further at the conceptual design phase, to include a detailed comparison of the thermal and mechanical loadings on the principal engine components.

Once the basic design parameters, such as combustion system, bore, stroke, rated speed, and mean effective pressure, have been determined and the project moves to the definitive design phase, the application of much more complex performance simulation programs can be considered. Such programs take account of valve geometry, timings, manifold design, firing orders, and many other factors; these programs are used to optimize valve sizes and timings and manifold sizes, estimate accurate pressure diagrams, obtain thermal loading data, and study the influence of injection.

Design analysis

Following twenty years' experience the company now have an extensive library of computer programs to assist the design analyst. These programs are used to calculate component loadings, stresses, vibrations, and dynamic motion. The programs can mostly be run in a relatively short time and are, therefore, applicable in the early stages of a project when a number of iterations may be necessary in order to obtain an optimum layout.

The main programs cover critical aspects of the crankshaft, bearings, connecting rod, timing drive, cylinder head and block, valve train (Fig. 5.29) and flywheel, vibrations, and combustion.

In the case of the crankshaft, development of the analytical programs (for bending and torsional stress, natural frequency, and damper response) has been in progress for many years (Fig. 5.30), and improvements are continuing. For crank stressing, 24-order torsional vibration analysis has been automated and is now a fairly routine procedure: the results of a typical calculation can be seen in Fig. 5.31.

Stress concentrations at crank fillets, oil holes, and bolt-head recesses are often critical, and efforts have been made to improve the accuracy of such predictions. The development of new methods has been backed up by experimental programs using the photo-elastic technique or strain gauges, fatigue testing, and correlation with finite element analyses.

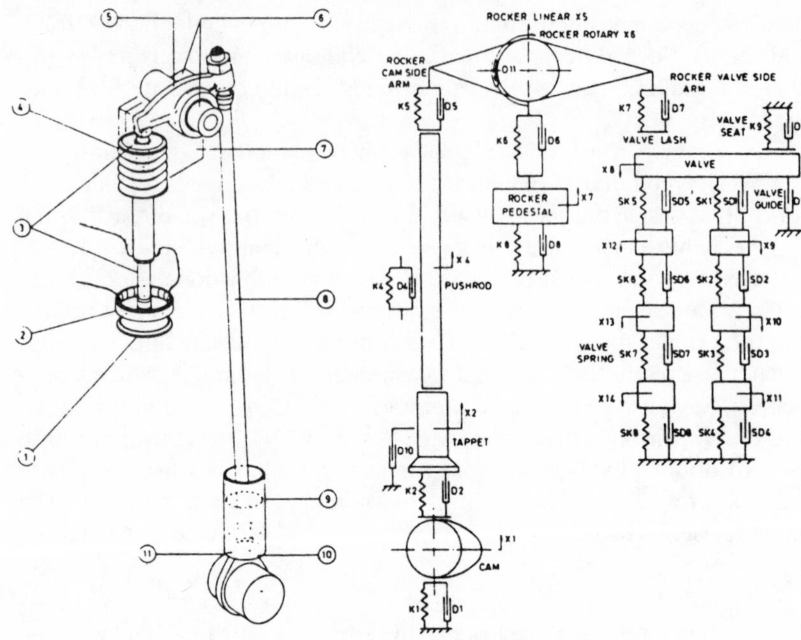

Fig. 5.29 Main areas of valve train investigation

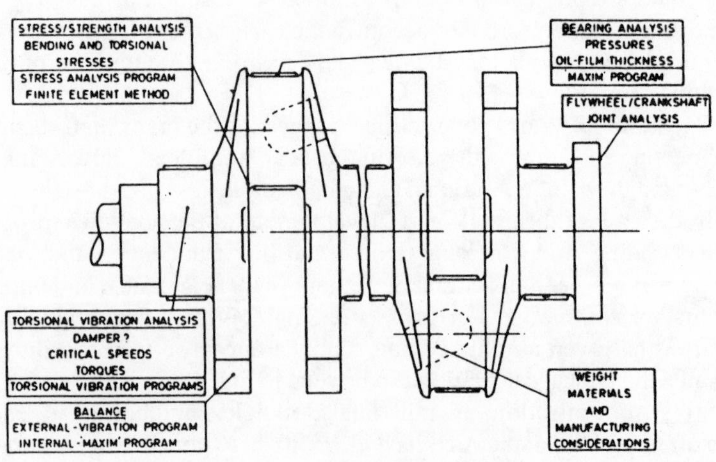

Fig. 5.30 Math model of valve train

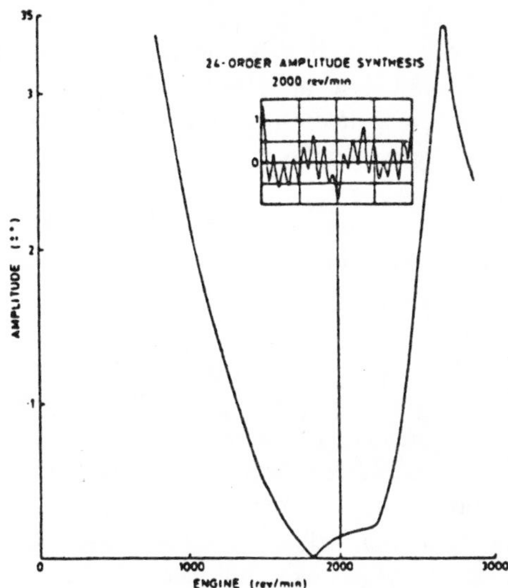

Fig. 5.31 Amplitude at crankshaft nose from torsional analysis

The interpretation of design calculations is, however, an important aspect to which considerable effort has been devoted. It is only too easy to devise new methods of analysis, the results of which give little indication of the safety or otherwise of the components in question. This occurs because criteria for safety, minimum oil films, stress limits, and so on, can often take many years to establish. In many instances even the basic fatigue properties of the material being used may be difficult to determine.

Finite element analysis

These techniques are employed to determine distortions, stresses, temperature distributions, and vibration characteristics.

The routine applications of the method are: confirmation of new designs once the layout has been optimized, determination of the basic stiffness of main engine structure, investigation of problems due to deformation, movement, or overstressing, and optimization of metal sections in components subject to combined thermal and mechanical loads.

Although the finite element method is widely used, its application to engine components demands specialist knowledge and experience if the results are to be valuable. The main phases in any analysis are as follows. (1) Establishing a model of component to be analysed: identification and location of loads, simplification of geometry – 2D/2½D/3D, selection of element type and generation of mesh, considering boundary conditions and constraints. Then comes (2), the analysis: individual load conditions, combined load results, and graphical presentation. Finally (3), interpretation of results: identification of critical areas, consideration of material properties, fatigue prediction, and recommendations for modifications.

Determining the basic model to be analysed can often be the most critical task. Over-simplification of the model may invalidate the whole analysis, raising questions such as; are thermal effects significant to cylinder block distortions and stresses, does the seating for the cylinder liner remain flat, is the tension in each retaining bolt equal, does movement of the bearing housings significantly influence crank stresses? Past experience will often indicate which factors are likely to be significant and should be modelled. The modelling of a joint such as the big-end eye of a connecting rod also needs careful consideration. Production tolerances may play a dominant role in determining peak stresses and this may not be accounted for in the basic model.

Experience will also guide the analyst when selecting his element mesh. He needs to mesh critical areas finely enough to ensure accurate predictions of high stress concentrations, but must also economize on the total number of elements employed in order to contain the costs of operating the program. For complicated components it may be more economic to have two models, a coarse mesh of the whole item and a fine mesh of the critical area or areas: the first model is used to determine boundary conditions for the second.

For components which have a well-defined family geometry, such as fly-wheels, straight-cut connecting rods, and valves, it is possible to establish a standard mesh and automate the analysis.

FEM for the engine structure

J. W. Cornforth of Ricardo reports[16] that many structures requiring three-dimensional analysis can be treated in some degree as assemblages of plates, and special plate or 'shell' elements have rotational degrees of freedom at their nodes to compensate for the fact that, unlike solid elements, they have no 'real' thickness. Cylinder blocks and crankcases are good candidates for this type of modelling and the use of shell

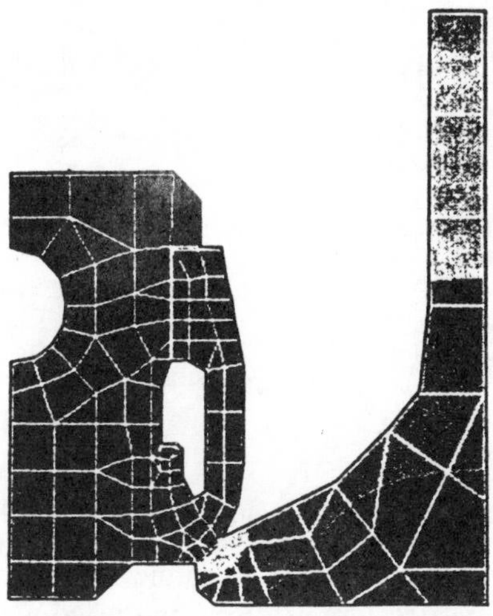

Fig. 5.32 Optimizing cooling passage for a valve seat

elements not only reduces the size of the model but also makes it very much easier to construct. Although the finite element method is in essence a linear method, it can be adapted to solve non-linear problems. This is done by using iterative techniques, with the inevitable consequences that the analysis is much more expensive.

Most finite element programs include a code for calculating heat flow, and there is no doubt that the finite element method is the most practical way of calculating temperature distributions in complex components. A good example of its use in optimizing the size and position of cooling passages for water-cooled valve sets (Fig. 5.32) where it is necessary to cool the seat sufficiently while avoiding over-cooling of the port. Heat flow analysis is always less expensive in computer time than stress analysis on an equivalent model, mainly because there is only one degree of freedom (temperature) at each node; so provided that a two dimensional or axisymmetric representation of the geometry can be used it is a very economic method of design optimization. Transient heat flow can also be analysed, although the equations are more complex because heat storage must be accounted for, and the solution itself is incremental and, as with the analysis of geometric non-linearity, it is therefore expensive. The finite element method can be used in a rather different way to solve

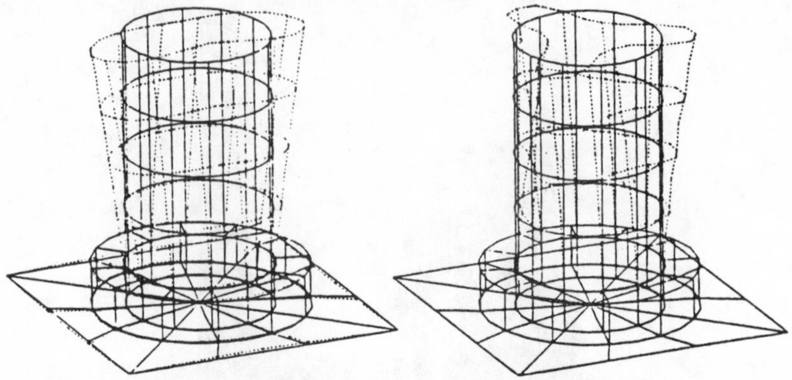

Fig. 5.33 Dynamic analysis of a cylinder liner

structural vibration problems. In dynamic finite element analysis the element stiffness matrices are set up in the usual way, but the solution involves finding the eigenvectors and eigenvalues of the structural equations of motion, which give the mode shapes and frequencies of the natural modes of vibration (Fig. 5.33).

Even in structures as complex as a cylinder block (Fig. 5.34) it is possible to predict mode shapes with good accuracy, and the results of this

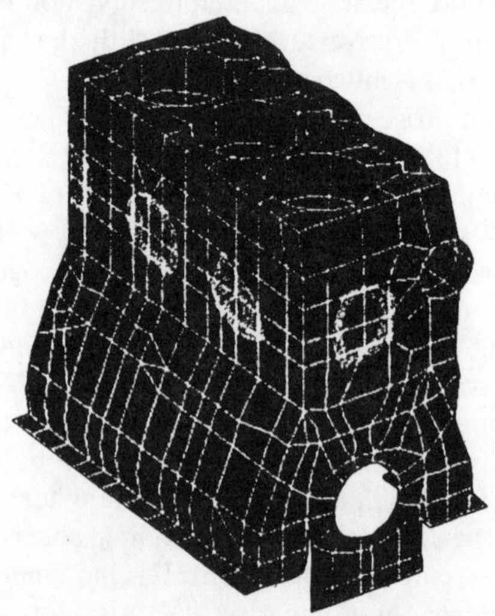

Fig. 5.34 Stress distribution for block and crankcase assembly

type of analysis can be very illuminating. Mode shapes do not give any information about the magnitude of vibration under cyclic or transient loading, but this can be calculated if the structural damping is known. The aim of work at Ricardo in this field is to reduce the time taken to develop low noise engines by improving the company's ability to *design* for low noise. There are many obstacles in the path to this goal – for example, the fact that at present structural damping can only reliably be found by experiment – but good progress is being made and the company are confident of success.

FIRE 1000: FIATS CIM ENGINE

Computer integrated design and manufacture (CIM)[39] has been used to the full in Fiat's new engine series of around one litre capacity for which almost the complete design process has involved CAD. Substantial specific fuel-economy gains (15 per cent) have followed (306 gm/kwh on urban cycle) computer control of the manufacturing process – especially in precision casting and core-making by new techniques – resulting in close dimensional tolerance cylinder-to-cylinder and engine-to-engine for inlet/outlet ducts and combustion chamber (Fig. 5.35). (It has been said on certain volume-built engines that up to 10 per cent discrepancy in compression ratio, cylinder-to-cylinder, and even higher differences in air–fuel charge deliveries, have been found.)

Manufacturing economy (production rate is to be 2500/day) is also found in the Fire 1000 engine, which has 30 per cent less components (273 against 368) and a unit weight of only 69 kg, now producing 33.5 kw and a potential power of 'twice that figure'. The maximum torque of 82 Nm, developed at 2750 r/min, is compared by Fiat with its 78 kg 903 cm^3 engine developing 68 Nm at 3000 r/min. Component savings have mainly been made in direct belt drive to the overhead camshaft, a crankshaft-driven oil pump, camshaft-driven petrol-pump and distributor, and a water pump volute machined in to the cylinder block. Key dimensions for the unit are bore × stroke of 70 × 64.9 mm, 77 mm cylinder centre distance and 9.8 : 1 compression ratio. Cast iron is used extensively: for cylinder block crankshaft, con-rods, camshaft and flywheel (Fig. 5.36).

Computer-aided design has involved the use of graphics to represent low or high stressed areas of components as light or shaded areas. Dynamic stress analysis has also been perfected for design use, the company say, at least for low frequencies. The result of careful structural

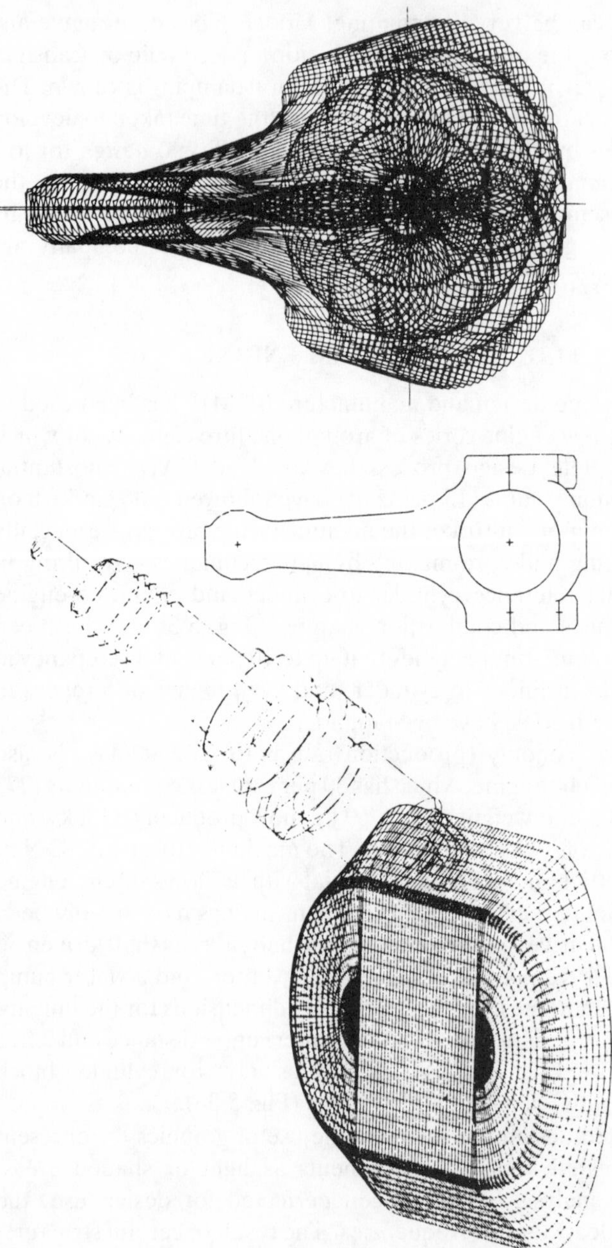

Fig. 5.35 Two interesting aspects of CAD are the computer plot of con rod motion (*right*) and combustion chamber profile (*left*)

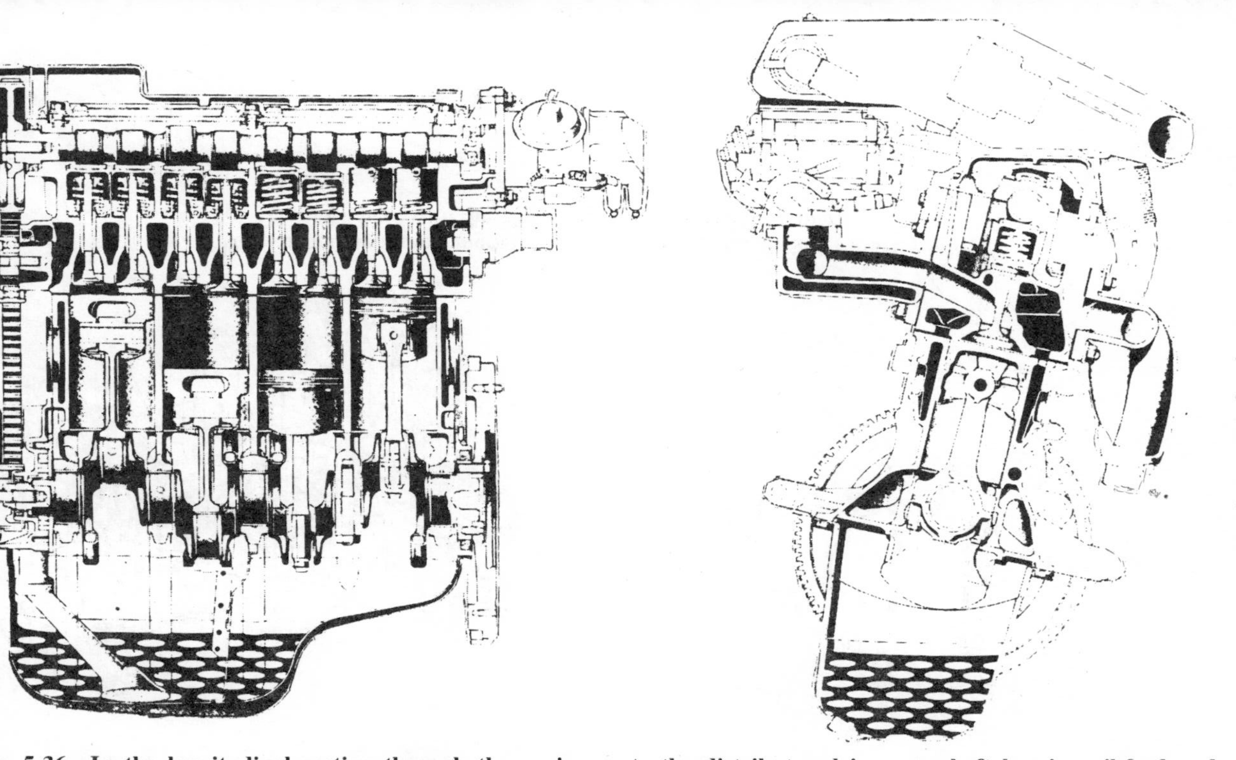

Fig. 5.36 In the longitudinal section through the engine, note the distributor drive, camshaft bearing oil-feed and oil-pump/front-cover assembly. The cross section reveals the cut-out bridge members used to support the main bearings also the deep section head with centre wells for valve spring assemblies

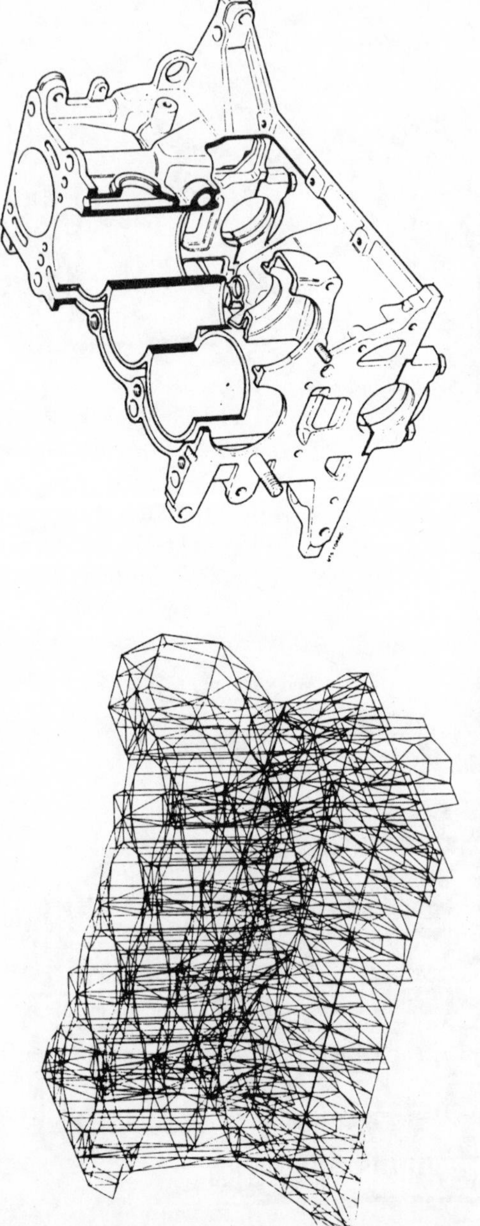

Fig. 5.37 The mathematical model (*left*) used to design the cylinder block (*right*) is shown here with the 'detached' layout of the 'wet liners'

design is particularly important on the cylinder block which weights 18 kg against its 24 kg predecessor. Side wall thicknesses are down to 4 mm maximum and CAD has particularly helped in main-journal bridge support design, which employs large cut-outs in the transverse diaphragm. Cylinder liners are not connected to the block, two walls of which house the 'wet' sleeves being restricted to the length of piston ring travel (Fig. 5.37).

The outer surfaces of the cylinder liner walls are used as references for machining the bores so that uniform thin wall thicknesses can be used. Head clamp-down screws align with the centre lines of the liner wall section. The liner water jacket is tapered to provide minimum temperature gradient from top to bottom of the liner. Friction between piston and cylinder has been precisely measured by a single cylinder test rig in which the crankcase walls are replaced by load cells.

The new casting system has been developed by Teksid; it involves a single pack system of cores for the cylinder head, which are made by cold-compaction, and three for the block. The cast crankshaft weighs 5.93 kg (25 per cent less than that of the 903) and has rolled crank pins. The deep upward-skirted cylinder head has central wells to accept the valve springs. Combustion chambers are as-cast finished, but tolerance on depth from the head face is within 0.2 mm which in turn holds clearance volume, cylinder to cylinder, within one per cent. Teksid have also developed a new casting process for the inlet duct involving a lost-polystyrene method.

All engine covers are diecast for minimum machining and, beneath the tappet cover, steel tubes are used in the head to feed the camshaft bearings, again to save machining. Toothed wheels for camshaft drive, and those in the oil pump, are in pressed sintered steel. The engine front cover also doubles as housing for oil pump, filter, and pressure control valves.

Installed in a *Uno*, Fiat say the engine would give 4 l/100 km at 90 km/h, surprising for a bath-tub chamber engine, but the design is said to be particularly tolerant of lean air/fuel ratios. The Weber 32 TLF carburettor is a new design notable for being secured to the inlet manifold by vertical screws that pass through the full depth of the body. It is suitable for alcohol/petrol mixtures. Within the manifold a mixture distributor is fitted at the confluence of the four branch pipes and the lower surface is channelled to improve fuel flow in cold starting. Inlet valves have 30.5 mm head diameters and open to 7.1 mm lift; maximum gas speed is 83.6 m/s, at maximum power, reducing to 41.8 m/s at maximum torque.

APPENDIX
Article sources used in compilation

Chapter One

(1) **S. S. Tresilian** (Rolls Royce), 'The piston engine', Parts 1 to 7b, *Automot. Des. Engr*, 1965, **4**(2)–**4**(10).

Chapter Two

(2) **W. T. Lynn** (Lucas CAV), 'Combustion fundamentals: the basic thermodynamic limitations of petrol and diesel engines', *Automot. Des. Engr*, 1964, **3**(7), 54–58.

See also:

M. Dargauthier, 'Evolution du moteur à combustion intern', *Ingénieurs de l'Automobile*, 1962, **35**(1).

C. R. Ball, 'The new Cummins V-type diesel engines', SAE preprint 553B, 1962.

R. C. Schmidt, 'Cummins diesels for stop and go services', SAE preprint 626A, 1963.

K. Schnauffer, 'Ein schnellaufender, unaufgeladener 450-PS-MAN-M Dieselmotor, MTZ, 1962, **23**(4).

M. A. Elliott *et al.*, 'The composition of exhaust gases from diesel, gasoline and propane powered motor coaches', *Air Pollution Control*, 1955, **5**(2).

J. D. Caplan, 'Causes and control of automobile emissions'. I.Mech.E. preprint, April 1963.

A. E. W. Austen and **W.-T. Lyn**, 'Relation between fuel injection and heat release in a d.i. engine and the nature of combustion processes, *Proc. Instn mech. Engrs*, 1960–61, **174/175**.

W. J. D. Annand, 'Heat transfer in the cylinder of reciprocating internal combustion engines', *Proc. Instn mech. Engrs*, 1963, **177**, 973–996.

Overbye '*et al.*, 'Unsteady heat transfer in engines', *Trans SAE* 1961.

M. H. Edson and **C. F. Taylor**, 'The limits of engine performance—comparison of actual and theoretical cycles', SAE preprint 633E, 1963.

W.-T. Lyn, 'Calculations of the effect of rate of heat release on the shape of cylinder pressure diagram and cycle efficiency', *Proc. Instn mech. Engrs*, 1960–61, **174/175**.

J. Beck, Discussion to paper on 'Some steps towards calculating diesel engine behaviour', *Trans SAE*, 1962.

C. N. Davies *et al.*, 'Fuel injection and positive ignition a basis for improved efficiency and economy', *Trans SAE*, 1961, **69**.

J. P. Soltau, 'Cylinder pressure variation in petrol engines', *Proc. Instn mech. Engrs*, 1960–61, **174/175**.

W.-T. Lyn, 'An experimental investigation into the effect of a fuel addition to intake air on the perforamce of a compression-ignition engine', *Proc. Instn mech. Engrs*, 1954, **168**, 265–279.

L. D. Derry *et al.*, 'The effect of auxiliary fuels on the smoke-limited power output of diesel engines', *Proc. Instn mech. Engrs*, 1954, **168**, 280–297.

P. H. Schweitzer *et al.*, 'Fumigation kills smoke', *Trans SAE*, 1958, **66**.

(3) **E. R. Norster** (Cranfield IoT), 'Chemistry of combustion', *Automot. Des. Engr*, 1964, **3**(4), 49–53.

See also:

E. S. Semenov, 'Studies of turbulent gas flow in piston engines', in *Combustion in turbulent flow*, 1963 Israel Program for Scientific Translation. Distributed by Oldbourne Press.

G. A. Harrow, P. L. Orman and **G. B. Toft**, 'The effect of engine operating variables on the time of flame propagation in spark-ignition engines', *J. Inst. Petrol.*, July 1963.

B. Karlovitz, *Combustion processes.* 1956, Princeton University Press.

Alcock and **Watts**, 'The combustion process in high-speed diesel engines', 1959, CIMAC.

Y. M. Paushkin, *The chemical composition and properties of fuel for jet propulsion*, 1962, Pergamon Press.

B. Lewis and **G. van Elbe**, *Combustion, flames, and explosions of gases*, 1951, Academic Press.

(4) **C. G. Lucas** and **E. H. James** (Loughborough University), 'Flame propagation and cyclic dispersion in the spark ignition engine', *Automot. Des. Engr*, 1970, **9**(5), 36–41.

See also:

R. Vichnievsky, 'Combustion in petrol engines', Joint Conference on Combustion, IMechE and ASME, 1955.

E. H. James, *A computer simulation of combustion in a spark ignition engine*', MSc thesis—Loughborough University, 1969.

J. H. Johnson, **P. S. Myers** and **O. A. Uyehara**, 'End-gas temperatures, pressures, reaction rates and knock', SAE Paper No. 650505, May 1965.

J. C. Salooja, 'Studies of combustion processes leading to ignition in hydrocarbon', *Combust. Flame*, June 1960, 117.

E. S. Starkman, F. M. Strange and **T. J. Dahm**, 'Flame speeds and pressure rise rates in spark ignition engines', SAE Paper No. 83V, August 1959.

J. S. Clarke, 'Initiation and some controlling parameters of combustion in the piston engine', Proc. IMechE, Paper No. 5, 1960–61.

S. Curry, 'The relationship between flame propagation and pressure development during knocking combustion', SAE Paper No. 647B.

G. M. Rassweiler, L. Withrow and **W. Cornelius**, 'Engine combustion and pressure development', *Trans. SAE*, 1940, **46**(1).

W. J. D. Annand. Private communication, 1969.

H. Rabezzana, S. Kalmar and **A. Candelise**, 'Combustion: an analysis of burning and expansion in the reaction zone', *Automot. Engr*, October 1939.

S. Curry, 'A three-dimensional study of flame propagation in a spark ignition engine', SAE Paper No. 452B, January 1962.

J. T. Wentworth and **W. A. Daniel**, 'Flame photographs of light load combustion point the way to reduction of hydrocarbons in exhaust gas', SAE Technical Progress Series, Vol. 6.

C. F. Marvin and **R. D. Best**, 'Flame movement and pressure development in an engine cylinder', NACA Report No. 399 : 1931.

H. Rabbazana, S. Kalmar and **A. Candelise**, 'Combustion: an analysis of burning and expansion in the reaction zone', *Auto. Engr*, Nov. 1939.

D. J. Patterson, 'Cylinder pressure variations: a fundamental combustion problem', *SAE Trans*, 1967, **75**, 621.

J. P. Soltau, 'Cylinder pressure variations in petrol engines', Proc. IMechE, Paper No. 2, 1960–61.

G. A. Harrow and **P. L. Orman**, 'A study of flame propagation and cyclic dispersion in a spark ignition engine', ASAE Symposium on combustion in engines, July 1965.

E. S. Semenov, 'A device for measuring the turbulence in piston engines', *Pribory i Tekhnika Eksperimenta*, 1958, No. 1.

K. I. Schlekin and **K. Troshin Ya**, 'Gas dynamics of combustion', NASA TT F–231, October 1964.

C. L. Bouchard, C. F. Taylor and **E. S. Taylor**, 'Variables affecting flame speed in the Otto-cycle engine', *SAE Jnl.*, 1937, **40–41**.

R. J. Ellison, G. A. Harrow and

B. M. Hayward, 'The effect of tetra-ethyl–lead on flame propagation and cyclic dispersion in spark ignition engines', *Jnl Inst. Pet.*, 1968, **54**(537).

(5) **J. A. Newlyn** (Leicester Polytechnic), 'Geometric limitations on performance of four valve per cylinder SI engines', *Automot. Engr*, 1984, **9**(4), 16–18.
See also:
C. F. Taylor, *The internal combustion engine in theory and practice. Vol. 2: Combustion, fuels, materials, and design*. The MIT Press.
H. Barnes-Moss, 'A Designer's view point', *Passenger car engines*, 1973, IMechE, London, pp. 133–147.
W. Annand and **G. Roe**, *Gas flow in the internal combustion engine: power performance, emission control and silencing*, Foulis.
S. Yagi, A. Ishizuya and **I. Fujii**, 'Research and development of high-speed, high performance, small displacement Honda engines, SAE Paper 700122.
C. Scheffler, 'Combustion chamber surface areea, a key to exhaust hydrocarbons', SAE paper 660111.
W. Daniel, 'Flame quenching at the wall of an internal combustion engine', *Sixth symposium on combustion* (Yale University), Reinhold, 1956.
W. Daniel, 'Engine variable effects, on exhaust hydrocarbon composition, (A single cylinder engine study with propane as the fuel)', SAE Paper 670124.
C. F. Taylor, *The internal combustion engine in theory and practice. Vol. 1: thermodynamics, fluid flow, and performance*, The MIT Press.
I. Fukutani and E. Watanabe, 'An analysis of the volumetric efficiency characteristics of a 4-stroke cycle engine using the mean inlet Mach Index', SAE Paper 790480.

(6) **S. F. Smith** (London Transport), 'Torque control for part-load economy', *Automot. Engr*, 1977, **2**(5), 41–42.

(7) **C. G. Lucas, E. H. James** and **R. Chrast** (Loughborough University), 'Exhaust pollution control', *Automot. Des. Engr*, 1970, **9**(2), 34–40.
See also:
P. S. Myers and **O. A. Uyehara**, 'Many possibilities exist for eliminating major pollutants from exhaust gas', *SAE Jnl*, March 1969.
T. A. Huls and **H. A. Nickol**, 'Influence of engine variables on exhaust oxides of nitrogen concentration from a multi-cylinder engine', SAE Paper No. 670482, 1967.
T. A. Huls and **H. A. Nickol**, 'Engine variables influence nitric oxide concentration in exhaust gas, *SAE Jnl*, August 1968.
E. S. Starkman, 'Fundamental processes in nitric oxide and carbon monoxide production from combustion engines', 12th International Congress of FISITA, Barcelona, 1968.
R. D. Kopa, 'Control of automotive exhaust emission by modifications of the carburation system', SAE Paper No. 660114, 1966.
E. S. Starkman, 'Various component gases of engine generated pollution pose differing health hazards', *SAE Jnl*, March 1967.
J. D. Benson, 'Reduction of nitrogen oxides in automobile exhaust', SAE Paper No. 690019, 1969.
G. J. Nebel and **M. W. Jackson**, 'Some factors affecting the concentration of oxides of nitrogen in exhaust gases from spark ignition engines', *Air Pollution Assoc. Jnl*, November, 1958.
A. S. Leah, 'Notes for short course on exhaust pollution, University of Leeds, Unpublished, 1969.
W. F. Deeter *et al.*, 'An approach for controlling vehicle emissions', SAE Paper No. 680400, 1968.

E. S. Starkman *et al.*, 'An investigation into the formation and modification of emission precursors', SAE Paper No. 690020, 1969.
R. C. Lee and **D. B. Wimmer**, 'Exhaust emission abatement by fuel variations to produce lean combustion', SAE Paper No. 680769, 1968.
E. S. Starkman and **H. K. Newhall**, 'Characteristic of the expansion of reactive gas mixtures as occurring in internal combustion engine cycles', SAE Paper No. 650509, 1965.
T. A. Huis *et al.*, 'Spark ignition engine operation and design for minimum exhaust emission', SAE Paper No. 660405, 1966.
D. L. Sutton, 'Engine tuning to minimize exhaust emissions', IMechE Auto. Div. Symposium, November 25th–26th, 1968, Paper No. 8.
D. F. Hagan and **G. W. Holiday**, 'The effects of engine operating and design variables on exhaust emissions', SAE Paper No. 486C, 1962.
E. Bartholomew, 'Potentialities of emission reduction by design of induction system', SAE Paper No. 660109, 1966.
C. L. Bailey, 'The influence of motor gasoline characteristics upon carbon monoxide emissions under engine idle conditions'. IMechE Auto. Div. Symposium, November 25th–26th, 1968, Paper no. 16.
P. L. Dartnell and **P. V. Lamarque**, 'The effect of combustion chamber shape and other engine design factors on exhaust emissions'. IMechE Auto. Div. Symposium, November 25th–26th, 1968, Paper No. 4.
L. A. McReynolds *et al.*, 'Hydrocarbons emissions and reactivity as functions of fuel and engine variables', SAE Paper No. 650525, 1965.
J. H. Freeman Jr and **R. C. Stahman**, 'Vehicle performance and exhaust emission, carburation versus timed fuel injection', SAE Paper No. 650863, 1965.
L. Eitinge *et al.*, 'Emission control requires good carburation and other engine modifications', *SAE Jnl*, September 1968.
M. W. Jackson *et al.* 'The influence of air–fuel ratio, spark timing, and combustion chamber deposits on exhaust hydrocarbon emissions', SAE paper No. 486A, 1962.
W. A. Daniel, 'Engine variable effects on exhaust hdyrocarbon composition (a single cylinder engine study with propane as the fuel)', SAE Paper No. 670124, 1967.
E. W. Beckman *et al.*, 'Exhaust emission control by Chrysler—the cleaner air package, SAE Paper no. 660107, 1966.
A. E. Dodd and **J. W. Wisdom**, 'Effect of mixture quality on exhaust emissions from single-cylinder engines'. IMechE Auto. Div. Symposium, November 25th–26th, 1968, Paper No. 17.
H. P. Davis *et al.*, 'The effects of knock on the hydrocarbon emissions of a spark ignition engine', SAE Paper No. 690085, 1969.
Ch. E. Scheffler, 'Combustion chamber surface area, a key to exhaust hydrocarbons, SAE Paper No. 660111, 1966.

(8) **D. J. Picken, H. A. Soliman** and **M. F. Fox** (Leicester Polytechnic), 'Inlet manifold requirements for low emissions', *Automot. Engr*, 1978, **3**(4), 49–61.
See also:
D. W. Hughes, J. R. Goulburn, N. R. Beale and **D. Hodgetts**, 'Engine induction developments—economic and reduced exhaust emissions, *Proc. Instn mech. Engrs*, 1976, **190**, 1–21.
R. Lindsey, A. Thomas and **J. L. Wilson**, 'Mixture quality, gasoline vaporisation and the Vapipe', *Power plants and future fuels*, 1975, IMechE, London, pp. 169–178.
G. G. Lucas and **E. H. James**, 'A computer simulation of spark ignition

engines', SAE Paper No. 730053, 1973.

V. Panduranga, 'Hydrocarbon emissions from automotive engines', *Indian J. Technol.*, 1973, 110.

R. A. Philips and **P. L. Orman**, 'Simulation of combustion in gasoline engines using a digital computer', in *Advances in Automobile Engineering*, Part 4, 1966, Pergamon Press.

P. H. Schweitzer, 'Control of exhaust pollution through a mixture optimiser', SAE Paper No. 720254, 1972.

D. B. Spalding, 'Theory of particle combustion at high pressure', *ARS Jnl*, November 1959.

(9) **I. C. Finlay, D. J. Boam** and **J. L. K. Bannell** (National Engineering Laboratory), Computer model of fuel evaporation in air valve carburettors', *Automot. Engr*, 1979, **4**(6), 51–56.

See also:

H. L. Yun, R. S. Lo and **T. Y. Na**, 'Theoretical studies of fuel droplet evaporation and transportation in a carburettor venturi', SAE Paper No. 760289, 1976.

W. A. Sirignano and **C. K. Law**, 'Transient heating and liquid/phase mass diffusion in fuel droplet vaporisation', in *Advances in Chemistry Series 166, Symposium on Evaporation–Combustion of fuels* (Edited by J. T. Zung), 1978, American Chemical Society, Washington DC, pp. 3–26.

W. E. Ranz and **W. R. Marshall**, 'Evaporation from drops', *Chem. Eng. Prog.*, 1952, **48**(3), 141–146; **48**(4), 173–180.

J. C. Slattery and **R. B. Bird**, 'Calculation of the diffusion coefficient of dilute gases and of the self-diffusion coefficient of dense gas', *AIChE Jnl*, 1958, **4**, 137–142.

C. R. Wilke, 'Diffusional properties of multi-component gases', *Chem. Eng. Prog.*, 1950, **46**(2), 95–104.

R. D. Ingebo, *Vaporization rates and drag coefficients for iso-octane sprays in turbulent air streams*, NACA Tech. Note 3265, 1954, National Advisory Committee for Aeronautics.

S. Nukiyama and **Y. Tamasawa**, 'Experiment on the atomisation of liquid by means of an air stream', *Trans. Soc. Mech. Engrs*, (*Japan*), 1939, **5**(18), 68–75.

D. J. Picken, H. A. Soliman and **M. F. Fox**, 'Inlet manifold requirements for low emissions', *Automot. Engr*, 1978, **3**(4), 59–61.

(10) **E. H. James** (Loughborough University), 'Combustion modelling in SI engines', *Automot. Engr*, 1984, **9**(3), 29–33.

B. Ahmadi-Befrui and coworkers, SAE Paper No. 810151, 1981.

S. F. Benjamin and co-workers, *Stratified charge automotive engines*, 1980, IMechE, London, pp. 92–100.

N. C. Blizard and **J. C. Keck**, SAE Paper No. 740191, 1974.

P. Blumberg and **J. T. Kummer**, *Combust. Sci. and Technol.*, 1971, **4**, 73–95.

A. A. Boni, SAE Paper No. 780316, 1978.

C. Borgnakke and co-workers, SAE Paper No. 800287.

F. V. Bracco, SAE Paper No. 741174, 1974.

G. C. Davis and **C. Borgnakke**, SAE Paper No. 820045, 1982.

J. N. Mattavi and **C. A. Amann**, *Combustion modelling in reciprocating engines*, pp. 231–264; 69–129; 1–35; 331–343. 1980, Plenum Press.

E. H. James, 1980, SAE Paper No. 800458, 1980.

B. E. Launder and **D. B. Spalding**, *Mathematical models of turbulence*, 1972, Academic Press.

T. Morel and **N. N. Mansour**, SAE Paper No. 820040, 1982.

J. I. Ramos and **W. A. Sirignano**, *18th International Symposium on Combustion*, 1981, The Combustion Institute, pp. 1825, 1835.

T. Singh and **K. Surakamol**, SAE Paper No. 790354, 1979.
L. J. Spadaccini and **W. Chinitz**, *Trans Am. Soc. mechn Engrs*, 1972, **94**, 98–108.
R. J. Tabaczynski and co-workers. SAE Paper No. 770647, 1977.

(11) **J. Fenton** (Editor), 'Ford's Dunton R and D centre', *Automot. Engr*, 1979, **4**(2), 88–89.

(12) **M. T. Overington** (Ricardo), 'High compression ratio gasoline engines and their impact on fuel economy', *Automot. Engr*, 1982, **7**(1), 26–27.
See also:
W. M. Scott, 'A passenger car diesel engine for America', *Proc. Instn mech. Engrs*, 1981, **195**(2).
M. L. Monaghan, 'The high speed direct injection engine for passenger cars', SAE Paper No. 810477.
M. G. May, 'Lower specific fuel consumption with high compression lean burn spark ignited four stroke engines', SAE Paper No. 790386.

(13) **C. R. Stone** and **E. K. M. Kwan** (Brunel University), 'Variable valve timing for ic engines', *Automot. Engr*, 1985, **10**(4), 54–58.
See also:
G. A. Torazza, 'A variable lift and event device for piston engine valve opration', 14th FISITA Congress, 1972, Paper 2/10, pp. 59–67.
D. A. Parker and **M. A. Kendrick**, 'A cam shaft with variable lift–rotation characteristics; theoretical properties and application to the valve gear of a multicylinder piston engine', 15th FISITA Congress, 1974, Papers B-1-11, pp. 224–232.
C. Paulmier, 'Variable valve timing for poppet valve internal combustion engines', *Ingens de L'Auto*, June/July 1974, p. 442.
J. Kerr, 'Variable valve timing to boost any engine', *The Engineer*, July 1980, pp. 28–29, 54.
N. Beresford and **K. Ruggles**, 'An investigation of induction ramming', project report 2/84, 1984, Department of Mechanical Engineering, Brunel University.
C. R. Stone, *Introduction to internal combustion engines*, 1985, Macmillan.
J. H. Tuttle, 'Controlling engine load by means of early intake-valve closing', SAE Paper No. 820408, 1982.
G. B. K. Meacham, 'Variable valve timing as an emission control tool', SAE Paper No. 700673, 1973.
P. H. Smith, *Valve mechanisms for high speed engines*, 1967, Foulis, pp. 108–109.
G. Zappa and **T. Franca**, 'A 4-stroke high speed diesel engine with two-stage supercharging and variable compression ratio', 13th CIMAC Congress, Vienna, 1979, session B3, Paper D19.

(14) **H. Forster** (Daimler Benz), 'The development potential of spark ignition engines', *Automot. Engr*, 1980, **5**(2), 23–26.

(15) **K. Radermacher** (BMW), 'The ETA engine concept', *Automot. Engr*, 1982, **7**(3), 21–24.

(16) **J. Fenton** (Editor), 'Engine analysis: dynamic, volumetric, and heat flows', *Automot. Engr*, 1985, **10**(3), 12–17.
See also:
R. Munro, W. Griffiths and **T. Cowell**, 'Performance predictive techniques for pistons', Associated Engineering Ltd Symposium, 1978.
J. Cornforth, 'Finite element analysis of engines', Ricardo Report DP 84/1359 (Unrestricted). 1984.
I. Finlay, **G. Gallacher**, **J. Bingham** and **J. Orrin**, 'Distribution of air mass flowrate between the cylinders of a carburetted automotive engine', SAE Paper No. 850180.
J. Bingham and **G. Blair**, 'An improved branched pipe model for multi-cylinder automotive engine calculations, *Proc. Instn mech. Engrs*, 1985, **199**(D1).
T. Priede, **J. Dixon**, **E. Grover** and **N.**

Saleh, 'Experimental techniques leading to the better understanding of the origins of automotive engine noise', in *Vehicle Noise and Vibration*, 1984, IMechE, London, pp. 141–160.

(17) **A. Dye** (Epicam), 'New approach to combustion analysis', *Automot. Engr*, 1985. **10**(1), 32–35.
See also:
C. G. Lucas and **E. H. James**, 'Flame propagation and cyclic dispersion in the spark ignition engine', *Automot. Des. Engr*, 1970, **9**(5), 36–41.
G. M. Rassweiler and **L. Withrow**, 'Motion pictures of engine flames correlated with pressure cards', SAE Paper No. 800131.
J. N. Mattavi, 'The attributes of fast burning rates in engines', SAE Paper No. 800920.
F. Matekunas, 'Modes and measures of cyclic variability', SAE Paper No. 830337.

Chapter Three

(18) **J. Hartley** (Consultant), 'Engine design series, *Automot. Des. Engr*, 1975, **14**(1)–*Automot. Engr*, 1978, **3**(3).
See also:
C. F. Taylor, *The Internal Combustion Engine in Theory and Practice*, Vol. 2, The M.I.T Press, Massachusetts.
R. J. Harker, *Determination of Engine, Crankshaft and Connecting Rod Loading*, SAE publication.
C. Lipton, *The Practical Design of Crankshifts and Connecting Rods*, SAE publication.
The Structural Durability of Crankshafts, Sulzer Tech. Rev., No. 2, 1943.
Hasselgruber and **Kroch**, *Calculation of Form Strength Factors for Crankshafts*, MTZ, Aug. 1960.
A. Buske, *Design and Production of Nodular Cast Iron Crankshafts*, MTZ, April 1961.
H. W. Barnes-Moss, 'A Designer's viewpiont', in *Passenger car Engines*, 1973, IMechE, London, pp. 133–147.
U. A. Kogan, 'Determination of the Torsional Oscillations and Parameters of oscillation Dampers for New Engines, *Avtom. Prom.*, USSR, May, 1961.
J. C. Clayton, 'Cast Iron Camshafts in Car Production', *Design and Components in Engineering*, 1971. Council of Iron-foundry Associations reprint.
S. D. Apsley, 'Iron Castings in the Motor Industry', *J. Automot. Engng*, Oct. 1971. Council of Ironfoundry Associations reprint.)
Engineering Data on Nodular Cast Iron, Council of Ironfoundry Associations.
Hunt and **Russell**, 'Crankshaft Forging: an evaluation of continuous grain flow', *Proc. Instn mech. Engrs*, 1967.
W. Egger, 'Fillet-rolling of Diesel Engine Crankshafts, *Machinery*, Nov. 1960.
J. C. Clayton, 'Engine Castings' in *Passenger Car Engines*, 1973, IMechE, London.
Gassner and **Schulz.** *Loading of Crankshafts,* MTZ, 1961.
Finnern and **Krzyminski,** *Methods of Increasing Bending Fatigue of Crankshafts by materials selection and heat treatment*, MTZ, June 1976.
Crandall, 'Oldsmobile's Computer Application to V-8 Crankshaft Design', SAE, 1966.
Hafner, *The Calculation of Torsion in Crankshafts for Piston Engines by Means of Electronic Computers*, MTZ, Oct. 1964.
J. P. Pirault, 'Location of bearing oil feeds', *J. Automot. Engng*, Aug. 1973.
W. V. Appelby, 'The BMC 'A' Series', *Design of small engines for mass-produced motor cars*, 1963, IMechE, London, pp. 1–9.
L. Kuzmicki *et al.*, 'The Hillman 875 cm^3 engine', *Design of small engines*

for mass-produced motor cars, 1963, IMechE, London, pp. 47–56.

L. Kuzmicki and **J. G. Haig**, 'Avenger Engine; a design concept for manufacture', *Passenger Car Engines*, 1973, IMechE, London.

J. A. Morgan and **D. J. Stojek**, 'A New Range of 4 cyl In-line Single OHC Engines by Ford of Europe', SAE Paper No. 710148.

D. E. Larkinson and **B. R. Jewsbury**, 'A Diesel Engine for Light Duty Applications' (*Perkins Engine Co.*), SAE Paper No. 750333.

R. A. S. Worters, 'The Design of Ford of England's 1000–1500 cm^3 engine family', *Design of small engines for mass-produced motor cars*, 1963, IMechE, London, pp. 25–35.

E. J. Murray, 'The design of pistons for petrol engines', Associated Engineering symposium, 1970.

G. Longfoot, The designs of pistons for diesel engines', Associated Engineering symposium, 1970.

J. E. Robinson, 'The design and development of pistons for automobile engines', *Piston Technology*, Hepworth and Grandage Ltd.

M. H. Howarth, *Design of High Speed Diesel Engines*, 1966, Constable and Co., Ch. 12.

P. E. Vickery, 'Small end lubrication—application of hydrodynamic lubrication theory', *Engineering*, March 1972; 'Evolution of the cast Armasteel con rod', SAE National Automobile Week, Detroit, March 1962.

C. A. Perkins and **J. M. Conway-Jones**, *Reciprocating engine bearing design*, Associated engineering symposium, June 1970.

D. D. Parker and **P. E. Vickery**, 'Engine bearing analysis on a small computer', *Tribology* 1971.

S. M. Robinson, 'Use of interference fits in thinwall bearing installations', *JAE 4*, June 1973.

(19) **B. Law** (Perkins), 'Computer prediction of engine imbalance', *Automot. Engr*, 1984, **9**(2), 44–46.

(20) **A. J. Reed** (Dunlop Polymer), 'Vehicle engine mountings', *Automot. Des. Engr*, 1963, **2**(1), 40–43.

(21) **J. Fenton** (Editor), Exhaust system analysis', *Automot. Dev. Engr*, 1964, **3**(4), 40–43.

(22) **J. Fenton** (Editor), 'Diesel vehicle noise control', *Automot. Engr*, 1979, **4**(1), 23–25; **4**(3) 15.

(23) **J. Fenton** (Editor), 'Silencer development', *Automot. Engr*, 1979, **4**(2), 43–45.

(24) **W. A. Pullman** (Lanchester Polytechnic), 'Gas dynamics applied: exhaust pulses and engine performance', *Automot. Dev. Engr*, 1964, **3**(4), 43–46.

See also:

B. N. Cole and **B. Mills**, 'Theory of sudden enlargements applied to the poppet valve with special reference to exhaust-pulse scavenging', *Proc. Instn mech. Engrs*, 1952–53. **1B**, No. 8, 364.

F. J. Wallace and **R. W. S. Mitchell**, 'Wave action following the sudden release of air through an engine port system', *Proc. Instn mech. Engrs*, 1952–53, **1B**, No. 8, 343.

A. W. Hussmann and **W. A. Pullman**, 'Pressure fluctuations in multi-cylinder manifolds', *Amer. Soc. mech. Engrs*, 1957, Paper 57-A-196.

F. K. Bannister and **G. F. Mucklow**, 'Wave action following sudden release of compressed gas from a cylinder', *Proc. Instn mech. Engrs*, 1948, **159**, 269.

F. K. Bannister, 'Induction ramming of small high-speed air compressor'. *Proc. Instn mech. Engrs*, 1959, **174**(13), 375.

F. J. Wallace and **M. H. Nassif**, 'Air flow in a naturally aspirated two-

stroke engine, *Proc. Instn mech. Engrs*, 1954, **168**(18), 515.
A. W. Hussmann and **W. A. Pullman**, 'Diesel exhaust blowdown energy', International Congress of Combustion Engines (CIMAC), Paris, 1959, p. 663.
P. H. Schweitzer, 'Research in exhaust manifolds', *Amer. Soc. mech. Engrs*, 1952, **74**, 517.
A. W. Hussmann and **W. A. Pullman**, 'Formation of pressure pulses by exhaust blowdown', *Amer. Soc. mech. Engrs*, 1959, Paper 58-A-145.
H. D. Carter, 'The loop scavenge diesel engine', *Proc. Instn mech. Engrs*, 1946, **154**, 386.
R. Feiss, 'Increase of the capacity of diesel engines of the four-stroke principle by means of the exhaust scavenging impulses', International Internal Combustion Engine Congress (CIMAC), Paris, 1955, p. 393.

(25) **J. C. Morrison** (Glasgow University), 'Exhaust system for IC engines', *Automot. Dev. Engr*, 1967, **6**(12), 38–42.

Chapter Four

(26) **D. Anderson** (Cranfield IoT), 'Engine cooling – a systematic approach', *Automot. Dev. Engr*, 1968, **8**(2), 38–41; **8**(3), 38–40.
See also:
M. G. Paish and **W. R. Stapleford**, 'A study to improve the Aerodynamics of Vehicle Cooling Systems (First Report)', MIRA Report No. 1966/15.
H. J. Hannigan, 'Coolant performance at higher temperatures', SAE Paper No. 680497.
J. C. Brabetz and **D. S. Pike**, 'Engines like to be warm', SAE Paper No. 891A, Aug. 1964.
A. V. Kostrov and **B. M. Kunyavski**, 'Influence of cooling water temperature on thermal operating conditions for vehicle engine components', *Moscow Automech. Inst.*, 1965, No. 9.
C. N. Moore and **H. L. Hohenstein**, 'Heat transfer and engine cooling—aluminium versus cast iron', SAE Paper No. 494A, March 1962.
C. F. Taylor, *The internal combustion engine in theory and practice*, Vol. 1, 1980, John Wiley and Sons, New York.
E. Gehres, 'An analysis of engine cooling in modern passenger cars', SAE Paper No. 660C, March 1963.
M. Jacob, *Heat transfer*, Vol. 1, John Wiley and Sons, New York.
W. M. Kays and **A. L. London**, *Compact heat exhangers*, 2nd Ed., 1964, McGraw-Hill.
D. G. Stratton, 'Engine cooling system design and development', *Proc. Instn mech. Engrs*, 1965–66, **180**. Part 2A.
E. R. Klinge, 'Truck cooling system airflow', SAE Paper No. 15–19/7/59.
M. G. Paish and **W. R. Stapleford**, 'A rational approach to the aerodynamics of engine cooling system design', *Proc. Instn mech. Engrs*, 1968–69, **183**, Part 2A.
A. P. Fraas an **M. N. Ozisik**, *Heat exchanger design*, 1965, John Wiley and Sons, New York.
M. L. Bell, *A study of factors affecting engine cooling*, ASAE thesis, 1968.
A. L. Longdon and **R. K. Shak**, 'Offset rectangular plate-in surfaces – heat transfer and flow friction characteristics', *Trans ASME*, July 1968.
S. F. Hoerner, *Fluid-Dynamic Drag*, 1965, published by the Author.

(27) **C. French** (Ricardo), 'Heat flow in the IC engine', *Automot. Dev. Engr*, 1963, **2**(2), 42–46.
See also:
G. Eichelberg, 'Temperaturverlauf und Wärmespannungen in Verbrennungsmotoren', *V.D.I. Forschungsheft*, 1923, 263.
G. Eichelberg, 'Investigations on internal combustion engine prob-

lems', *Engineering*, Oct. 1939, p. 463.
N. D. Whitehouse *et al.*, 'Method of predicting some aspects of performance of a diesel engine, using a digital computer', *Proc. Instn mech. Engrs*, 1962, **176**(9).
J. F. Alcock *et al.*, 'Distribution of heat flow in high-duty internal combustion engines', CIMAC Conference, Zürich, 1957.
C. F. Taylor, *The internal combustion engine in theory and practice*, Vol. 1, 1980, John Wiley and Sons, New York.
W. H. McAdams, *Heat transmission*, 3rd edition, 1954, McGraw-Hill.
J. F. Alcock, 'Heat transfer in diesel engines', ASME-IMechE International Heat Transfer Conference, Boulder, Colorado, and London, 1961.
D. R. Pye, *The internal combustion engine: Vol. II. The aero engine.* 1934, Oxford University Press, Chapter VII.
P. V. Lamarque, 'The design of cooling fins for motor cycle engines', *Proc. I.A.E.*, 1942, **37**, 99.
National Advisory Committee for Aeronautics, *Index of Technical Publications*, 1915–1947, pp. 256–261.
J. Mackerle, *Air cooled motor engines*, 1961, Cleaver-Hume Press.
A. Nagel, 'The transfer of heat in reciprocating engines', *Engineering*, Jan. 1929, p. 59.

(28) **S. Hurford** (Delanair), 'Heat flow in the coolant heat exchanger', *Automot. Dev. Engr*, 1963, **2**(3), 43–47.

(29) **S. Hawes** (Airscrew Howden), 'Modern fan technology', *Automot. Engr*, 1975, **1**(3) 15–17.

(30) **G. Bennisgen** (Reintz), 'Gasket by design', *Automot. Dev. Engr*, 1968, **7**(10), 106–107.

(31) **G. Stahl** (Reintz), 'An analysis of gasket joints', *Automot. Dev. Engr*, 1973, **12**(10), 75–81.

Chapter Five

(32) **I. Dabney** (Gen. Rad.), 'Computer-Aided-Engineering: the integrated approach to design test and manufacture', *Automot. Engr*, 1982, **7**(4), 66–68.

(33) **R. Southall** (SDRC), 'Engine components by CAE', *Automot. Engr*, 1984, **9**(5), 16–20.

(34) **J. Fenton** (Editor), 'Designing for finite life', *Automot. Engr*, 1983, **8**(1), 14–18.

(35) **J. Fenton** (Editor), 'Design of engine lubrication systems', *Automot. Engr*, 1984, **9**(1), 24–25.

(36) **F. A. Davis** (Brunel University), 'Plain bearing wear in IC engines', *Automot. Engr*, 1981, **6**(4), 31–36.

(37) **R. T. C. Harman,** 'Predicting fuel consumption', *Automot. Engr*, 1985, **10**(5), 46–48.
See also:
'ECE 15 driving cycle', from Annex III, Type I Test, in *Official J. European Communities*, L197, vol. 26, July 1983.
'EPA Urban and Highway driving cycles', in *Code of Regulations, 40, Protection of Environment*, 1979, US Government Printing Service, Washington DC.
'NZ Urban and Suburban driving cycles', in *Driving patterns of private vehicles in New Zealand*, 1984, LFTB publication LF2039.
'Electric vehicle test procedure' – SAE J227(a), *SAE Handbook*, 1981, section 2709.
C. S. King, 'A car for the nineties. BL's energy conservation vehicle', *Proc. Instn mech. Engrs*, 1983, **197**, Paper No. 64.
'Citroen, the road ahead', *Autocar*, November 1984, pp. 50–53.
O. Lindstrom, 'Prospects for electric vehicles', *Electric Vehicles Developments*, 1984, No. 18, pp. 4–7.
R. T. C. Harman and **D. J. Byers**, 'An AC induction motor electric

vehicle', Proceedings of the Electric Vehicle Exposition, *EVE 80*, Adelaide, August 1980, pp. 469–488.

'EVs and hybrids', *Automot. Engr*, 1984, **9**(2), 48–50.

R. T. C. Harman, 'The efficiency potential of electric and fuelled vehicles', *Trans Instn Prof. Engrs N.Z.*, 1985.

(38) **J. Fenton** (Editor), 'Engine systems analysis, structural design, and draughting by CAD', *Automot. Engr*, 1985, **10**(1), 40–42.

(39) **J. Fenton** (Editor), 'Fire 1000; Fiat's CIM engine', *Automot. Engr*, 1984, **9**(6), 66–67.